PLANETS and PLANETARIANS:

A History of Theories of the
Origin of Planetary Systems

PLANETS and PLANETARIANS:

A History of Theories of the Origin of Planetary Systems

STANLEY L. JAKI

A HALSTED PRESS BOOK

JOHN WILEY & SONS
New York – Toronto

Published in the U.S.A. and Canada by
Halsted Press, a Division of John Wiley & Sons, Inc., New York

ISBN 0 470-99149-6

Library of Congress Cataloging in Publication Data

Jaki, Stanley L.
    Planets and Planetarians: A History of Theories of the
        Origin of Planetary Systems

    "A Halsted Press book."
    Includes index.
    1. Planetary Systems, Formation of.   2. Cosmogonies.
    3. Extraterrestrial Life.   4. Principle of Plenitude.
QB601.J34   521'.54'09   77-4200
ISBN 0-470-99149-6

Printed in Great Britain

CONTENTS

INTRODUCTION

When ten years ago man stepped on the moon, even the sceptics must have sensed that the step was only the start of man's march into the realm of planets. An extraordinary realm, to be sure, which unlike things and processes on earth, appeared to man since ancient times as the realm of exactness. The regularity and precision of the motion of planets became ultimately, in the hands of Newton, the clue to a physics which has not ceased unfolding more and more of the exactness that after all rules this globe of ours as well. One of the great triumphs of that physics was the demonstration by Laplace that the system of which our earth is a principal member has a very large measure of stability, though not an absolute one. The solar system, like anything else, is a child of time. It was born and has grown into its present frame which is destined to collapse in the long run. Indeed, it was a famous hypothesis submitted by Laplace that first assured scientific respectability to an evolutionary approach to the system of planets. In almost two centuries following the first publication of Laplace's nebular hypothesis, and especially during the last three generations, much effort has been spent on elucidating the history of the planetary system. That history is still to be traced out in an essentially satisfactory manner.

Such may seem relatively little progress but even less has been made in what should seem a far easier task, namely, the historiography of those efforts aimed at tracing out that history, or the history of theories of the origin of planetary systems. In fact, there seems to exist on this topic no monograph sufficiently detailed and steeped in the reading of the original documents. While some essay-reviews of the state of the art covering the last few decades are of distinct help to the historian for a primary orientation, he should be rather suspicious of accounts concerning the first three decades of this century. Suspicion should yield to distrust as he tries to orient himself from texts and histories of astronomy about the 19th-century part of the story. Finally, he should not be unwilling to take the attitude of plain disbelief with respect to most secondary information about what had been really proposed during the 18th and 17th centuries on the question. This sad record is largely the making of those who in proposing theories on the evolution of the solar system offered a review of what had been said previously, and of those, also mostly scientifically trained men, who drafted histories of such theories. Their cavalier handling of the facts of scientific history may seem a small matter. Yet a reasonably correct presentation of those facts is of paramount importance with respect to the cultural impact of science. A scientific explanation influences culture in the measure in which it is believed, rightly or wrongly, to be the crowning of past developments. As such, it is apt to be invested with the aura of being the long-sought solution to an age-old problem.

It seems unlikely that there should result beneficial cultural effects from a long chain of unfounded assertions that the final solution had been secured. The many theories already proposed on the evolution of the solar system form such a chain. It hardly ever happened over almost four centuries that a theory on the origin of planets was proposed with a genuine touch of diffidence. This diffidence should have been all the more proper as all theories were held together not so much by solid scientific considerations as by the wish to have *the* solution. No wonder that all theories were plagued either by the dubious technique of assuming what was to be explained or by a baffling oversight of the bearing on the question of some basic laws of physics. That first-rate scientists could be guilty on either or both counts should give pause to anyone still fancying that a scientific mind, or any mind using the scientific method, is easily safe from errors, let alone from some elementary mistakes in plain logic.

The search for a satisfactory explanation of the evolution of the system of planets has for its most obvious motivation the urge to know. The system of planets is like an extraordinary mountain in the vast array of the phenomena of nature and it exerts as much an overpowering attraction for scientists as do high peaks for mountaineers. The parallel can be extended further. The conquest of a mountain turns it into a well explored terrain, into one of the various types mountaineers contend with. The explanation of the origin of the system of planets also implies finding in it a typical feature. Typically enough, the preference has always been in favour of finding an explanation which provides assurance that systems of planets are typical, that is, regularly occurring features of the universe. Theories, on the basis of which our own planetary system had to be viewed as an exceedingly rare product of the forces of nature, produced, as a rule, considerable uneasiness.

This difference in appraisals derives from an instinctive commitment to the idea of plenitude, according to which living forms of all grades should be present everywhere in the universe. Since planets are the only known possible abodes for life, theories which promised planets in great numbers found not only ready acceptance, but were also, time and again, embellished with copious remarks, even with lengthy sections, on denizens of other planets. Whether this close tie between theories on the evolution of planetary systems and wishful vistas about planetarians helped or obstructed rigour in scientific reasoning should not be difficult to guess. At any rate, as long as such theories are being constructed with an *a priori* desire to assure high frequency for planetary systems, the chances remain meagre for an objective evaluation of the possibility that a planetary system like ours may be after all an extremely rare phenomenon, a product of a long chain of interactions of very small probability. A careful survey of the history of the question, and particularly of its most recent phase, may suggest precisely this, especially in view of the fact that although a very wide range of mechanisms have been tried, none of them worked.

This survey contains only a few technical details, always relegated to the notes. The story told here is intended for a wider range of readers interested in the fate and fortune of scientific ideas. These can be set forth without mathematics and Laplace himself served classic evidence, and in connection with our topic, that this can be done in a highly informative manner. Indeed, it was not until the beginning of this century that mathematics began to play a noticeable part in the presentation of theories of the origin of planetary systems. Yet even during that recent phase of our story, proponents of theories provided convincing examples that what had been first proposed in a forbiddingly mathematical form can also be presented in ordinary words. However, this last phase of our story also strengthened a pattern already evident in its previous phases. The pattern consists in the fact that the truth of theories did not depend so much on the amount of mathematics used as on the plain application of basic laws of physics to data embodied in the planetary system itself. The failure or success of theories to cope with those data are facts that require no mathematics in order to be spelled out in a manner understandable for everyone. Emphasis on technical details can easily distract from crucial assumptions and problems, and create the impression that it is meaningful to discuss the merits of a superstructure which lacks solid foundation. In fact, the presentation and discussion of many recent theories has been beset with the illusion that the glitter of mathematical details can dispense with a thorough probing into basic presuppositions and motivations.

To see all this in a convincing fashion one need only take a careful look at the original documents, recent and old. A principal aim of this book lies in presenting the chain of theories as they were proposed by their authors and not as they were distorted or embellished in the filters of second-hand accounts. A sustained look at the originals will at the same time unravel something of the human feature of science, a deep awareness of which is indispensable if scientific theories, and cosmology in particular, are to play a beneficial role in the human quest for an ever more reliable interpretation of the universe we live in.

That quest constitutes culture. The impact of cosmology on culture is incalculable and is most naturally noted when discussion turns to the origin of the system of planets. 'The advances of astronomy,' as was stated in a typical popularisation of mid-20th-century knowledge of the solar system, 'make their impact felt sooner or later on all sciences, help man both to understand nature better and to dominate it for the good of mankind. An important phase of astronomy for man will open on the very day when one discovers in a corner of the sky among the ever changing forms of matter new planets in their process of formation.' While this is certainly true, it is an even more important truth that constructing theories on the origin of planetary systems is not equivalent to proving their occurrence elsewhere, let alone in great numbers. In fact, even a good theory is not equivalent to a good prospect of ever observing a planetary system in formation. A good theory, to quote words immediately preceding the passage cited above, should be produced by 'a conscientious study of the sky, by setting forth the essential phenomena, by a demand of a rigorous agreement between theory and observation, by the confrontation of the results of the various sciences'.[1] If a conscientious attitude was needed in 1952 when eminent scientists still could call intercontinental ballistic missiles a dream never to come true, the same attitude, which always means caution as well, is even more necessary in this age of space probes. The new wealth of information about the planetary system that accrued through manned flights to the moon and through space probes sent to and around other planets, weakens not a whit the need for that caution which is a chief lesson of a survey of man's so far futile attempt to cope with the origin of our system of planets. Many of the new data have already imposed drastic revision of well-established notions and raised more than one grave question mark about the early history of planets and their system. The extensive documentation of this book is offered to those who are not reluctant to face up to the fact that in science, and especially in cosmology, there are more stubborn puzzles than welcome answers.

[1] G. Bruhat and E. Schatzman: *Les Planètes* (Paris: Presses Universitaires, 1952), p. 281.

CHAPTER ONE

THE GLITTER OF GEOMETRY

Pre-Socratics, planets, and purpose

Wanderers the Greeks called them, for this is what the word 'planets' ($\pi\lambda\alpha\nu\hat{\eta}\tau\epsilon\varsigma$) means. As any few vagrants would be among thousands of stable men, the planets, too, were inevitably conspicuous in a sky of fixed stars. Together with the motion of the moon and of the sun, the wanderings of Mercury, Venus, Mars, Jupiter, and Saturn amidst the stars were the prime target of the attention of ancient Greek astronomers. They traced the intricacies of planetary motions to an astonishing degree of precision.[1] The question of the origin of the planets, and of their system, was a different matter. Ancient Greeks made only a few utterances which qualify as vague anticipation of modern discussions of planetary origins. No less intriguing should appear the fact that those utterances did not come from astronomers among ancient Greeks, but from their *physikoi*, or natural philosophers.

In view of the proverbial acumen of ancient Greeks, their reticence on the question of planetary origins may very well justify probings into this point, although it should be kept in mind that not every important or interesting question should necessarily be formulated and developed at the very outset, dramatic and fascinating as this may be. The shunning by Greek astronomers of the topic of planetary origins is not an overly conjectural matter. Ultimately, it is closely related to the origin of the most haunting and challenging aspect of speculations on planetary origins, the question of planetarians, or beings with a sense of purpose. The question as related to earthlings forms in a sense the second great watershed of ancient Greek thought, the *Phaedo*, Plato's account of Socrates' heroic stand in support of the perennial validity of certain value judgments. To prove his principal point, the survival of his personal identity and moral responsibility, Socrates felt impelled to make a frontal attack on the purely physical interpretation of the whole gamut of human experience by early Greek natural philosophers, whose rise during the 6th century B.C., marked the first watershed in ancient Greek conceptual development.

There is no defending the effort of Socrates, who in order to vindicate eternal purpose for any person advocated the assigning of purpose to each and every process in the physical universe. Socrates' claim that everything in the external world happened for the best in a purposeful sense was as far-fetched as was the contention of pre-Socratic *physikoi* that all reality consisted in the perennial transformation of matter. Socrates was, however, very much to the point as he referred to the notion of vortex ($\delta i\nu\eta$),[2] which, according to th *physikoi* and Anaxagoras in particular, explained the stability of the earth in the centre of a great cosmic whirl or vortex. The modern question of planetary origins beckoned from the wings, but the centre stage was, in a genuinely Socratic manner, occupied by a far deeper question. It concerned the meaning of human purpose, on which many centuries later doubts were cast by theories of planetary origins all the more powerfully, the more plausible they became.

One could only wish that Socrates had discussed in detail the thought of Anaxagoras,[3] who in all evidence was at one with some other pre-Socratic natural philosophers in picturing the universe in an evolutionary framework of explanation. Its emergence had from the very start as its hallmark the emphasis on 'spontaneous' necessity in the formation of the whole and parts of the cosmos. Aristotle, who always wrote with an eye on the pre-Socratics, showed no hesitation in making the generalization that all who advocated a spontaneous cosmic evolution rested their case on the spontaneous emergence of vortices.[4] To this telling remark Aristotle added no specific information about the ways in which pre-Socratic *physikoi* imagined the evolution of the earth, planets, and stars into one system in a huge vortex.

Very likely, Aristotle was short on words in this respect, because supporters of cosmic evolution through vortices did not themselves seem to go into detail. The only extant pre-Socratic fragment which may rank as a statement on planetary origins contains no reference to vortex motion. The fragment is ascribed to Anaximander[5] according to whom once the basic pair of opposites, the hot and cold, began to be differentiated from the eternal substratum, there developed a sphere of flame around the cold earth and air. Initially, that sphere of flame surrounded the earth-air core as closely as is the case of the bark around a tree. Later, however, the sphere of flame moved out toward the periphery and became enclosed in several rings corresponding to the sun, the moon, and the stars. The planets as such were not mentioned in the context of this model on which Anaximander grafted for obvious reason the *ad hoc* proviso that the actual appearance of the sun and the moon was due to a circular hole in their respective rings filled with fire.

The extant evidence of Leucippus' reliance on vortices for cosmological purposes is very meagre, and similar is the case for his successor as the spokesman of atomism, Democritus.[6] Yet it is most likely that both attributed a chief role to vortices in the constant rise and decay of innumerable worlds in some of which, according to Democritus, there was no sun or moon. Democritus was, in fact, most explicit on the point that spontaneous evolution through random combination of atoms was not to produce similar worlds in an infinite universe. Very numerous, though also confusing in details, are the references in later tradition to Anaxagoras' emphasis on vortices.[7] At any rate, the universal popularity in late-5th-century Greece of vortex motion as an all-purpose physical mechanism is put beyond any doubt by some lines in Aristophanes' *Clouds*, a play dealing with the cultural crisis which prompted Socrates' reaction to the purely relativistic and utilitarian doctrine of the Sophists who eagerly unfolded the ultimate implications of the doctrine of the *physikoi*. It was Socrates' all-consuming concern that old Strepsiades anticipated as he exclaimed with great indignation: 'What! Vortex? That is something I own. I knew not before that Zeus was no more, but Vortex was placed on his throne!'[8]

Plato, planets, and divine principles

A generation later and under the impact made on him by Socrates, Plato was busily at work on restoring, if not Zeus, at least the 'divine principles' to a commanding role in cosmological discourse[9] to save thereby the validity of ethical values. For Plato this meant that, in line with traditional and pantheistic Greek religiosity, a divine and permanent character had to be attributed to everything from the moon to the stars. In place of spontaneous emergence came, therefore, a world view stabilized along geometrical patterns. Thus almost at the very outset of the fashioning of the world by the demiurge, Plato described him in the *Timaeus* as proceeding to the arithmetic subdivision of the 'Whole', or the universe animated by the world-soul. The lengths marked off on the 'Whole' corresponded to the sequence 1, 2, 3, 4, 8, 9, 27, that is, to the series of numbers formed by unity and by the squares and cubes of its double and triple measures. The procedure, cryptic at first sight, is revealed a little later as being the allocation of the moon, the sun, and the planets at proper intervals below the sphere of the fixed stars, 'the Same and Uniform' frame of the universe: 'And he gave the supremacy to the revolution of the Same and Uniform; for he left that single and undivided; but the inner revolution he split in six places into seven unequal circles, severally corresponding with the double and triple intervals, of each of which there were three. And he appointed that the circles should move in opposite senses to one another; while in speed three should be similar, but the other four should differ in speed from one another and from the three, though moving according to ratio'.[10]

This partitioning of the space between the earth and the sphere of the fixed stars into segments, the size of which corresponded to a specific numerical sequence, was in line with the Platonic tenet that the structure and laws of the heavenly region could only be understood in terms of numbers and geometry. This tenet was valid to a lesser degree about the sublunary regions, but even there Plato wished to have geometry play an important role. According to him the tetrahedron, cube, octahedron, and icosahedron, or four of the five perfect solids, generated the four sublunary elements, the fire, earth, air, and water, respectively. The dodecahedron generated the material of the starry realm, the incorruptible ether moving forever in a circle. All this was but a 'probable' theory in Plato's words, as he put it in the *Timaeus*.[11] What could not be questioned was the absolute perfection of the heavenly regions and their one-to-one correspondence with the perfect geometrical figure, the sphere. Perfection was akin to harmoniousness which in turn was best reflected in the pleasing tonalities of certain musical intervals. Since these represented arithmetic ratios, it was all too natural to search for their replicas in the periods of celestial movements, and especially of the movements of the planets. This was the Pythagorean doctrine of the harmonies of the spheres which Plato wholeheartedly endorsed in his *Republic*.[12] His dicta were to have one particularly notable echo as the republic of science advanced within striking distance to its modern watershed with a geometrical speculation on the origin of the system of planets.

If the Platonic motto was true that the chief activity of deity consisted in 'geometrising', then the geometrical approach must have certainly appeared best suited for the investigation of that region of the universe which by definition was eternal and divine. From this it also followed that in that region was no room for physical emergence, let alone for decay, of any sort. Consequently, an evolutionary theory of planets could be of no use whatever for Plato. He even found it highly improper to call some stars 'wanderers'. The Platonic heaven was the embodiment of perfect motion, which meant motion in a circle. 'Wandering' meant odd, irregular routes, the abandoning of the right path. To attribute such 'misbehaviour' to 'the great gods, the sun, the moon and some other stars', was blasphemy in Plato's eyes. For him the name planet (wanderer) implied a brazen lie on the lips of all Greeks. He made this remark in *The Laws*[13] where he laid down a long list of precepts for sound social living. One of them was a reshaping of the study of astronomy. A chief point in this respect consisted in planting the conviction in the youth that each of the planets 'moves in the same path – not in many paths, but in one only, which is circular, and the varieties are only apparent'.[14] Actually, this educational manifesto aimed at restoring an old belief undermined by 'that other doctrine about the wandering of the sun and the moon and the other stars', which was 'not the truth but the very reverse of the truth'.[15]

The representatives of that other doctrine were the Ionians, Anaxagoras, and the atomists, especially Democritus. The latter is best known through the relentless criticism to which his advocacy of atoms and of infinitely numerous 'worlds' in endless rise and decay was subjected by Aristotle. Had Democritus taken up in detail the question of planetary origins, he would have certainly come into explicit strictures on the part of Aristotle. Democritus' ideas on atoms and infinity were for Aristotle the worst possible aberrations from the path of 'right' thinking. A special discourse on the chance formation and dissolution of the system of planets would not have fared any better with the one who built in his *On the Heavens* the divinity of the spherical superlunary world into a painstakingly argued, though sadly misleading, cosmological system. In the Aristotelian heavens everything was eternally permanent and also 'metaphysical', that is, lying beyond the reach of ordinary physics. No wonder that Aristotle's most influential work, the *Metaphysics*, culminated, in a sense, in the enumeration of the circles, principal and auxiliary, which were needed to 'explain' the uneven motion of planets embedded in a set of concentric spheres.[16]

Planets in Epicurus' heavens

Clearly, if the realm of planets was eternal and divine, no ordinary physical process could have ever been at work in establishing or maintaining it. Still one would look in vain for more than indirect hints on planetary evolution in the dicta of those who in the post-Aristotelian epoch advocated a cosmology in which the heavenly regions enjoyed no special status. They were the Stoics and Epicurus. The former believed in a finite world which contracted and expanded as the respective strength of the hot and cold alternated. It was in that process that the world, as pictured by Stoics, perished in periodic conflagrations only to be reborn again from the cosmic ashes.[17] The universe of Epicurus was infinite, like that of the atomists, in which there were infinitely numerous 'worlds'. These latter he defined as 'a circumscribed portion of sky, containing heavenly bodies and an earth and all the heavenly phenomena, whose dissolution will cause within it to fall into confusion: it is a piece cut off from the infinite and ends in a boundary either rare or dense, either revolving or stationary: its outline may be spherical or three-cornered, or any kind of shape'.[18] As the passage shows, Epicurus was no champion of world formations by large-scale vortices. He rejected as false the notion that a large vortex in big, empty space was a chief and typical means to produce a 'world': 'For it is not merely necessary for a gathering of atoms to take place, nor indeed for a whirl and nothing more to be set in motion, as is supposed, by necessity, in an empty space in which it is possible for a world to come into being, nor can the world go on increasing until it collides with another world, as one of the so-called physical philosophers ($\varphi v\sigma\iota\kappa\delta\varsigma$) says'. To this he added emphatically: 'For this is contradiction of phenomena'.[19] The real meaning of this last phrase will become clear shortly. Reference to vortex was absent in the kind of world-formation which Epicurus submitted as the correct one. It was based on the chance event when 'seeds of the right kind have rushed in from a single world or interworld, or from several: little by little they make junctions and articulations, and cause changes of position to another place, as it may happen, and produce irrigations of the appropriate matter until the period of completion and stability, which lasts as long as the underlying foundations are capable of receiving additions'.[20] The spherical shape of our heavens (world) seemed to him to be a special case and with a touch of inconsistency he traced it to the circumstance that a vortical motion was *assigned* to the stars within our world.[21]

This phrasing, almost indicating an outside or general cause, was hardly in line with Epicurus' assertion that every formation was due to the random swerving and whirling ($\delta\iota\nu\eta\sigma\iota\varsigma$) of atoms.[22] It was the collisions and combinations of whirling atoms which, according to him, resulted in the formation of larger bodies, including the sun, the moon and the stars. By the latter he most likely meant the planets as well. But if the sun and the moon were exactly as large as they appeared to be, to recall Epicurus' astonishing claim,[23] then how big did he picture the planets? Probably as small as peas, for in the scientific method of Epicurus, who argued with marvellous similes the existence of non-observable atoms, observation did not play a systematically critical role. The universe of Epicurus was incompatible with the regularity, quantitative order, and uniform evolution either on the microscopic or on the macroscopic level. He decried the 'slavish artifices of astronomers',[24] and their 'foolish notions',[25] because they implied uniform, cosmic causality, which in his eyes constituted the principal threat to human freedom and happiness.[26] A hundred years after Socrates the problem of reconciling science with human purpose was still begging a satisfactory solution. While Socrates nipped in the bud some good scientific starts by overemphasising the purposeful over the quantitative, Epicurus vitiated the scientific enterprise by discarding the notion of universally valid laws and patterns. With that he also thwarted the possible utilisation of his atomistic world view for speculation on planetary origins.

Most of these features and foibles were echoed in Lucretius' *De rerum natura*, the great cosmological poem celebrating Epicurus' ideas. The 'beginning' of a particular world consisted, according to Lucretius, in a 'sort of strange hurly-burly'.[27] Epicurus was again echoed when Lucretius referred to the innate desire of things to join with their like, leading to further differentiation. In the process the heavy fragments squeezed out the light ones, in particular the ethereal particles, causing the formation of the spherical skies: 'In this way therefore at that time the light and expansive ether, with coherent body bent around on all sides, and expanded widely on all sides in every direction, thus fenced in all the rest with greedy embrace. This was followed by the beginnings of sun and moon, whose globes revolve in the air between the two [the earth and the heavenly sphere]'.[28] Lucretius spoke of the vortex of heavens (turbo coeli) not in connection with the evolution of this or that heavens or world, but in connection with the varied motions of celestial bodies. Like Epicurus, he insisted that there was no uniform explanation for these differences. Yet he also recalled the 'venerable judgment of that great man Democritus', according to whom the force of the celestial vortex increased with distance from the earth and thus the sun was constantly falling behind the motion of the signs of the zodiac.[29] In addition to the sun and the moon, Lucretius mentioned as heavenly bodies only the 'bright constellations' and the 'shining stars of the universe'.[30] The planets as a separate class were not listed by him even when he named internal force, currents of air, and desire for food, as three possible explanations of the movement of the stars. Astronomical learning fared badly in the book of Lucretius who insisted that the wheel of the sun could not be much larger than seen by the naked eye.[31] In such an outlook there could hardly arise any special interest in the much smaller five wanderers of the sky.

If there was a saving grace in the cosmological dicta of Epicurus and Lucretius, it consisted in the absence of fantasy and astrology. What fantasy could do about planets was well illustrated in *Concerning the Face which Appears in the Orb of the Moon*, a work of Plutarch (fl. 100 A.D.). While he asserted the earthlike character of the moon, he also questioned the sphericity of the earth, gave the size of the moon as that of the Peloponnesus, and described the moon as the Elysian field of the virtuous souls and the earth as the place of punishment for those who had misbehaved on the moon.[32] Scientific details were completely lacking in *A True Story* of Lucian (fl. 160 A.D.) who acquainted his readers with the fantastic-looking inhabitants of the moon, the sun, the morning star, the dog star, and even of the Milky Way.[33]

Planets and the three faces of Ptolemy

As to astrology, its rank abuse of the realm of planets was codified in Ptolemy's *Tetrabiblos*, which found an eager readership not only among the Arabs and the Medievals but also among the moderns of our times. In Ptolemy's account Saturn was a fearsome figure that 'caused long illnesses, consumptions, withering . . . rheumatism, and quartan fevers, exile, poverty, imprisonment, mourning, fears and deaths, especially among those advanced in age'.[34] This last proviso spoke for itself. Saturn was the harbinger of similar ills for dumb animals and the cause of all kinds of bad weather. Ptolemy credited Jupiter with good things, both material and spiritual. He traced to Mars 'wars, civil faction, capture, enslavement, uprisings, the wrath of leaders, and sudden deaths arising from such causes; moreover, fevers, tertian agues, raising of blood, swift and violent deaths, especially in the prime of life; similarly, violence, assaults, lawlessness, arson and murder, robbery and piracy'.[35] Mars also brought about sudden shipwrecks, foul air, and devastation by locusts. With Venus it was all sunshine. She 'brings about results similar to those of Jupiter, but with the addition of a certain agreeable quality',[36] which Ptolemy relished giving in detail. To the most insignificant of planets, Mercury, was left the role of doing all that the others did. When in conjunction, say, with Venus and Mars, Mercury had to be both benevolent and destructive, a feat that was not beyond the range of possibilities in that realm of sheer fantasies, astrology.

Ptolemy is more willingly remembered as the codifier of the specifically Greek approach toward the science of the heavens. That science was based on the principle that the task of astronomy was restricted to the construction of geometrical devices which enabled one to predict the position of the sun, the moon, and of the planets, that is, 'to save the phenomena'.[37] This programme meant the elimination of questions bearing on the physical causes of the motion of heavenly bodies, and on their physical origination and evolution into one system. In his Almagest,[38] the classic epitome of the analysis of planetary motions for well over a thousand years, Ptolemy proved himself the supreme master of epicycles, extants, eccentrics, and deferents. The predictive value of Ptolemy's system, not surpassed by even that of Copernicus, reflected in full force the power of geometry. Yet for Ptolemy and for the great majority of ancient Greek astronomers those geometrical devices had nothing to do with the heavenly reality. As a result, there is nothing in the Almagest about the dynamics of planetary motions. Ptolemy touched on this topic in a much smaller work, Planetary Hypotheses, in which he compared the co-ordinated motion of planets to that of a group of dancers and to a unit of soldiers drilling with their weapons.[39] In line with the Socratic-Aristotelian tradition, Ptolemy thought of the planets as permeated by some sense of purpose so that they could trace out unerringly their intricate loops in the sky. In such an outlook it had to appear simply stupid and sacrilegious to place the earth, that huge inert lump of matter, among the planets. Such was Ptolemy's unsparing stricture of Aristarchus of Samos (fl. B.C. 250) who put the sun in the very centre of planetary orbits and let the earth move around the sun with the rest of the 'divine' planets.[40]

About the Planetary Hypotheses it was discovered only very recently[41] that its full text contained an evaluation of planetary distances, which from the 9th century on was inserted with increasing frequency in Arabic, Jewish, and Latin astronomical and cosmological texts. In the *Almagest* Ptolemy calculated only the distances of the moon and of the sun.[42] His calculations were in part based on the evaluation of the earth's radius and circumference by Eratosthenes during the closing decades of the 3rd century B.C. But long before Eratosthenes, Aristotle had already referred to geometrical methods which showed that the sun was much larger than the earth.[43] The methods in question are not known, but it is certain that geometry must have been at a high stage of development in Aristotle's days, because a generation later Euclid was able to systematise an impressive body of geometrical theorems. That geometry was a most powerful beacon in showing some basic features of the system of planets must have appeared even more convincing when Aristarchus of Samos devised an ingenious method of measuring the relative and absolute distances of the sun and the moon.[44] Although his calculation of about 360 earth-radii for the sun's distance fell short of the true value of about 23,400 earth-radii by a factor of about 65, it certainly gave some taste of the vastness of the planetary system.

This vastness took on a direct concreteness in the *Planetary Hypotheses*, where Ptolemy illustrated with numerical data for all planets the meaning of what he called 'the nesting of the spheres'.[45] This meant that the interior radius of a spherical shell, whose width was given by the difference of the greatest and smallest distance of a planet from the earth, was taken to be equal to the exterior radius of the spherical shell within which the planet immediately below the former planet moved around the earth. Ptolemy calculated the distances for Mercury and Venus by starting from the distance of the moon, which together with the distance of the sun could be directly evaluated. The distances of Mars, Jupiter, and Saturn were based on the greatest distance of the sun. As Ptolemy admitted, the procedure gave no explanation of the fact that the greatest distance of Venus was shorter by almost 100 earth-radii than the smallest distance of the sun, or 1,160 earth-radii. He should have rather marvelled that the difference was only so small! As a firm advocate of geocentrism, Ptolemy found nothing wrong with the case of Mars whose greatest distance in the geocentric system was seven times larger than its smallest distance. Within the framework of the 'nesting of the spheres' Ptolemy naturally assumed that the spherical shell of the orbit of Saturn was adjacent to the sphere of the fixed stars. 'In short', Ptolemy summed up his reasoning and calculations, 'taking the radius of the spherical surface of the earth and the water as the unit, the radius of the spherical surface which surrounds the air and the fire [the least distance of the moon] is 33, the radius of the lunar sphere is 64, the radius of Mercury's sphere is 166, the radius of Venus' sphere is 1,079, the radius of the solar sphere is 1,260, the radius of Mars' sphere is 8,820, the radius of Jupiter's sphere is 14,187, and the radius of Saturn's sphere is 19,865.'[46] The glitter of geometry was rarely more deceiving.

Medieval vision and Renaissance vagaries

During the 9th and 10th centuries several Arab astronomers made their own calculations of planetary distances on the basis of the Ptolemaic model of the 'nesting of the spheres'. One of them was al-Fargani, whose *Elements of Astron-*

omy became the chief source in this connection for the medievals. Medieval respect for al-Fargani's figures (64, 167, 1,020, 1,220, 8,876, 14,405, 20,110)[47] received its best documentation when they obtained a hallowed place in Dante's *Convivio*.[48] Whatever the admiration of the medievals for Greek science, their Christian faith in a wholly transcendental and personal Creator made them reject unhesitatingly the pantheistic Greek view of the divinity of the heavenly regions. It should, therefore, be of no surprise that this radically new outlook on the cosmos issued before long in some all-important insights. One of these was expressed with striking conciseness by Oresme, the learned bishop of Lisieux. While sharing the traditional concept of a spherical universe, he saw it as the product of the Creator who brought it out of nothing as a perfect mechanism, or clockwork, to be specific.[49] The universe of planets could, therefore, be pictured as persisting in its motion according to autonomous mechanical laws set by the Creator. Underlying this concept was not only the rudimentary notion of the conservation of momentum (impetus), so indispensable for future developments in physics, but, more importantly, a confidence that personal, human values had a meaningful assurance in a universe which was the handiwork of a personal God. Within that framework it then became ultimately possible to think of the world mechanism as something evolving along scientific (mechanistic) laws anchored in the rationality of the Creator, as will be seen in detail in the next chapter.

Another crucial contribution of the medieval reform of the Greek world picture is best seen in the *De docta ignorantia* (1440) of Nicholas of Cusa, undoubtedly the most influential work for cosmological thinking until the early decades of the 17th century.[50] Prompted by his Christian faith in the personal and transcendental Creator, the future Cardinal made a resolute departure from the Greek dichotomy between a divine or superlunary and an ordinary or terrestrial region by treating the earth, the planets, and the stars on equal footing. In view of the infinite power of the Creator, he also argued that the extent of the universe should be indefinitely large. He likened it to a sphere, the circumference of which was nowhere and the centre everywhere, a proposition out of which he explicitly inferred the relativity of position and motion.[51] He also struck a very modern note as he insisted that the universe should appear much the same when viewed from widely distant points. He was less fortunate in some particulars, though his resolve to remain consistent with his basic postulates should be an object of admiration. He attributed to *all* cosmic bodies the same three-layered structure, namely, an earthlike core surrounded by the sphere of water and air, which in turn was enveloped by a shell of fire that radiated its heat and light only outward. His motivation for making this particular proviso was also of a piece with his belief in the Creator whose infinite power implied, according to him, an immense number of abodes for intelligent beings. Such was a bold vision, which demanded that the earth be placed within the moon's air-water layer,[52] but which, with the idea of a personal Creator in the immediate background, still exuded a natural trust in human purpose and values, the immensity of the universe notwithstanding.

This trust began to dissolve into vague verbalisations as soon as the idea of an infinite universe with infinitely numerous abodes for life was proposed with a distinctly pantheistic touch. The first to be mentioned is the *Zodiacus vitae* written by Marcellus (Stellatus) Palingenius in 1534.[53] The Renaissance magus best remembered in this respect is, of course, Giordano Bruno. His scientifically valuable scattered dicta owed much to his reading of Nicholas of Cusa, but he certainly did not share the latter's admiration for geometry. Bruno's *La cena de le ceneri* (1584), the first programmatic defence of the Copernican system, is a classic example of the obscurantist distrust in what Bruno called 'the file of geometry'.[54] In Bruno's pantheistic universe, subject to perennial cyclic transformation, stars were the divine and eternal entities,[55] a tenet which left Bruno's dicta on planets somewhat unclear. Within each star (and planet) everything had to become everything, and in order to secure physical or scientific support to this contention Bruno distorted beyond recognition two of the four motions which Copernicus assigned to the earth. According to Bruno, the earth not only performed a daily rotation and an annual orbiting, but also pitched, yawed, and rolled in some arcane ways so as to permit each of its parts to occupy successively the position of any other part. While this served perfectly Bruno's pantheism, its 'evolutionary' character lacked even the veneer of science. It certainly prompted him to no speculation about the evolution of planetary systems, although he claimed to know, in contrast to Copernicus, the physical explanation of celestial processes.[56] The vision Bruno offered about infinite realms of stars and planets was a morbid one if it is to be judged in its own context and not through the distorting lenses of selective retrospect. In view of that context it should appear highly ironical that Bruno strictured Copernicus for not seeing beyond the solar system.[57]

Unlike Bruno, Copernicus made painstaking efforts to come to grips with the geometrical or quantitative problems of the true ordering of planets. Strengthened by his faith in geometry, the work-pattern of the Supreme Artificer,[58] Copernicus made but feeble attempts to offer a physical explanation for the physical problems posed by the motion of the earth. Still his contribution to mastering fundamental prerequisites for a meaningful discussion of planetary origins was crucial. Recognition of the earth's motion could not ultimately fail to focus attention on questions of planetary dynamics. Also, Copernicus made it clear that from the heliocentric revolution of planets one could derive their relative distances in terms of the distance of the earth from the sun. He not only spoke of the 'wonderful commensurability' in the universe and of the 'sure bond of harmony for the movement and magnitude of the orbital circles'.[59] He also gave numerical values for the distances of Mercury, Venus, Mars, Jupiter, and Saturn in terms of the earth-sun distance. His data were in a remarkably close agreement with the modern data given in parentheses: 0.37 (0.38), 0.72 (0.72), 1.52 (1.52), 5.22 (5.20), 9.17 (9.54).

Keplerian obsession: geometry as the womb of planets

That the foregoing sequence might conceal a specific progression remained hidden even to a Kepler talented though he was to spot pleasing correlations in masses of apparently unconnected numerical values. The frustration in this connection can particularly be felt on studying the details of that famed system of the five perfect solids in which he believed to have found the *raison d'être* of the cosmic system, or the system of planets. He told it in great detail in his *Mysterium cosmographicum*,[60] the last explicit and extensive effort to understand the origins of the planetary system along lines drawn two thousand years earlier by Plato. Kepler was also the first and the last major scientist who revealed in minute details the psychological and motivational framework in which he arrived at his great discoveries. Of these he considered the most pivotal the model of the planetary system set forth in the *Mysterium cosmographicum*, although he was not reluctant to point out its countless errors as he sent to press its second edition.[61] He had, however, never come to doubt his conviction that the truth of the planetary system and of the physical world must be a geometrical truth.

The planetary system which Kepler championed was, of course, the Copernican one, and Kepler became its first outspoken and convincing advocate by devoting the first chapter of his book to a defence of the advantages of the heliocentric arrangement of the planets over the Ptolemaic one. As Copernicus' ideas were still a brash innovation half a century after the publication of his *De revolutionibus*, Kepler felt the need to give, as he put it, 'a novel foundation to Copernicus' hypothesis about the new world'.[62] What was really novel in the new explanation given by Kepler was the unbounded trust by which he used geometry as the common idiom to speak about God and cosmos. Many before Kepler praised the Pythagoreans' faith in numbers and quoted Plato's dictum that God is a geometer. It is quite possible that some before Kepler raised the question of why there were only six planets and not three, twenty, or a hundred. After all, the progression given in the *Timaeus* seemed to imply that there could only be six measures between the sphere of the fixed stars and its centre, the earth.

When in the summer of 1595 Kepler was seized with his vision, Jean Bodin's (1530–96) *Universae naturae theatrum*, just to be printed, carried a speculation on the 576 earth-diameters which were believed to separate the earth from the sun, in terms of the five perfect solids. 'It seems to us', he wrote in a chapter on the celestial bodies and their number, motion, and size, 'that the number 576, or the square of 24, expresses not only the distance of the sun from the earth but also the [sum of] orthogons that are in the perfect solids; for the tetrahedron contains 24, the cube as many, the octahedron 48, the dodecahedron 360, the icosahedron 120, which altogether make 576. Again, the distance from the earth to the moon, which is 16 earth-diameters or the square of 4, contains, I believe, other measures as we see that the Creator in his incredible wisdom determined everything with most certain numbers, weights, and measures'.[63]

Bodin was only one of the many who in the late 16th century engaged in such numerological speculations. The material of his book was not original. His field was political science, economics, and religious philosophy, not geometry and astronomy. Kepler undoubtedly knew about such numerological speculations that were more obscurantist than enlightening. That he nevertheless produced a highly appealing speculation was due a great deal to the fact that for him admiration for geometry was not merely a fashionable pastime, but a total commitment to its truthfulness about physical reality. Most pertinent evidence of this comes from a letter of his to his former teacher in Tubingen, Michael Maestlin, who a year or so earlier saw through print Kepler's manuscript: 'May God make it happen that my happy speculation (the *Mysterium cosmographicum*) should fully exert everywhere among reasonable men the effect which I tried to achieve by the publication, namely, that belief in the creation of the world be strengthened through this incidental aid, that the Creator's thought be recognised in its very nature, and that his inexhaustible wisdom may shine each day more brightly. Then will man at long last measure the power of his mind on the true scale and recognise that God has created in terms of quantities everything in the whole world, including the human mind, so that it may comprehend all such things. For just as the eye is created for colours, and the ear for sounds, so man's mind is fashioned not for the understanding of anything but for the understanding of quantities. The mind grasps a thing all the more correctly, the closer it is to sheer quantities, that is, to its origin. The farther is anything from quantities, the greater is its share of darkness and errors'.[64]

On the face of it there were but quantities in Kepler's model of the size and structure of the universe, or the system of the planets. Its origins went back, as Kepler himself recorded it,[65] to his immediate fascination with Copernicus' theory to which he was introduced by Maestlin soon after he had matriculated in Tubingen. Right there he felt the urge to 'establish with physical, or rather metaphysical demonstrations',[66] what Copernicus proved mathematically. In particular he wanted to show 'metaphysically', that is, to derive *a priori*, 'the number, size, and periods of the planetary orbits'.[67] But, as he recalled it, during his years in Tubingen the ambitious programme did not go beyond a public defence by him of the truth of Copernicus' views.

His first academic year in Gratz, where he took in April 1594 the post of the mathematician of the province, left him with no time to concentrate on his favourite project. With the summer of 1595 the long awaited opportunity had finally come. First he tried to fit planetary distances and periods to arithmetic progressions. He got nowhere. Then he assumed the existence of a planet between Jupiter and Mars and of another between Venus and Mercury, thinking wishfully that they were too small to be seen. This trick yielded some regular sequence but also implied patent arbitrariness with the actual world. As he noted, 'anyone who discusses the fashioning of the world should not derive his arguments from numbers which obtained some dignity only from things posterior to the creation of the world'.[68] The number six, standing for the orbits of the six then known planets, seemed to have precisely such a preeminence over the number eight and, therefore, the latter was not to be relied upon. Even in its most euphoric flights Kepler's mind was to keep a subtle tie with the actual world.

As the days of summer went by, so grew his frustration. But on the 19th of July fortune struck, which he devoutly ascribed to Providence. As he was giving a lecture on the great conjunctions of Saturn and Jupiter he marked the places of the successive conjunctions on a circle representing the zodiac. Since these points are separated from one another not by 120 but by about 111 degrees, lines connecting the points of three successive conjunctions do not form a closed triangle. Actually, one must draw forty triangles, each starting 120 minus 111 or 9 degrees behind the previous one, to reach once more the original starting point. Kepler duly completed the series of those forty 'triangles' which looked like a net with a large circular hole in the middle [*Illustration I*]. He immediately saw that the circular hole corresponded to a circle inscribed into an equilateral triangle, while the circle representing the zodiac was a circle circumscribed to that triangle. This meant that the radius of the inscribed circle was half of the larger circle. All of a sudden Kepler's deepest mental interest was drawn to the forefront of his consciousness. Since he was lecturing on Saturn and Jupiter, the vision of the two circles could not fail to remind him of a similar proportion between the orbits of the two planets. From his mind steeped in geometrical mysticism, the primacy of triangle among all geometrical figures was evoked with natural ease. And so was he accustomed to think of Saturn and Jupiter as being the primary planets. 'Immediately', he recalled, 'I tried to fix with a square [and a circle inscribed to it] a second distance, that between Mars and Jupiter, a third with a pentagon, and a fourth with a hexagon. . . . But it would be endless to mention all the details'.[69]

It is not clear from his account whether he did this 'immediately' by interrupting his lecture. At any rate, he soon had to realise that hexagons could be followed by heptagons, octagons, and so forth to no end. From plane geometry he could not get a set of figures that admitted only six numbers, corresponding to the six planetary distances. At this crucial juncture it was the obvious three-dimensionality of the world of planets that led him out of the impasse. He saw, again 'immediately', that in the realm of solid bodies there were only five perfect ones, that is, bounded with identical sides. These five, when arranged concentrically and a sphere between each, could yield the six and only six specific distances. What could indeed be more natural than to do in three demensions what he had already tried, but in vain, on a plane?

His was on that day an amazing rash of fortune. The five perfect solids could be arranged in 120 different sequences, but he picked at the first try the one which alone yielded a good approximation of the planetary distances given by Copernicus. To the sphere representing the earth's orbit he circumscribed a dodecahedron; the sphere enveloping this was the orbit of Mars. Around it he fitted a tetrahedron, with a circumscribed sphere representing Jupiter's orbit. Over the sphere of Jupiter came a cube with a circumscribed sphere giving the orbit of Saturn. The sphere of the earth circumscribed an icosahedron, inscribed into which was the sphere of Venus' orbit. This in turn enveloped an octahedron with its inscribed sphere standing for the orbit of Mercury [*Illustration II*].

This model of the world pivoted on the planetary orbits Kepler unabashedly offered as truth itself. According to him, God Himself had to conceive the world along the perfect lines of perfect solids, because His nature admitted only perfection. Consequently, beside the route taken by Copernicus, who argued from the phenomena and 'like a blind man leaned on a stick', there was another, far more excellent avenue, the *a priori* reasoning from the perfection of geometry and metaphysics.[70] Nothing shows more tellingly Kepler's confidence on this score than the manner in which he was to criticise his very hero, the canon from Frauenburg. The perfect model had to fit reality in a perfect manner and Kepler boldly went about tailoring reality to the *a priori* model. Hurdles presented themselves as soon as he considered the fact that in Copernicus' system the deferent circles of planetary orbits were not concentric. Thus instead of ideally thin spheres he had to adopt spherical shells of various thickness. It remained to be seen whether the spacing left between the shells would still accommodate the perfect bodies. For the tetrahedron, dodecahedron, and icosahedron, generating the orbits of Mars, earth, and Venus, the data of Copernicus closely fitted the theory. But for the cube and octahedron generating the orbits of Jupiter and Mercury the facts clashed with its *a priori* verities. The real nature of *a priori* reasoning now gave itself away. With patent arbitrariness Kepler dismissed the difficulty about Jupiter with the remark that its orbit was too far to make the discrepancy troublesome for our perception.[71] As to Mercury, he replaced the octahedron with a cube, a step which struck at the very foundation of his own *a priori* reasoning. Kepler was on more secure ground as he called attention to the unreliability of some of the data that found their way into the various planetary tables. He had, however, no justification for accusing Copernicus of deliberately altering some data in connection with the motion of Mercury[72] to fit facts to theory. The shoe was on the other foot.

Once the *a priori* derivation of the number and distances of planets was vindicated by talking away the difficulties, Kepler could turn to the third point which he had wanted for several years to clear up about the planets. It consisted in showing that, for some reason, they could not have periods other than they actually had. Formulating the problem was a gem of originality. The solution he gave fell far short of the *a priori* and geometrical one he hoped to formulate. He said nothing new by recalling that in the Copernican ordering of planets their periods increased with distance from the sun. Originality was, however, in evidence as Kepler looked for a physical explanation of this regularity. That the traditional 'intelligences' moving each planet grew more 'imbecile' with increased distance from the sun was a possibility that did not appeal to Kepler. He preferred the weakening of a single force emanating from the sun.[73] This was a halting start in his speculations in celestial dynamics where, as will be seen shortly, he did not reason along purely geometrical patterns. He never seemed to realise what potentialities lay for celestial dynamics in his felicitous reasoning, published in 1604, that the intensity of light decreased with the square of distance from a point source.[74]

By 1604 he had already been for three years the successor of Tycho Brahe in the post of imperial mathematician in Prague, and also heir to his observations on planetary motions that far exceeded in precision anything of the kind. Sending a copy of the *Mysterium cosmographicum* to Tycho was very instrumental in securing for Kepler an assistantship with him, as Tycho highly enjoyed Kepler's mystical bent of mind and *a priori* speculations. Kepler's *a priori* derivation of the system of planets was also praised unreservedly by Maestlin before the Tubingen Senate,[75] while Georg Limnaeus, astronomer in Jena, congratulated Kepler for resurrecting the old Platonic method of reasoning about nature.[76] Galileo did not deign to comment upon a work of which he requested two copies from Kepler, in spite of the fact that Kepler begged him to express his views on it in a 'very long letter'.[77] As to Tycho, he was also a hardy observer who brooked no fiddling with well established data of observation. Kepler must have sensed this, for already before he arrived in Prague in early 1600 the distant influence of Tycho, the uncompromising observer, began to tell on him.

Kepler's letters from 1599 reveal a subtle apprehension that Tycho's data may play havoc with the planetary system of the *Mysterium cosmographicum*. On the surface everything was exuberant confidence. His words—'Come Tycho and be not late!'—were a paraphrase of Isaiah's longing for the presence of the Messiah. He wrote this in his letter of February 26 to Maestlin noting that Tycho's data prompted him to have no doubts about the truthfulness of the world picture based on the five perfect solids. Kepler also boasted about his plan to wrest Tycho's riches from him and with a sense of superiority he described Tycho as the immensely rich man who did not know what to make of his own wealth.[78] In April, in a letter to Herwart von Hohenburg he praised geometry and arithmetic as the only information that man could truthfully apprehend, because the human mind was created by God for precisely such a purpose. In the same letter he urged that 'everyone should fall silent and listen to Tycho, who had already spent thirty-five years in observations and who saw more with his eyes than many others with the acumen of their minds'.[79] Was Kepler beginning to have reservations about the infallibility of his own intuition? In May, in a long letter to Herwart von Hohenburg, he criticised Reimarus Ursus (von Baer), Tycho's predecessor in Prague, for ignoring the limits of the usefulness of geometry in astronomy: 'In geometrical matters our knowledge is certain, or as you well know, there you either know an axiom or you know that you don't know it. But in astronomy there is room for opinions, as far as hypotheses are concerned. In geometry nobody's invention is rejected, but in astronomy some ought to be dismissed. Therefore it happens that he who trusts [too much] his own ingenuity by which he explores geometry, may readily reject and neglect all astronomers and this is what Ursus seems to do'.[80] Kepler failed to realise that he himself was soon to outgrow, and rather painfully, somewhat similar illusions. This meant that an affirmative answer was called for by the incredulous question which he let slip three months earlier into his letter to Maestlin in connection with his *Mysterium cosmographicum*: 'Am I paying the penalty for youthful temerity and affectated pompousness?'[81]

Years of sobering up

He was to pay heavily, though indirectly, as if he were to undergo over the next years a kind of lobotomy in small steps. He did not seem to notice that he was shedding, slowly but surely, that most cherished part of his mental world, the belief in the possibility of an *a priori* genesis of the features of the universe, or the system of planets. How could he notice it indeed! Soon after his arrival in Tycho's castle in Benatek, near Prague, he was given the task to recompute the orbit of Mars on the basis of Tycho's data. In no time he got so obsessed with them as to go almost out of his mind. Well, he was given to eccentric statements. On getting the assignment he first boasted that he would do it in eight days. When eight long years later he presented to the public the fruit of his labours, the *Astronomia nova* on the motion of Mars, the new book resembled only in one respect the *Mysterium cosmographicum*. In both works Kepler gave an ample glimpse to his reader of the mental labyrinths he had to explore in order to reach an unexpected goal.

Compared with the *Mysterium cosmographicum* the *Astronomia nova* must have appeared far less ambitious. Instead of the whole world it was now the question of the orbit of one single planet. Kepler could have stated and proved over two dozen quarto pages that the orbit was an ellipse (the Second Law), and that the radius vector swept out equal areas in equal times as the planet advanced in its orbit (the First Law). Happily for posterity he recalled all his conjectures, mistakes, erroneous beliefs and lucky guesses scattered over 337 pages. He got the First Law through the combination of two errors that almost miraculously cancelled one another. As to the Second Law, only a chance glimpse at his trigonometrical tables led him to perceive that the pattern chosen by nature was an ellipse.

That nature was choosing her own patterns regardless of the subjective preferences of the speculative mind was a recognition which represented a watershed[82] not only in Kepler's mental development but also in intellectual history. The data of observation could no longer be manipulated, however cleverly or surreptitiously, to fit *a priori* theories. Kepler himself was fully aware that the triggering device toward entirely new vistas was an observation of Tycho. It differed by eight minutes of an arc from the theoretical edifice which he first built for the explanation of the motion of Mars on the basis of the axiom shared by all earlier astronomers that the basic form of a planetary orbit must be a circle. 'Those eight minutes', Kepler wrote, 'showed the road toward a complete reformation of astronomy'.[83] For this to happen a price was to be paid, as Kepler put it. The ancient edifice, which he wanted to put on perfect, *a priori*, and computational foundation, had to be discarded.[84]

The ancient edifice was in an immediate sense the allegedly circular orbit of Mars and of other planets. In a deeper sense it meant the construct of the *Mysterium cosmographicum* with its thin spherical shells tucked among the five perfect solids. Its renunciation meant the shattering of hopes in an *a priori* codification of the shape, size, and periods of the planetary system, or its abstract genesis. Characteristically, the *Astronomia nova* contained but few references to the *Mysterium cosmographicum*. This is not to suggest that Kepler was forgetting about it. As will be seen shortly, it remained for the rest of his life his favourite brainchild. Still he could not help feeling that the *Mysterium cosmographicum* had no natural place in a treatise where he first saw geometry without a divine glitter. 'It must be recognised', he wrote, 'that even in patterns which appear to the mind as the most reliable ones of all possibilities, there seems to be contained some geometrical uncertainty'.[85] The admission came at the end of his lengthy efforts to give a 'physical' explanation of the orbiting of planets around the sun.

The contrast could not have been greater with the encomiums which opened the *Mysterium cosmographicum* on the divine truth of geometry in general and of its one-to-one correspondence with the structure of human mind and of physical reality in particular. The *Astronomia nova* opened with a detailed discussion of the point that Sacred Scripture did not contradict the physical truth of the earth's motion. Groping for physical truth was the keynote of the work's full title which announced in large Greek letters an 'inquiry into causes' (ΑΙΤΙΟΛΟΓΗΤΟΣ) leading to a 'celestial physics'. But if the physics was 'celestial' it was such only in the sense that Kepler postulated a 'non-material substance' (species immateriata) emanating with infinite speed and in straight lines from the sun and rotating with it.[86] It was that structure, resembling the spokes of a wheel, which pushed around the planets that were tied to it by the magnetic fibres permeating them. Kepler now could argue that the strength (density) of the magnetic spokes of the sun diminished with distance and that the farther a planet was from the sun, the slower its orbital motion had to be. Yet, although he insisted in the same context that the diffusion of the sun's magnetic force was, like everything else in the physical world, subject to the laws of geometry, and although he referred to his law of the diminution of the intensity of light from a point source, he failed to state that the dynamic influence of the sun on the planets diminished with the square of distance.[87]

Compared with this enormous misfortune for the development of true celestial physics, of small value should appear the interesting hypothesis which Kepler offered as explanation of the slower motion of planets at aphelion than at perihelion. According to him the planet kept its magnetic axis always perpendicular to the line joining the perihelion and aphelion points of its orbit.[88] Consequently, at these points the two magnetic poles of the planet were at equal distance from the sun, and thus there was no net magnetic force along the orbit either to decelerate or to accelerate the planet. He also compared [*Illustration III*] the process to the action of a rudder,[89] but his discourse, burdened with many arbitrary elements, carried no conviction. Peter Crueger, astronomer in Danzig, chided him for introducing 'strange speculations, which belong to physics not to astronomy'.[90] Neither Crueger, who at least took note of the elliptical orbit of planets, nor others seemed to perceive the possible significance of the phrase in which Kepler summed up, in writing the Introduction to the *Astronomia nova*, his physical explanation of the motion of planets. The phrase referred to the rotation, 'in analogy of an extremely swift vortex',[91] of that non-material substance emanating in straight lines from the sun which itself was rotating due to its magnetism. It was only decades later that the notion of vortex became the catchword of intelligibility in science and the starting point of physical theories on planetary origins. By then Kepler had already departed from the scene without ever suspecting of having struck the keynote of the immediate future. He remained straddled on the watershed which he was the first to reach. What kept him there was not only the lure of Aristotelian animism but also the glitter of geometry.

Moments of relapse

Kepler's erstwhile obsession with an *a priori* use of geometry in astronomy was a mental weakness the curse of which was not without some momentary relapses. The most significant of these was touched off when on April 8, 1610, Kepler received a little book, the *Sidereus nuncius* or *Starry Messenger*,[92] which Galileo had just published on his startling discoveries made with the telescope. The details of that relapse were jotted down by Kepler himself in the book-length commentary which he penned in great haste during the latter half of that month. In it he tells us that those discoveries were already reported to him in substance in mid-March.[93] Reading the book itself made it clear to him that the new stars seen by Galileo were not stars but satellites revolving around Jupiter. The presence of new moons around a planet meant an immediate end to the quiet enjoyment in which Kepler spent the months following the publication of his *Astronomia nova*. All of a sudden his favourite brainchild, the *Mysterium cosmographicum*, came back to the centre of his attention as something to be defended. The chief lesson of the *Astronomia nova*, the arduous tracing out of an elliptical orbit for Mars, did not kill all the strains of Kepler's attachment to a planetary world built around the five perfect solids.

The new challenge to the old dream was twofold. One was real, the other hypothetical. The former concerned the task to accommodate the newly sighted moons of Jupiter in the spacing provided by the model he set forth in the *Mysterium cosmographicum*. Kepler not only felt that this could be readily done, but he also claimed that the same model afforded room for a moon around Mercury and Venus, for two around Mars and for six or eight around Saturn.[94] He made no secret of his hopes that if the world of planets were generated by the five perfect solids the number of moons around each planet had to show some arithmetic progression. Some old dreams again got the best of him. Only a dreamer could now claim that Pythagoras, Plato, and Euclid connected the world of planets with the five solids. He now ascribed not only to himself but also to Copernicus the discovery of the proper sequence of the five solids.[95] Only a dreamer forgetful of hard-learned lessons, such as the elliptical orbit of Mars, could now describe himself as one of those thinkers 'who intellectually grasp the causes of phenomena, before these are revealed to the senses', and who 'resemble the Creator more closely than those others, who speculate about the causes after the phenomena have been seen'.[96]

The hypothetical challenge concerned the possibility of planets around each and every star. This was the challenge of Bruno's infinitely numerous worlds which he deeply detested and feared. His first remark retained its soundness and validity even to the day: 'No moons have yet been seen revolving around them. Hence this will remain an open question until this phenomenon too is detected by someone equipped for marvellously refined observations'.[97] But almost immediately he also tried to vanquish the challenge on the basis of the *a priori* arrangement of the five solids. Planets around other stars either followed that perfect pattern or not. If they did, then there were infinitely numerous, useless duplications in the universe to the smallest details. In an infinitely large number of planets there would be, to quote Kepler's words, 'as many Galileos observing new stars in new worlds as there are worlds'.[98] If the arrangement of planets around other stars differed from ours then they formed systems with an infinite variety of imperfections including their denizens. Kepler did not allow even the moons of Jupiter to be pictured as a small-scale replica of the planetary system based on the five perfect solids: 'Let the Jovian creatures, therefore, have something with which to console themselves. Let them even have, if it seems right, their own four planets arranged in conformity with a group of three rhombic solids'.[99] As to the existence of denizens, inferior to humans, on Jupiter and other planets, Kepler had no doubt. Nor did he doubt that once the art of flying had been mastered, 'settlers from our species of man will not be lacking' on other planets and their moons.[100]

In the midst of this rapture of *a priori* speculations the great and hard work on Mars receded to the background. Its elliptic orbit was not mentioned at all as Kepler referred to his 'revising the orbits and motions of Mars, Earth, and Venus on the basis of Brahe's observations'.[101] This was all the more revealing as at the very outset of his comments on Galileo's book, Kepler mentioned how he was thinking in early March about making his book on the motion of Mars known to Galileo, with whom his correspondence came to an end twelve years earlier. If Kepler could be so inattentive to the real gem in his book on Mars, why should Galileo have paid special attention to it? Galileo, who might have found powerful support in the *Astronomia nova* for his relentless criticism of the Aristotelian world picture and physics, went on to ignore the elliptical orbit of planets for the remaining thirty-two and most crucial years of his life. He never realised how ironical it was to defend the heliocentric ordering of planets and still retain that pivotal point of the Aristotelian system, the necessarily circular motion of everything in the heavens.

Galilean sidelights

Galileo had some justification in turning away with disgust from the often morbid and psychic style of Kepler. Unlike Kepler, Galileo had no wish to entertain flights of fancy about Jovians or other planetary beings. In his 'Third Letter on Sunspots', written in 1613, Galileo branded 'as false and damnable the view of those who would put inhabitants on Jupiter, Venus, Saturn and the moon, meaning by "inhabitants" animals like ours, and men in particular'.[102] He claimed he could prove that other planets were without inhabitants of any kind, though he added in the same breath: 'If we could believe with any probability that there were living beings and vegetables on the moon or any planet, different not only from terrestrial ones but remote from our wildest imaginings, I should for my part neither affirm it nor deny it, but should leave the decision to wiser men than I'.[103] By these he meant some high Church authorities who soon began to show their concern. In March 1615, Father Giovanni

Ciampoli, an official in the Curia and a friend of Galileo, referred in gentle advice to him not only to ecclesiastical apprehensions but also to the readiness to jump from investigations of the system of planets to the topic of their alleged inhabitants: 'Your opinion regarding the phenomena of light and shadow in the bright and dark spots of the moon creates some analogy between the lunar globe and earth; somebody expands on this, and says that you place human inhabitants on the moon, the next fellow starts to dispute how these can be descended from Adam, or how they can have come off Noah's ark, and many other extravagancès you never dreamed of'.[104]

On this point Galileo heeded the advice, as he was hardly in sympathy with the efforts of Bruno who put people everywhere in the universe. Galileo had no sympathy at all for the animistic views of either Bruno or Kepler. The latter came in for criticism in Galileo's *Dialogue concerning the Two Chief World Systems* for his animistic fantasies and not for praise about the elliptic orbit of Mars. It was explicitly mentioned in the résumé of the contents of the twenty-eighth of the seventy chapters in the *Astronomic nova*, drawn up no doubt for the benefit of those who did not wish to read the book itself. But even if Galileo had read at least the reference in the résumés to the orbit of Mars, it did not strike a responsive chord with him. As a result, there was to come in Galileo's *Dialogue* a futile page on how the planets obtained their circular orbits. It was sprinkled with references to the law of the free fall of bodies. Of his discovery of that law Galileo could be rightly proud. Different was the case with his claim that calculations fully vindicated his model of the mechanical evolution of the system of planets. He presented the model 'to illustrate a Platonic concept'.[105] It meant to show that all planets were created in one place, away from the centre, the sun, and by descending towards it to various distances, they acquired the different orbital velocities which they had. Galileo skirted the question of what turned the linear descent of each planet into an orbital motion, but he felt certain that 'one may determine, from the proportions of the two velocities of Jupiter and Saturn and from the distance between their orbits, and from the natural ratio of acceleration of natural motion, at what altitude and distance from the centre of their revolutions must have been the place from which they originally departed'.[106] He said the same about Mars, the earth, Venus, and Mercury, 'the size of whose orbits and the velocities of whose motions agree so closely with those given by the computations that the matter is truly wonderful'.[107]

The actual situation was far from being such. Galileo not only failed to offer anything specific about those calculations, but later in the *Dialogue* he did not recoil from using only half of the previously given value of the constant of acceleration in computing the time of fall from the moon to the earth.[108] At any rate, the orbits were elliptical, but of this Galileo could not learn anything from Kepler's little book in which the latter's chief contention was to show that Galileo's telescope embodied Keplerian theories on the combination of lenses. Kepler completed the manuscript of his comments on April 19 adding the next day a Postscript, which ended with a reference to a book on 'Harmonics' he planned to write.

Obsession revisited

When finally published in 1619, Kepler's *Harmonice mundi* represented a fantastic potpourri of mostly discarded speculations on various 'harmonious' relations among measures and proportions in the macrocosmos as well as the microcosmos. The 250 or so quarto pages of the book contained only one 'harmonious proportion' that turned out to be truly 'delightful' to men of science. It is known as Kepler's Third Law which states that the ratio of the squares of the periods of any two planets equals the ratio of the cubes of their mean distances from the sun.[109] Kepler stumbled on this all-important gem unexpectedly in the course of a search which no longer was dictated by a primitive, *a priori* confidence in the value of some specific geometrical pattern. The lessons of his arduous work on Mars did not fail, after all, to make a lasting impact on his mind. They did not, however, cure it completely. It was not the Third Law which gave him the greatest satisfaction for wasting so much energy and time in exploring in vain countless avenues for 'harmonies'. He waxed ecstatic over the harmonies which he found in the smallest and greatest values of angular velocities of each planet and of pairs of these. From these harmonies science was not to profit at all. In line with his customary exuberance Kepler declared that his great lifelong ambition had been achieved by unveiling some of the divine music which is the Creator's delight in his handiwork and which now can delight man as well.[110]

Kepler's own delight with cosmic 'harmonies' almost immediately found a striking expression as he published in 1622 the Fourth Book of his vast summary of the Copernican system, the *Epitome astronomiae copernicanae*.[111] The Book which proudly carried the subtitle, 'Celestial Physics', contained amplifications on the magnetic cause of planetary motion[112] together with a more systematic diagram of the process [*Illustration IV*]. The genesis of the planetary system was not, however, to be 'physical'. Although Kepler refrained from reproducing the famed diagram in the *Mysterium cosmographicum*, he had the figures of the five perfect solids printed individually to support his claim that the world, or the system of planets, was in substance at least generated by geometry. The additional factor was the 'harmonies' which, according to Kepler, remedied the discrepancy between the actual values of planetary distances and the ones predicted by his favourite geometrical model. In his words, 'the archetype of the movable world is constituted not only of the five regular [solid] figures – by which the chariots of the planets and the number of the courses were determined – but also of the harmonic proportions with which the courses themselves were attuned, as it were, to the idea of celestial music or of a harmonic concord of six voices'.[113] He left the determination of the actual numerical form of these harmonies to the 'freedom of the composer', that is, the Creator.[114]

Clearly, this freedom could not be constrained by human choice among various geometrical models, 'perfect' and attractive as they might have appeared. Kepler fully recognised this as he prepared, in 1621, a second edition of the *Mysterium cosmographicum* with copious notes almost matching in length the original.[115] The notes represented his 'Retractations'. His critical sense left the arguments of hardly any chapter untouched. He now knew that man could not trace the true contours of the universe out of his mind's recesses. He now knew that mystical contemplation of perfect bodies and sundry ratios would not give him a magic clue to the genesis of the world of the planets, to the question of why there could only be so many planets, at such and such distances and with periods they actually have. He no doubt said the truth that the *Mysterium cosmographicum* was the happiest, most significant book on the topic,[116] a great first in short. Paradoxically, no other book of Kepler received an even remotely similar warm reception. Contrary to what Kepler claimed in the Preface of the second edition, interest in the work was not due to the 'truth' of its contents. He himself put it better in his comments to the crucial Twenty-first Chapter on the discrepancy between observation and theory. Twenty-five years earlier, blinded by the glitter of geometry, he blandly talked away those discrepancies. Now he acknowledged that wherever there was a close agreement, it was merely fortuitous. Still he cherished all details in the book as pleasant memories. 'They remind me', he wrote, 'of my meandering, of my hitting many walls, as I approached through the darkness of ignorance the shining door of truth'.[117]

The last metaphor was Keplerian exaggeration. The three major gems in his works on astronomy lay in a vast field of errors, of irrelevant data and details, of mystical fantasies, of useless speculations, of morbid detours of self analysis, and, last but not least, of an organismic and animistic conception of the world and of its processes. This is why Mersenne strictured him repeatedly.[118] Mersenne had high hopes that experimental and quantitative science would serve as a powerful antidote against magic, cabbala, and astrology, the natural allies of an animistic world view. No trained astronomer, Mersenne failed to see the importance of Kepler's Three Laws. He had shown greater sympathies for tracing out the harmonious proportions in music as well as in the physical world, as shown in his last major work, the *Harmonie universelle* (1636–37). Its inspiration was as scientific as it was Pythagorean. But Mersenne, whose first great work was a massive teatise on creation, was fully aware that an *a priori* approach to nature clashed with the utter contingency of the world, implied in the belief in the Creator. It was in that sense that he turned the table on Kepler by arguing that the actual measures in the world, in particular the sizes of planets, their distances and the like, were the actualisation of only one of infinitely numerous possibilities. Since those data could not be deduced from any known law of nature, they were to be viewed as being contingent on the sovereign act of the Creator.[119] Mersenne's reasoning might in some specifics have gone too far,[120] but it is useful to remember that those data about the planets to say nothing about data concerning the universe as a whole, are still waiting for a scientific explanation.

When Mersenne's argument saw print in 1630, Kepler was already a largely forgotten figure caught up in the last turmoils of a life beset with endless reversals. It was almost a replay of Keplerian fate that death claimed in 1641, at the age of twenty-two, the young genius, Jeremiah Horrocks, before his lengthy defence of Keplerian astronomy, *Astronomia Kepleriana defensa et promota*, could be printed. Horrocks' work, published long after his death,[121] was conspicuous not only for his perspicacity of what was really valuable in Kepler's astronomical work. Its contents at the hands of Newton served later as an indispensable stepping stone to the formulation of the true law of gravitation and of celestial dynamics. In Horrocks' work conspicuous also was the absence of any reference to the *Mysterium cosmographicum*. This was all the more telling as the other major works of Kepler, including his *Tabulae rudolphinae*, were duly appraised in the lengthy Prolegomena which Horrocks wrote to his work. He seemed to have unerringly sensed that a chief achievement of Kepler's super-human labours was to discredit the glitter of geometry which turned out to be a false prospect for tracing out, as if by introspection, the genesis of the system of planets.

NOTES TO CHAPTER ONE

1. The chief appeal of Copernicus' work lay not in greater accuracy but in conceptual simplicity. Great progress in accuracy came only through Kepler and Newton. The best general discussion of Greek astronomy is still to be found in the more than half-century-old works of J. L. E. Dreyer, *History of the Planetary Systems from Thales to Kepler* (1905), reprinted under the title, *A History of Astronomy from Thales to Kepler*, with a Foreword by W. H. Stahl (New York: Dover, 1953), and of Sir Thomas Heath, *Aristarchus of Samos, the Ancient Copernicus: A History of Greek Astronomy to Aristarchus* (Oxford: Clarendon Press, 1913). Their usefulness has not been superseded by O. Neugebauer's monumental *A History of Ancient Mathematical Astronomy* (New York: Springer Verlag, 1975).

2. *Phaedo* 99b.

3. For an excellent reconstruction and documentation of Anaxagoras' speculations, see *Anaxagoras and the Birth of Physics* by Daniel E. Gershenson and Daniel A. Greenberg (New York: Blaisdell Publishing Company, 1964).

4. *Physics* 196a. See also *On the Heavens* 295a.

5. For a translation of the passage, see Heath, *Aristarchus of Samos*, pp. 26–27.

6. See G. S. Kirk and J. E. Raven, *The Presocratic Philosophers: A Critical History with a Selection of Texts* (Cambridge: University Press, 1957), p. 132.

7. 'The tradition concerning Anaxagoras' use of rotational motion is one of the most confused and contradictory of all traditions about his doctrines', wrote Gershenson and Greenberg, *op. cit.*, p. 349, and this seems to hold true in general of the whole gamut of ancient Greek dicta on the role of vortices in the physical world, and especially in cosmogonical processes.

8. *The Clouds* 380–84, in *Aristophanes*, translated by B. B. Rogers (Loeb Classical Library; Cambridge, Mass.: Harvard University Press, 1960), vol. 2, p. 301.

9. Such was the authoritative estimate of Greek posterity as voiced by Plutarch (fl. 100 A.D.). See *Plutarch's Lives*, translated by B. Perrin (Loeb Classical Library; Cambridge, Mass.: Harvard University Press, 1916), vol. 3, p. 291.

10. *Timaeus* 36C–D. Quotation is from *Plato's Cosmology: The Timaeus of Plato*, translated with a running commentary by Francis M. Cornford (London: Routledge & Kegan Paul Ltd., 1937), p. 74.

11. *Ibid.*, p. 223 (56B).

12. See Book X (617A–B).

13. See *The Dialogues of Plato*, translated into English with Analysis and Introduction by B. Jowett (3d rev. ed.; London: Oxford University Press, 1895), vol. 5, p. 204 (821A).

14. *The Dialogues of Pluto*, vol. 5, p. 205 (822A).

15. *Ibid.*

16. *Metaphysics* 1074b.

17. See S. Sambursky, *Physics of the Stoics* (New York: Macmillan Company, 1959), pp. 106 and 113.

18. From his 'Letter to Pythocles'. Quotation is from *Epicurus: The Extant Remains*, with short critical Apparatus, Translation and Notes by Cyril Bailey (Oxford: Clarendon Press, 1926), p. 59.

19. *Ibid.*, p. 61.

20. *Ibid.*

21. *Ibid.*, p. 63.

22. *Ibid.*, p. 25.

23. *Ibid.*, p. 61.

24. *Ibid.*, p. 63.

25. *Ibid.*, p. 79.

26. *Ibid.*, pp. 59 and 65.

27. *Lucretius, De rerum natura*, with an English translation by W. H. D. Rouse (3d rev. ed.; Cambridge, Mass: Harvard University Press, 1937), p. 371.

28. *Ibid.*, p. 373.

29. *Ibid.*, p. 385.

30. *Ibid.*, p. 379.

31. *Ibid.*, p. 381.

32. See *Plutarch's Moralia, Vol. XII*, with Notes and English translation by H. Cherniss and W. C. Helmbold (Loeb Classical Library; London: William Heinemann, 1957), pp. 73, 121, 205 and 211.

33. See *Lucian*, Greek text with an English translation by A. M. Harmon (Loeb Classical Library; London: William Heinemann, 1913), vol. 1, pp. 261–69.

34. Ptolemy, *Tetrabiblos*, edited and translated into English by F. E. Robbins (Loeb Classical Library; Cambridge, Mass.: Harvard University Press, 1964), p. 181.

35. *Ibid.*, pp. 183–85.

36. *Ibid.*, p. 185.

37. The classic treatment of the history of the role of this principle in astronomy is the work by Pierre Duhem, *To Save the Phenomena: An Essay on the Idea of Physical Theory from Plato to Galileo*, translated from the French [1908] by Edmond Doland and Chaninah Maschler, with an Introductory Essay by Stanley L. Jaki (Chicago: University of Chicago Press, 1969).

38. See the English translation by R. Catesby Taliaferro in *Great Books of the Western World, Volume 16, Ptolemy, Copernicus, Kepler* (Chicago: Encyclopaedia Britannica Inc., 1952), pp. 1–478.

39. See the German translation by L. Nix of Book II in *Ptolemaeus Claudius: Opera astronomica minora* (Leipzig: B. G. Teubner, 1907), pp. 119–21.

40. *The Almagest*, Book I, chapter 7, *transl. cit.*, pp. 11–12.

41. Bernard R. Goldstein, 'The Arabic Version of Ptolemy's Planetary Hypothesis', *Transactions of the American Philosophical Society* (Philadelphia) New Series – Volume 57, Part 4, 1967.

42. Book V, chapters 13 and 15.

43. *Meteorologica* 345a–b.

44. Heath, *Aristarchus of Samos*, pp. 394–95, or Proposition 13 of Aristarchus' work 'On the Sizes and Distances of the Sun and Moon', and Table of distances on p. 350.

45. Goldstein, 'The Arabic Version . . .' p. 7.

46. *Ibid.*

47. See the critical edition by Romeo Campani of the medieval Latin translation used by Dante, *Alfragano, Il 'Libro dell' aggregazione delle stelle'*, (Citta di Castello: Casa Tipografico-Editrice S. Lapi, 1910), pp. 146–47.

48. See the English translation, *The Convivio of Dante Alighieri* (London: J. M. Dent, 1908).

49. Nicole Oresme, *Le livre du ciel et du monde*, edited by Albert D. Menut and Alexander J. Denomy, translated with an Introduction by Albert D. Menut (Madison: University of Wisconisn Press, 1968), p. 289.

50. Nicolaus Cusanus, *Of Learned Ignorance*, translated by Fr. German Heron, with an Introduction by D. J. B. Hawkins (London: Routledge & Kegan Paul, 1954).

51. *Ibid.*, p. 111.

52. *Ibid.*, pp. 112–13.

53. The work enjoyed great popularity for the next two generations. Its English translation by Barnabe Googe (1576) was widely used as a textbook. See *The Zodiake of Life by Marcellus Palingenius*, with an Introduction by Rosemund Tuve (New York: Scholars' Facsimile and Reprints, 1947).

54. See the critical edition of the text by Giovanni Aquilecchia (Torino: Giulio Einaudi Editore, 1955), p. 227, or its English translation, *The Ash Wednesday Supper*, with Introduction and Notes by Stanley L. Jaki (The Hague: Mouton, 1975), p. 165. Somewhat less obscurantist was Bruno's cosmology in his *De l'infinito universo et mondi* (1584), available in English in *Giordano Bruno: His Life and Thought, with annotated Translation of his Work On the Infinite Universe and Worlds*, by Dorothea Waley Singer (New York: Henry Schuman, 1950), pp. 225–378. Bruno subsequent publications were, however, wholly dominated by his interest in magic, cabbala, mnemotechnique, and necromancy, as means to bring about the Hermetic 'reformation' of the world.

55. *The Ash Wednesday Supper*, p. 149. The pantheistic, cyclic world of Bruno is discussed in detail in my *Science and Creation: From Eternal Cycles to an Oscillating Universe* (Edinburgh: Scottish Academic Press, 1973), pp. 262–66.

56. *The Ash Wednesday Supper*, pp. 63, 133 and 143.

57. *Ibid.*, pp. 56–57.

58. See Chapters 9 and 10 of Book I of his *On the Revolutions of the Heavenly Spheres*, translated by Glenn Wallis, in *Great Books of the Western World, Volume 16, Ptolemy, Copernicus, Kepler*, especially pp. 521 and 529.

59. *Ibid.*, p. 528

60. Published in 1596. References are to the critical edition in vol. 1, of *Johannes Kepler Gesammelte Werke*, edited by Walther von Dyck and Max Caspar (Munich: C. H. Beck, 1938–)

61. *Werke*, vol. 8, pp. 1–128.

62. *Werke*, vol. 1, p. 23.

63. First published in 1596. A French translation followed in 1597 together with a new edition of the Latin original. See p. 596 in the latter (Frankfurt: apud heredes A. Wecheli, C. Marnium & I. Aubr., 1597).

64. *Werke*, vol. 13, p. 113

65. *Mysterium cosmographicum*, Preface, in *Werke*, vol. 1, p. 9.

66. *Ibid.*

67. *Ibid.*

68. *Ibid.*, p. 10.

69. *Ibid.*, p. 12.

70. *Ibid.*, p. 26 (cap. 2).

71. *Ibid.*, p. 48 (cap. 13).

72. *Ibid.*, p. 62 (cap. 18).

73. *Ibid.*, p. 70 (cap. 20).

74. In his *Ad Vitellionem paralipomena quibus astronomiae pars optica traditur* (1604), in *Werke*, vol. 2, p. 22.

75. See Maestlin's letter to the pro-rector of the University of Tubingen in Kepler's *Werke*, vol. 13, pp. 84–86.

76. In a letter of April 24, 1598, in Kepler's *Werke*, vol. 13, p. 207.

77. In a letter of Oct. 13, 1597, in *Werke*, vol. 13, p. 146.

78. *Werke*, vol. 13, p. 292.

79. *Ibid.*, p. 305.

80. *Werke*, vol. 13, p. 345.

81. *Ibid.*, p. 290.

82. This point was aptly made by Arthur Koestler in Part IV, 'The Watershed', dealing with Kepler's discoveries, of his *The Sleepwalkers: A History of Man's Changing Vision of the Universe* (New York: Macmillan, 1959). The unquestionable merits of Koestler's work are marred by far-fetched analogies, curious insinuations, and careless quotations, such as when he uses three different letters of Kepler as if they were one (see pp. 278 and 581, notes 16 and 17).

83. *Astronomia nova*, in *Werke*, vol. 3, p. 178.

84. See the concluding words of Chapter 21, of Part Two of the *Astronomia nova*, in *Werke*, p. 187.

85. *Ibid.*, p. 362.

86. *Ibid.*, p. 242.

87. *Ibid.*, p. 241.

88. *Astronomia nova*, pp. 254–62 and 348–64.

89. *Ibid.*, pp. 349, 353, 356 and 360.

90. In a letter of July 1, 1622, to Philipp Müller; quoted in W. von Dyck and M. Caspar, 'Nova Kepleriana, No. 4,' in *Abhandlungen der Bayerischen Akademie der Wissenschaften. Mathematisch-naturwissenschaftliche Abteilung*, vol. XXXI, Abh. 1 (1927), p. 107.

91. *Astronomia nova*, p. 34.

92. See the English translation by Stillman Drake, in his *Discoveries and Opinions of Galileo* (Garden City, N.Y.: Doubleday, 1957), pp. 21–58.

93. See the English translation by E. Rosen, *Kepler's Conversation with Galileo's Sidereal Messenger* (New York: Johnson Reprint Corporation, 1965), p. 9.

94. *Ibid.*, p. 14.

95. *Ibid.*, p. 37. The five perfect solids are discussed in Book XIII of Euclid's *Elements* [of geometry] but without any reference to the physical world, let alone to the system of planets.

96. *Ibid.*, p. 38.

97. *Ibid.*, p. 39.

98. *Ibid.*, p. 44.

99. *Ibid.*, p. 46.

100. *Ibid.*, p. 39. Kepler's posthumous *Somnium*, a small essay on astronomical phenomena as seen from the moon, originally composed in 1609 and greatly expanded by him with notes during the next two decades, contains several conjectures on the size and habits of the moon's denizens. See *Kepler's Somnium: The Dream, or Posthumous Work on Lunar Astronomy*, translated with a commentary by E. Rosen (Madison: The University of Wisconsin Press, 1967), pp. 27 and 129.

101. *Kepler's Conversation*, p. 47.

102. *Discoveries and Opinions of Galileo*, p. 137.

103. *Ibid.*

104. *Ibid.*, p. 158.

105. *Dialogue concerning the Two Chief World Systems*, translated by Stillman Drake (Berkeley: University of California Press, 1953), p. 29. This concept or model of the formation of the system of planets is 'Platonic' only in the sense that it readily lends itself to a geometrical presentation. It is again ascribed to Plato in Galileo's *Dialogues concerning Two New Sciences* (1638), translated by Henry Crew and Alfonso de Salvio, with an Introduction by Antonio Favaro (1914; New York: Dover Publications Inc., n.d.), p. 261. For further details, see W. R. Shea, *Galileo's Intellectual Revolution: Middle Period, 1610–1632* (New York: Science History Publications, 1972), pp. 126–28.

106. *Dialogue concerning the Two Chief World Systems*, p. 29.

107. *Ibid.*

108. *Ibid.*, pp. 223–24. Galileo's calculation, giving the time in question as 3 hours 22 minutes 4 seconds, instead of 4 days and about 20 hours, was also faulty because he did not know that the rate of acceleration was not constant.

109. *Harmonice mundi*, in *Werke*, vol. 6, p. 302.

110. *Ibid.*, p. 328. In the opening part of the corresponding section (Book V), Kepler stated that the geometrical model of the planetary system as given in the *Mysterium cosmographicum* was valid only inasmuch as it determined the number of planets, but not their distances, and he referred his reader to Book IV of his *Epitome astronomiae copernicanae* soon to be published for more definite statements on the matter (pp. 291 and 299).

111. This part of the *Epitome* is available in English translation by Charles Glenn Wallis, in *Great Books of the Western World, Volume 16, Ptolemy, Copernicus, Kepler*, pp. 845–960.

112. *Ibid.*, pp. 898–99.

113. *Ibid.*, p. 871.

114. *Ibid.*

115. See *Werke*, vol. 8, pp. 1–128.

116. *Ibid.*, p. [9] of Dedication.

117. *Ibid.*, p. 120.

118. See R. Lenoble, *Mersenne ou la naissance du mécanisme* (Paris: J. Vrin, 1943), pp. 154–55 and 367–70.

119. See Mersenne's *Questions rares et curieuses, théologiques, naturelles, morales, politiques et de controverses résolues par raisons tirées de la philosophie et de la théologie* (Paris: chez Pierre Billaine, 1630), pp. 72 and 121.

120. Within two generations it became clear that contrary to Mersenne's claim (see Lenoble, *op. cit.*, p. 402) there was no need to make the elliptical orbit of planets contingent on a special act of the Creator.

121. First by John Wallis in 1673 and again in 1678 as part of *Jeremiae Horrocii opera posthuma* (London: apud Mosem Pitt, 1678), pp. 1–239. See especially the Prolegomena, pp. 1–21.

CHAPTER TWO

THE SPELL OF VORTICES

Cartesian aspiration and preliminary sketch

The first copies of the *Mysterium cosmographicum* were just leaving the printing shop in Tubingen, when René Descartes was born on March 31, 1596, in far-away Touraine, Brittany. Twenty-three years later, in 1619, Kepler's *Harmonice mundi*, containing his last nostalgic vision about geometrical patterns in the cosmos, was published in Linz. In the early autumn of the same year young Descartes, coming from the coronation festivities of Emperor Ferdinand in Frankfurt, reached the Danube and decided to stay, probably in Neuburg, a Bavarian town separated by a week's journey from Linz, where Kepler lived, farther down the river. Descartes needed rest and solitude. For some time his mind had been drawn deeper and deeper into the vortex of concentration. A year earlier he met, while studying military engineering in Breda, Isaac Beeck-man, the savant from Middelburg, and what he learned from Beeckman became a catalyst in his intellectual gropings. In Beeckman Descartes found the type of scholar whom he had already been seeking for some time but in vain. 'Apart from me', so Beeckman recorded the substance of Descartes' words to him, 'he has never found anyone who united very closely in his studies physics and mathematics'.[1]

The idea of a complete union between these two sciences was now taking an overpowering hold on Descartes' thinking. As winter approached he closeted himself away from the world in a 'stove-heated room', to recall his words.[2] He also recorded in another context[3] that he had not tasted wine for three months when in early November he felt a steady rise of temperature in his head, accompanied with a sense of exhilaration. Then came the evening of November 10, or the vigil of Saint Martin, a time of gay celebrations in France. On that night Descartes had three dreams. All three carried one and the same message about the fundamental role which geometry or mathematics should play in any meaningful discourse about the physical world. The role in question was not, of course, the one advocated in the *Mysterium cosmographicum* or even in the *Harmonice mundi*. Had Descartes visited Kepler in Linz, where he probably passed through with the Imperial troops in the spring of 1620 on his way to Hungary to fight the Turks, the meeting would have hardly been to his taste. For him the geometrical method meant that nature should be inter-preted as a machinery throughout, for a machine and its parts were obvious embodiments of geometrical pattern. While this had been noted before, the interactions among the mechanical parts of a world conceived as a machine still represented an area largely unexplored.

Implementing this programme of explaining the world as a mechanical process proved to be far more tedious and lengthy than a dream or two. The first steps, however, could hardly have been more auspicious. During the 1620's Descartes developed the rudiments of analytical geometry into a full-fledged science. His success convinced him that the essence of physical entities was extension itself, the regularities of which had now been unfolding through the new science in which geometry was reduced to algebra. To be sure, geometrical patterns were one thing, the shapes and interactions displayed by the actual physical world another. While the former now showed a one-to-one correspondence to mathematical formulas and their precision, the latter, especially in their details, seemed to be an extremely imprecise affair. Clearly, if one was to give convincingly the new, or mechanical explanation of the physical world, one had to confine oneself to its gross features. While an explanation resting on such qualification certainly fell short of the ideal, it could amply show the truth of a physics cast in the moulds of mechanics. Descartes must have felt so, if this is to be judged from his eagerness to implement the task he had assigned to himself.

Le monde ou traité de la lumière,[4] Descartes' ambitious account of the universe, or a treatise on light, as he called it, was nearing completion in 1632, when news reached him about the condemnation of Galileo. Living in Holland since 1629, Descartes had no reason to fear the Inquisition, but his sincere and lifelong loyalty to the Church made him decide against publishing the Le monde in which he endorsed the motion of the earth as a planet around the sun. Descartes, or rather his system, derived immense profit from this decision. The often sketchy pages of the Le monde would have at the outset discredited Descartes' efforts to give a new science of the physical world not only as it actually existed but also as it developed into its present form.

The originality of Descartes' attention to the idea of an orderly, cosmic evolution along the well defined lines of purely mechanical interactions cannot be emphasised enough. It was a giant step beyond the vague utterances of ancient atomists and especially beyond their picture of the universe as the result of a mere chance. It also represented a major advance over Oresme who had assigned an autonomous continuation of the mechanism of the world once it had been produced as a perfect clockwork and set in motion by the Creator.[5] Descartes now argued that the Creator's intelligence, power, and especially his immutability, were a sufficient, general guarantee that the universe would develop even from an original chaotic stage into its present form.[6] The special guarantee consisted in some basic laws of physics which could securely be formulated, so Descartes believed, from a careful consideration of the perfect rationality of God. The laws concerned inertial motion, conservation of the quantity of motion in impact, and centrifugal force as a form of impact. Of these he gave only the first correctly, while the third was a completely false idea. His postulating of the formation, through friction, of three kinds of matter, very fine, intermediate (globular), and coarse, which constituted the bodies of stars, the transparent ether, and the earth (together with planets and comets), respectively, also revealed something of the arbitrariness of his method. He wanted to make his mark by being systematic and consistent, but he offered no details on the all-important question of how the original motion and stirring of particles, fine, globular and coarse, would lead to the formation of an indefinitely large number of rotating 'heavens', that is, compartments of the universe each dominated by a star at its centre (*Illustration V*). He merely made the assertion that chaotic collisions among the three kinds of particles would somehow issue in a general rotational motion.[7] His was the first of many unconvincing modern efforts to derive the circular motion of the planetary system from a chaotic confusion of colliding particles.

On the formation of stars and planets the *Le monde* contained equally little. In the rotating 'heaven' filled with the ethereal fluid composed of globular particles, the fine particles, which he also identified as the element of fire, occupied the centre and formed the body of the sun. Pieces of the coarse matter, which he pictured as gradually growing to planetary sizes by being hooked to one another, took up positions away form the centre in proportion to their massiveness.[8] The author of the *Le monde* was still far from translating his vision of a dynamic or mechanical formation of the system of planets around the sun into a really expressive mechanical model. He seemed to be able to express in words only half of what might have been in his mind. Tellingly, the model he proposed was based on half circles. He spoke of two rivers touching one another at the point where both were sharply bending in opposite directions (*Illustration VI*). In rivers, he noted, the heavier objects floated along lines where the flow was fastest. Foam and light debris collected in the slower zones. Since in a sharply bending river the flow was speediest along the longer bank, a massive object, like a boat, moved along it, that is, along a greater circle. But the implied inference to a full circle was not developed in Descartes' model of two rivers. Instead, he argued that the boats coming from straight sections of the rivers into the bending would, because of the dynamics of the flow in that bend, be driven toward the point of junction and would keep on a practically straight course. This meant that the boats would pass from one river into the other.[9]

The boats represented such large bodies made of the coarse matter as planets and comets. This is not to suggest that Descartes envisaged the crossing of a planet from the 'heaven' of its star or sun into the 'heaven' of another. It was the formation of the earth-moon system that he seemed to base on such a crossing. He spoke of the arrival of the moon to the earth's orbit at a time when the earth was passing along there and then, without offering details on the process of how they joined into one system. He merely stated that there could be small 'heavens' within the 'heaven' of a star. As further illustration of this he referred to the spotting of moons around Jupiter and Saturn by 'new astronomers'.[10] He must have had in mind above all the observations of Galileo, who in addition to the 'Medici stars' around Jupiter noticed protrusions at two opposite points of Saturn, later identified as parts of its ring.

Descartes called the moon's orbit a circle[11] and drew it as such, although a later diagram of it showed an ellipse (*Illustrations VII and VIII*). This patent inconsistency and Descartes' reference to 'new astronomers' might have alerted any perceptive and well-informed reader of the *Le monde* to the fact that solid astronomical information was conspicuously missing from the work. If it displayed astronomical interest, it concerned the topic of comets. Descartes' model of two rivers mainly served his contention that comets were bodies similar to planets and passed continually from one 'heaven' into another. The text of the *Le monde* came to an abrupt end with a lengthy discourse on the tail of comets which were, according to Descartes, the effect of the refraction of the light of their heads in the ethereal fluid composed of compact layers of transparent globules, his second kind of matter. The reasoning provided at best a good example of the extent to which Descartes was willing to carry his basic assumptions. The length at times appeared 'very paradoxical' even to Descartes.[12] In his resolve to give a uniformly mechanical explanation of everything, he redefined light as pressure produced by the centrifugal tendency in the ethereal 'heaven'. From this it followed that there would still be some sunlight even if there were no sun, a body consisting of fine particles, or fire, in the centre of its 'heaven'. His was a commitment to a type of rigorous thinking that would have been obviously out of place if his intention had merely been, as he alleged, to write a 'Fable' about the world.[13]

The world of the learned got only a glimpse of that 'Fable' when in 1637 Descartes published his *Discours sur la méthode*. Once more he asserted that his intention was not to speak about the actual world, created ready-made by God, but about a 'new world' and about its evolution: 'I resolved', he summed up the cosmological contents of his still unpublished *Le monde*, 'to speak only of what would happen in a new world if God now created, somewhere in an imaginary space, matter sufficient wherewith to form it, and if He agitated in diverse ways, and without any order, the diverse portions of this matter, so that there resulted a chaos as confused as the poets ever feigned, and concluded His work by merely lending His concurrence to Nature in the usual way, leaving her to act in accordance with the laws which He had established'.[14] But 'new' as this world or system could be, there was, to quote his words, 'nothing to be seen in the heavens and stars pertaining to our system which must not, or at least may not, appear exactly the same in those of the system which I described'.[15] This was an exact echo of the claim in the *Le monde* that there could indeed be a science of the universe which would have been rightly called *a priori* by the Schoolmen.[16]

Complete theory from vortex motion

It took another seven years before Descartes presented, in 1644, the fully developed form of the 'new system'. Its title, *Principia philosophiae*,[17] bluntly conveyed Descartes' intention to break entirely new grounds. Its impact was instantaneous. It caught imagination by storm, as indicated by the publication within three years of a French translation.[18] In fact, Descartes rested his case largely on his ability to help the mind of his reader by constant appeal to analogies easily representable to one's imagination. The principal of these analogies was the vortex motion, or the spontaneous formation of large and small eddies in fast moving rivers. It was the vortex which carried much of the burden of proof on behalf of the new system. It served above all as a supreme reassurance for Descartes that the theological objection to the earth's motion was now removed by reference to objects kept motionless in the centre of vortices which themselves were drifting.[19] It seems that his long wait to make public his system was due to his resolve to gain full conviction about the usefulness of vortices in theological as well as scientific respect.

Very little is known about the genesis in Descartes' mind of the cosmic relevance of vortex, or *tourbillon* in the French translation, words which he did not use in print before the publication of the *Principia*. In view of what he wrote in the *Principia* it would be tempting to see his system of cosmic vortices surrounding each star behind his reference to the 'highest and most perfect science', which he felt ready to formulate, so he wrote in 1632 to his trusted friend, Father Mersenne: 'Although they [the stars] appear rather irregularly dispersed in the sky, I do not doubt that there is a natural order among them which is regular and determined; the knowledge of this order is the key and foundation of the highest and most perfect science that man can have about material things; all the more so, as by that science one could know *a priori* the different forms and essences of all bodies on earth, whereas without it we must rest satisfied with finding them *a posteriori* and through their effects'.[20]

Part of that *a posteriori* route would have consisted in studying various models and cases of vortex motion. Descartes did some experimentation with pieces of various solids in a rotating liquid, but whenever he spoke of his findings there was a striking connection between his confident tone and his readiness to gloss over problems. He was all too ready to dispose of them by making quick recourse to a convenient analogy. Thus in his letter of January 6, 1639, to Mersenne he declared that no motion in the 'subtle matter' (layers of globules) was different from the fact that in one and the same river the straight and curving streams could move at different speeds, although there was only one basic moving force at play.[21] In October of the same year he described again to Mersenne the behaviour of small lead globules mixed with large pieces of wood in a pen, when the latter was rotated. As Descartes interpreted the effect, the very small lead globules 'pushed' toward the centre the bigger wooden pieces. In the same way, he claimed, the 'subtle matter' pushed coarse or terrestrial matter toward the centre of rotation.[22] The next year, in a letter to Mersenne again, he argued that the stability of the earth in the centre of its vortex corresponded to the case when one threw pieces of iron and wood into a rotating pan filled with water. The pieces of wood (earth) moved toward the centre, so Descartes reported, but he said nothing about the respective positions of the iron particles and of water. He quickly brought the topic to a close by noting that in a letter one could not explain these things in detail.[23] In November 1641 he curtly remarked to Regius (Henry le Roy) that the reason why in vortices solid bodies occupied the centre was because the water tended toward the periphery.[24] In all these examples, intended for the explanation of gravitation by centrifugal force, impact, surface area, specific weight, solidity, friction, and any other physical parameter remained frustratingly intermingled.

With respect to a scientific account of the vortex, the *Principia* represented no real advance. There he declared that 'from the vortex alone one can most easily understand all the phenomena of the planets without any further manipulations'.[25] The proof of this consisted not in mathematics or geometry but in a quick reference to the formation of vortices in a bending river. In that reference Descartes spoke of straws moving in circles around the centre of a whirl, and moving the faster the closer they were to the centre. To this he added the telling proviso that the path of those straws was not a perfect circle but one 'with aberrations with respect to latitude and longitude. Thus we can picture without any difficulty all these [features] about the planets, and by this alone all their phenomena are explained'.[26]

This wholly qualitative argumentation deserves to be recalled for two reasons. One of them lies in the sharp contrast which Descartes' argument presents with his proud claim at the end of the Second Book of the *Principia* that he admitted 'no other principles in physics than those of geometry and abstract mathematics'.[27] When in 1637 Descartes published his *Météores*, or his explanation of atmospheric phenomena, he clearly conveyed his intention to discuss problems of physics within a distinctly geometrical framework.[28] But soon afterwards he found it necessary to defend himself against charges that he had fallen short of the lofty goal: 'If it pleases him,' wrote Descartes to Mersenne about a Mr. Argues, one of his critics, 'to consider what I have written on salt, snow and the rainbow, he will recognise that all my physics is nothing but geometry'.[29]

This reference of Descartes to his discussion of snow crystals is germane to the second of the two reasons that make his explanation of planetary motions by vortices noteworthy. In his discussion of snowflakes in the *Météores* Descartes had a notable predecessor, Kepler, who had published a special treatise on the 'hexangular snow' in 1611.[30] With this work of Kepler Descartes was fully familiar, as shown by his letter of March 4, 1630, to Mersenne.[31] Again, in his letter of November 1, 1635, to the older Huygens, Descartes referred to Kepler as superior even to Galileo and Scheiner in dioptrics, or the science of lenses.[32] Three years later, and a year after he published his *Dioptrique*, Descartes haughtily dismissed the charge that his discussion of the ellipse had been borrowed from Kepler. Yet in the same context Descartes also declared that Kepler was his principal master in optics.[33] Why is it then that in his specification of the path of planets as something differing from the perfect circle 'with respect to latitude and longitude' he did not simply speak of an ellipse?

The question is all the more perplexing as on one occasion, in connection with the moon's orbit around the earth, Descartes specified those differences from the circle as being 'in the manner of an ellipse'.[34] He might have, of course, never learned of Kepler's discovery of the elliptical form of planetary orbits in view of the relatively slow spreading of valuable details in Kepler's often rambling writings. This should, however, seem most unlikely because of Descartes' repeated acknowledgments of his debt to Kepler. In addition there is the remark by which Descartes brushed aside, two years after the publication of the *Principia*, the cosmological treatise of Roberval who put forward his advocacy of the earth's motion as being merely the idea of Aristarchus of Samos. Roberval 'wrote out everything', so Descartes stated, 'from Copernicus and Kepler'.[35] The remark seemed to imply that Descartes had already been familiar for some time with Kepler's synthesis of astronomy and cosmology, the *Epitome astronomiae copernicanae*, which contained repeated references to the elliptical orbit of planets. One is, therefore, compelled to conclude that Descartes ignored deliberately the ellipticity of planetary orbits possibly for the same reason for which he also ignored all the finesse in the works of Copernicus and Tycho Brahe, to whom he referred several times in the *Principia*.

Actually, Descartes' explanation of the world, and especially of the development of the world into its present system, was such as to encourage the neglect of practically all quantitatively exact details of astronomical lore. Of the various laws of physics, which he felt to have correctly enunciated, only the set of laws concerning impact had quantitative veneer. He must have realized that his physics offered no basis for quantitatively exact predictions about the cosmological results (among them the elliptical orbit of planets) of basic physical processes as he specified them. His distrust of Galileo's exact law of free fall and of the method by which it was obtained[36] evidenced his awareness of the inferiority of his own physics which allowed but generic statements and inferences. Descartes' discussion of the systems of Ptolemy, Copernicus, and Tycho Brahe, which introduced his cosmogony, contained no quantitative details. About 'eclipses, stations, and retrogradations of planets, about the precession of equinoxes, about the variation of the obliquity of the ecliptic and similar things', he noted that 'they contain nothing which is not easy for those who are somewhat familiar with astronomy'.[37] Such was his convenient excuse for not treating these and many similar problems of astronomy as part of his cosmology. The only quantitative details which Descartes gave about planets concerned their periods and distances. In listing the latter he revealed much of his mistrust of exactness by giving some traditional data only in round figures. According to him the distance of Mercury from the sun was 'more than two hundred earth-diameters', that of Venus 'more than four hundred', that of the earth 'six to seven hundred', that of Mars 'nine hundred to a thousand', that of Jupiter 'more than three thousand', and that of Saturn 'five to six thousand'.[38]

Unsatisfactory as such procedure may appear from the modern viewpoint, the cosmogony which Descartes offered in the *Principia* had both grandeur and originality. It encompassed the whole indefinitely large universe and presented the evolution of each part, the planetary system in particular, as intimately interconnected with the development of the whole. The process as outlined by him had the semblance of being purely mechanical throughout, that is, of being the successive unfolding of the consequences of some basic processes of motion, impact, rotation, and friction. At no step did it seem to imply something patently impossible, and at each step it made a plausible appeal to the visualising ability of the reader. It could readily create the impression in the mind of a non-expert that physics in its Cartesian version was the embodiment of easy-to-grasp intelligibility. Last but not least, the cosmogonical process unfolded by

Descartes had an unusually large measure of novelty. It must have struck each and every reader in an age which under the impact of the Reformation came heavily under the sway of biblical fundamentalism, a far cry from the way in which Augustin of Hippo, Gregory of Nyssa, Basil the Great, to mention only a few, interpreted the biblical account of creation.[39]

This fundamentalist trend not only led to a morbid preoccupation with biblical chronology in the case of Newton and many others, but also created the impression that the idea of a creation in full form of everything in the beginning was an integral part of Christian creed. Thus Descartes took pains to declare his adherence to the fundamentalist interpretation of the biblical account of the creation. The evolutionary account of the world remained a hypothetical case though one which, so Descartes emphatically asserted, gave a better insight into the rationality of the created whole and into God's immense power: 'Just as for the understanding of the nature of plants and men it is by far better to consider how they could slowly be born from seeds rather than how they were created in the very first origin of the world, in the same way, if we could figure out very simple and easy-to-understand principles from which, as if from some seeds, the stars and the earth and finally everything that we see in this visible world could demonstrably originate, although we know well that they have never originated that way, we would indeed much better explain their nature than if we merely described them as they are.' And since, Descartes added, 'it seems to me that I have found such principles, I shall present them briefly'.[40]

The presentation was not much shorter than the preceding ninety pages in much of which Descartes set forth the philosophical foundation of his physics and cosmogony. From man's self consciousness he boldly made his well known inference to the dualism of mind and matter, from there to extension as the chief feature of material entities, and from the world to its first cause, God. From God's immutability he inferred the conservation of motion, a principle which he tried to illustrate with his analysis of the various cases of impact between bodies. The infinite power of God served in turn for Descartes as a persuasive pointer to the existence of an indefinitely large world. That he did not attribute to the world actual infinity was due to his metaphysical sensitivity. Recognition of the scientific pitfalls of an actually realised physical infinity was still centuries away.[41] With the same metaphysical sensitivity Descartes also recognised the need for taking a close look at the question of man's standing in a world which contained an indefinitely large number of planetary systems. He warned against a naïve assignment of the existence of everything for man's benefit, while reasserting his belief in purpose in a religious sense.[42] His was certainly not a mind to be trapped in flights of fancy about denizens on other planets. He dismissed with a slight touch of sarcasm the merits of assuming the existence of such beings because of the indefinitely large extent of the world. One could just as well make the same inference, he remarked with tongue in cheek, from the distances which astronomers assigned to the world, that is, to the planets.[43]

The birth, life, and death of a star

Planets, or rather their evolution, formed a central part of the genesis of the world as presented by Descartes in the *Principia*. Its reader was given a clear hint of this in the very first step which, according to Descartes, constituted the start of the cosmogonical process. That first step consisted in the fact that all particles of matter (he pictured matter as divisible into all conceivable parts and actually divided into many small parts) had 'in some way circular motions'.[44] About this circular motion originally impressed by God on all particles of matter Descartes immediately stated that its total amount was equal to the total amount of motion now present in the world. His positing at the outset this rotational motion was a move that shall appear the more felicitous, the more chapters unfold in this study of theories on the origin of planetary systems. It does not seem that Descartes' procedure had to do with some possible misgivings on his part about the chaos as a starting point for the necessary emergence of rotational motion. He still maintained that the present form and dynamism of the world could be developed from a chaotic stage.[45] His abandoning of the chaos as a starting point was rooted in the sole consideration that an original parcelling of matter into equal parts was more consonant with the absolute perfection of God, the Creator.[46] The only 'inequality' which Descartes permitted, in order to satisfy obvious evidence, derived from the unequal separation of stars. He therefore postulated that originally God had also set up very large domains that slightly varied in diameter around the mean value of the distance separating the stars from one another. About these large domains he stated that they were also rotating around their centre and that their original number corresponded to the actual number of stars *and* planets.[47] Such was his hint that in his cosmogony planets were to be shown as having the same origin as stars.

Shortly afterwards, Descartes made another hint that he had novel ideas about the origin of planets. The hint came as he described the formation of the three forms of matter composing the actual world. Since he had defined material existence as synonymous with extension, his world was full. This meant that the large parts (domains) into which matter was originally distributed had to be irregular polyhedrons. From their close packing it logically followed that once they had been set in rotation by God exceedingly small particles or filings should break off at their edges and corners. These filings, due to their smallness, obtained very large velocities in the process and continued fragmenting, because the smaller they became, the larger was the ratio of their surface to the matter contained within.[48] The small parts in turn developed into perfectly spherical globules. Descartes now had two of his three kinds of matter or element. The first element consisted of the minute, speedy filings composing the sun and the stars. The second element consisted of the globules filling in the form of ethereal liquid each stellar domain with its interplanetary spaces. The third kind of matter, or element, composed 'of rather coarse parts', formed the earth, planets, and comets.[49]

Taken together with his previous remark that planets were former stars, this was a momentous hint not only about comets as being former stars, but also about the genesis of the third kind of matter as constituting the factor which transformed a star into either a planet or a comet. Details of this fascinating idea were to be given a 'little later'.[50] Actually, the explanation came considerably later, after the account of the formation of stars, a topic with many ramifications if not detours. First came the distribution of the large domains, or stellar vortices, into 'three heavens' (*Illustration IX*). The first 'heaven' was the vortex of the sun (S), the second 'heaven' was composed of the surrounding and 'very numerous' vortices (f, F, etc.). The vortices beyond (Y, X, etc.) formed the third 'heaven' about which Descartes pointedly said that it 'cannot in any manner be seen by us in this life'.[51] He showed no concern for the fact that only a few vortices could be drawn around the sun's vortex, or placed around it physically, if they were similar in size to it. He did not care to discuss the number of concentric layers of vortices forming the second or visible 'heaven' around the sun, nor did he pay attention to the very great number of visible vortices in the belt of the Milky Way. His sole and rather brief account of the Milky Way remained hidden for two more decades in the manuscript of his *Le monde*. Was his silence on the Milky Way due to his tacit realisation that its cosmic singularity clashed sharply with his postulate of the homogeneous parcelling of the heavens into equal domains or vortices?[52]

With the distribution of the universe into three heavens Descartes was ready to unfold the formation of a star in the centre of a large vortex or celestial domain. Primarily, the process rested on the centrifugal force which Descartes defined as a 'law of nature whereby all bodies that move around in an orbit tend, inasmuch as they can, to recede from the centre of their motion'.[53] The corresponding tendency of globules left, therefore, a spherical hole in the centre which was occupied by particles of the first element. The process explained the spherical form of stars, but could also appear 'paradoxical', as Descartes put it,[54] because the same process made the stars as sources of light to some extent superfluous. In this connection he merely repeated what he had already written in the *Le monde* about the consequences of defining light as the outward pressure of globules in the rotating 'heaven' or vortex. He saw nothing paradoxical in the spreading of that pressure or light not only in the plane of the ecliptic, the plane of rotation of the vortex, but also toward its poles. The cause of the equal spreading of the sun's light in every direction lay, according to him, in the tight packing of globules. Because of that he also felt justified in concluding that the globules could not flow from one vortex into another. To secure the relative stability of vortices, demanded by the obvious stability of the relative position of stars, he also specified that the planes of the rotation of neighbouring vortices were as far away from one another as possible. This meant that the pole of a vortex touched a neighbouring vortex closer to the latter's ecliptic than to either of its poles. Finally, and again to show that his system would readily satisfy the phenomena, namely, the differences in the light of stars, he further emphasised that it was most natural that the vortices should not be thought of as being strictly equal.[55]

With all this Descartes provided the general framework for the formation of a star. He could now turn to the first element itself of which the body of a star was composed. Due to their finesse, particles of the first element could move from vortex to vortex, but their principal course was a closed circuit. From the centre of the vortex, where they formed the star's body, they moved toward the periphery along the exceedingly small interstices formed by the contiguous globules. Once at the periphery, their route continued toward the poles where they entered the channel constituting the axis of the vortex and finally reached the centre, or the body of the star. Reflecting now on this general scheme Descartes specified several details to enhance the plausibility of his system. Thus he argued that since it was easier for the fine particles to move toward the centre in the axial channel than to pass through the globules, there always remained a sufficient quantity of the first element in the centre to form there the body of the sun.[56] Also, because the flow of fine particles in and out of a vortex varied both with respect to quantity and direction, Descartes elaborated in a rather pedantic manner on the continual slight changes in the shape of the sun, in its position in the centre of the vortex, and in the boundaries of the vortex itself.[57]

Since the principal component of a vortex consisted of the globules, Descartes hastened to make some specifications about them before continuing with his discourse on the first element, which, as will be seen shortly, contained the decisive clue to his cosmogony. Clearly, Descartes wanted to be sure that by the time the actual formation of the planets was broached, all particulars had already been carefully staked out so as to bar immediate doubts about the naturalness of the process envisaged by him. He claimed that the globules closer to the sun were smaller than those farther away, and also moved around faster than the more distant ones. This was, however, true only up to a certain distance from the sun. Beyond that distance all globules were supposed to be equal and the farther they were, the faster they moved. The justification he offered for this was that since the vortices were not perfect circles the path of globules closer to the boundaries was disproportionately longer. While this proposition seemed to be implied logically in the picture he had already unfolded about the vortices, it was otherwise with the determination of that 'certain distance'. It could only be established through observational evidence to be found in the passing of a comet from one vortex to another and in the motion of Saturn.[58] He postponed giving the proof of why that crucial distance was at the orbit of Saturn or somewhat below it [Illustration X].

It now remained to take a closer look at the first element and Descartes' was a close look indeed. His mental eyes discerned the formation of three spiral grooves on some particles of the first element as they obtained a 'triangular form both in depth and width' through their passage amidst the tightly packed globules.[59] He likened these particles to cochleae and called them striatae. To this he added a 'most worthy consideration'.[60] According to it the orientation of spiral grooves on the striatae depended on whether the particles of the first element moved to the periphery of the vortex within its northern or southern hemispheres. He noted that the explanation of magnetic force largely depended on the twofold orientation of the grooves. What he seemed to suggest was that one class of the *striatae* had right-handed spiral turns, whereas the turns of the other class were left-handed. After these dicta on magnetism he further specified that the spiral turns were strictly three in number on all *striatae*, and that only particles of the first element approaching in size the interstices between the globules did undergo this arcane, chiselling process.

Anyone who had willingly followed Descartes through more than half of the Third Book of the *Principia* devoted to the evolution of the physical world, or more specifically to the evolution of stars and planets, would hardly disagree with Descartes' claim that as the *striatae* approached the centre they began to lose speed rapidly. This meant that they became separated from the smaller and still very speedy particles of the first element. Because of that differentiation the *striatae* formed small groups and became connected with one another through their spiral grooves. Further accretion of these groups resulted in the formation of heavy, foamlike masses which, when reaching the sun's surface, appeared as dark blots.[61] This was Descartes' famous explanation of sunspots noticed by scientists about a quarter of a century earlier. Descartes now argued that the appearance of most sunspots along the equator of the sun was due to the rotation of its material. The disappearance of some sunspots was solved with a recourse to the analogy of boiling liquid in which patches of foam were continually forming and dissolving.[62] From the same analogy he also derived the turning of some sunspots into solar flares[63] and he even utilised the vapour emitted from boiling liquids. He spoke of a thick, ethereal atmosphere surrounding the stars which in the case of the sun extended to the orbit of Mercury and perhaps beyond.[64]

It could, however, also happen that the foamy patches, or agglomerations of *striatae*, would form within a star in such quantities as to cover its whole surface. As a result, the star's light was either dimmed or completely extinguished. This was Descartes' explanation of reports on the decrease in the light of some stars and on the complete disappearance of some of them.[65] (He did not try to reconcile this view with his previous 'paradoxical' assertion that there would be some sunlight even if there would be no sun. Possibly he felt that this held true only within one vortex.) On the other hand, the foamy cover could suddenly break up and dissolve, which meant that an extinct star could suddenly become visible again, a process illustrated, so Descartes believed, by novae. He referred to the nova of 1572 about which he also recalled that it disappeared two years later.[66] To explain this latter point he only needed to assume that the *striatae* became very copious once more in that star and covered it with a thick layer (*Illustration XI*).

Dead stars into planets

Descartes found in the analogy of foamy liquid a model which in all appearance provided a dazzlingly uniform and intelligible explanation of a great number of, and in part new, celestial phenomena. Its appeal must have appeared far superior to anything offered beforehand. He seemed to have felt elated about its persuasiveness because he exploited its possibilities with a will. Of his truly wilful elaborations on the behaviour of the foamy cover enveloping a star here only the one touching on the formation of a planet can be recalled. 'It may also happen', he turned suddenly to a long postponed topic, 'that a whole vortex, in which some such fixed star is contained, is absorbed by other surrounding vortices and its star drawn into one of those vortices turns into a planet or a comet'.[67] This was especially the case when a star was covered not by one but by several hardened layers of *striatae* as if by concentric barks.[68] Since the particles of the first element could not leave the body of such a star, it no longer contributed to the centrifugal pressure of its vortex. If in addition the vortex in question was noticeably smaller than the surrounding vortices and its boundaries markedly different from a sphere, it was bound to cease to exist as an independent vortex.[69] In the diagram [*Illustration IX*] vortex Q with star N at its centre illustrated this case.

Following the capture of star N by vortex H, N began its 'descent' as an opaque and hard body toward S, the sun. The 'descent' was through an area about which, or rather about the size and speed of globules there, Descartes had well in advance specified certain features, so that the turning of a former star into a planet or a comet would appear a most natural consequence of his laws of velocity transfer through collision. Thus if N was a very solid body, that is, if its quantity of *striatae*, as Descartes now defined solidity,[70] was very large, then it gained speed rapidly as it moved through the outer section of vortex H where the orbital speed of somewhat larger globules than the average was greater than in the vicinity of the critical distance, or the orbit of Saturn. If the speed of N was equal at that distance to the speed of globules, or to their minimum speed throughout the vortex, then N did not 'descend' any farther, because below that distance the speed of globules was increasingly faster, so that by any further 'descent' N could only gain in speed and thereby move farther away from the centre. In other words, a 'more solid' former star like N crossed the outer half of vortex H in a single curve and thus could be identified as a comet to which Descartes assigned the wavy route 1, 2, 3, 4, 5, 6, 7, 8 [*Illustration IX*] from one vortex to another.[71] Such was Descartes' explanation of the reputedly single appearance of a comet within a vortex along a curving path. The idea of the return of comets in highly elliptical orbits was still to be set forth in Newton's *Principia* and to be given further verification by the work of Halley.

If the former star N was not 'very solid', then its speed at the critical distance was less than the speed of globules there and thus it could continue its 'descent' into the layers of smaller and speedier globules. When and where its speed became equal to theirs, it settled in an equilibrium orbit, because above that orbit it could only lose some of its speed to the globules whose speed in the interior part of the vortex decreased toward the critical distance. Or as Descartes put it: 'If it [N] has less solidity, and therefore descends below that [critical] terminus, it stays there afterwards forever at a certain distance from the star, which occupies the centre of that vortex [H], rotates around it and is [now] a planet'.[72]

As was noted before, Descartes' scheme was at no point short on at least an apparent plausibility. Yet its share of artificiality was not small either. Indeed, Descartes almost immediately bogged down in further qualifications about solidity. It now became a function not only of the quantity of *striatae*, but also of the size and shape of the former star.[73] Furthermore, to make the transfer of velocity from globules to the new planet appear plausible, Descartes had to attribute to the globules a very high measure of solidity (he tried to disregard inelastic collisions as much as possible), while claiming at the same time a complete liquidity to their combined mass.[74] Almost three centuries of physicists followed him in his unsuspecting endeavour to conjure up the existence of a material, the ether, embodying contradictory qualities. He also spoke prolifically about the tail of comets in which he saw a phenomenon of refraction of the light of their heads at the critical distance where the speed of globules was a minimum. He had to expend many words to explain away the fact that planets and stars displayed no tails.

Descartes claimed, of course, that the difficulty was easy to answer.[75] As to the stars, he noted that no tail around comets was observable as long as the apparent diameter of their heads was no longer than a star. As to the planets, Saturn and Jupiter in particular, which appeared bigger and brighter than the fixed stars, he wrote: 'I have no doubt that they appear on occasion with such a tail . . . in countries where the air is very clear and pure; and I remember very well to have read somewhere that this had been observed, although I cannot recall the name of the author'.[76] No wonder that for someone trapped in such reasoning even the author of the highly misleading and by then heavily discredited *Meteorologica* was a welcome witness, although Descartes' chief aim consisted in supplanting the system of Aristotle to whom he referred but rarely: 'In addition to what Aristotle says in chap. 6 of the first book of the *Meteorologica*, namely, that the Egyptians have a few times observed such a tail around stars, this should, I believe, be understood of these planets and not of the fixed stars; and concerning what he says, that he himself saw a tail around one of the stars that are in the thigh of the Dog, this had to happen because of some extraordinary refraction in the air, or rather because of some indisposition in his eyes, for he added that this tail appeared the less clearly, the more intently he fixed his eyes on it'.[77]

Equally revealing, though in a different sense, was Descartes' discussion of the orbit of planets. He never referred to their elliptical orbit as he listed five causes why their orbits should be different from a perfect circle.[78] The first was the not exactly spherical shape of any vortex. Where its width was less than the average, the planet picked up speed and moved closer to the sun; the opposite was the case at points where the width was larger than the average. The second cause was the variability of the flow of the first element which moved at right angle to the orbit of planets on its course from the centre to the periphery of the vortex. The third cause was somewhat mysterious but in line with the arcane *striatae* responsible for it. Descartes believed that the axis of a planet was to line up parallel to the direction in which the *striatae* happened to flow through the vortex toward the periphery after they detached themselves on occasion from the surface of the central star. The fourth cause represented a rudimentary anticipation of the conservation of angular momentum and was illustrated by the behaviour of a small top. In spite of its smallness it could make, according to Descartes, two or three thousand turns in a few minutes before coming to a stop. Thus a star, destined to become a planet, could acquire at its formation a specific rotation which then would for a long time maintain itself against external influences. The fifth cause derived from the respective ratios of surface to volume (or rather to the amount of material contained within it) in the different planets. If the ratio was large, that is, if the planet was relatively small, the impact on its surface was, according to Descartes, more effective than in the case of a larger planet where the relatively larger mass presented more resistance to any change in its motion.

In all this Descartes carefully avoided quantitative exactness. Aiming at it was still to come in physics as a powerful antidote to mixing arbitrarily the interplay of various physical factors. Descartes gave his vote now to this now to the other, depending on what appeared more convenient in 'explaining' a particular feature of the question. While his 'explanations' could seem plausible with respect to this or that feature of the planetary system, the case was strikingly different when it came to the task of bringing together into one single system·fourteen celestial bodies. These were the sun, the six then known planets, the moon, four satellites of Jupiter, and two of Saturn, 'if two planets do indeed turn around it', Descartes added cautiously.[79] He based their coming together into one system on the undeniable though exceedingly tenuous possibility that the celestial space now occupied by the sun's vortex had formerly consisted of fourteen vortices. It was unquestionably true, to recall his words, that 'nothing prohibited' making this assumption. Yet one could still ask whether there was anything in his theory which suggested precisely this kind of original partition of the celestial space in question. His was only the first of many theories on planetary origins that presupposed what they were supposed to explain. Again, it was possible that some of these vortices, such as Jupiter's and Saturn's, were bigger than the others and that in some of the vortices the covering of stars with *striatae* had set in earlier than in the others, so that Jupiter and Saturn acquired their moons before being engulfed into the sun's vortex. Still all these 'possibilities' when taken together could only add up to a very high degree of improbability.

Telltale remarks

Such must have been, as will shortly be seen, the tacit verdict of most of the competent and judicious readers of Descartes' theory of the origin of the system of planets. The theory came to a close with three considerations, all of which anticipated, unwittingly though, the principal frustrations and stumbling blocks of all future theories. First came Descartes' remark that a proof of his theory was that planets closer to the sun were less solid than those farther away.[80] Mars seemed to form an exception, for as Descartes noted, it was smaller than the earth. His solution was that 'solidity' or quantity of matter, did not depend on size alone. This meant that Mars had to have a very high density. As was learned many years later, Mercury, Venus, and the earth had much the same high density, a peculiarity for which no planetary theory has so far offered a satisfactory solution.

With his second concluding remark[81] Descartes called attention to the agreement between his theory and the increasingly slower orbital motion of planets in proportion to their distances, as in his system the orbital speed of globules decreased with distance from the sun to Saturn's orbit. But in the same system the globules near the sun supposedly moved with the same speed as did a point on the surface of the sun. It was that surface which carried the sunspots, the triggering device of the origin of planets and of their system. The motion of those spots, making their round trip in 26 days, or about a month, contrasted sharply with the much faster orbital speed of Mercury. The latter, moving in an orbit 60 times larger than the sun's perimeter (so Descartes specified the matter), took only 3 months to complete its orbit. Even Saturn, the slowest of planets, seemed to move much faster than the solar spots. It completed in 30 years, or 360 months, its orbit which Descartes gave as being 2,000 times greater than the sun's perimeter. If Saturn's speed had been equal to that of the solar spots, then its orbiting would take 100 years, or 1,200 months, Descartes noted[82] in recognising the difficulty. Had he cared for quantitative precision or consistency, he should have given the last figure as approximately 2,000 months. His lackadaisical handling of quantitative data was also evidenced by his failure to correlate the foregoing figures for the relative sizes of planetary orbits with the actual figures he had earlier given for planetary distances. A little reflection might have made it clear to him that his argument implied for the sun a size that was much too small.

No detail would have been more germane to Descartes' second remark than Kepler's Third Law connecting the orbits and periods of planets, but its precision was patently incompatible with Descartes' resolve to be attentive only to generalities. Any reference, however veiled, to that Law would have merely alerted attention to the inherent weaknesses of his theory. While he could still ignore with some impunity Kepler's Third Law, about which and about Kepler's two other laws the learned world became largely informed only seven years later through Riccioli's *Almagestum novum*, the unduly great orbital speed of Mercury could not be glossed over because of the planet's relative vicinity to the solar spots. Descartes, therefore, quickly recalled his previous 'proof' of the existence of a very dense, turbulent cloud of ether around the sun extending to Mercury and somewhat beyond.[83] Once more a desired feature was ready in advance. Due to the braking effect of that dense ether the motion of Mercury was pictured to be somewhat slowed down to make it thereby appear plausible that the slow motion of sunspots was the result of a far more drastic braking process. It took two more centuries before it became recognised that the slow motion of sunspots evidencing the slow rotation of the sun represented a far more serious problem than Descartes imagined.

Descartes' third concluding remark concerned the earth-moon system. He repeated the substance of what he had already put in writing in the *Le monde* on why the moon always showed the same side to the earth and why the moon's orbit differed from a perfect circle. To explain the former point he assigned a greater solidity to the far side of the moon. In connection with the latter he remarked that the orbit 'rather approaches the figure of an ellipse'.[84] One of the diagrams (*Illustration VII*) of the *Le monde* appeared now in a subtly changed form, showing a perfect ellipse for the moon's orbit. In the same diagram the orbit of the earth was still a perfect circle. Descartes hastened to add that the actual situation was different: 'Finally, no one will be surprised [on finding] that the planets, although they always seem to display motion in a circle, don't ever describe a perfect circle, but in every way, both with respect to longitude and latitude, they slightly deviate [from it]. Since all bodies which are in the universe are contiguous and act on one another, the motion of any of them depends on the motion of all others and, therefore, varies in innumerable ways'.[85]

That bodies in the universe interacted with one another in a most complicated manner was certainly true. But undue emphasis on the complicated contours of nature invited an acquiescence into a method which restricted the scope of physical science to the gross features of configurations and processes. The gigantic efforts of Descartes to justify that method constituted in a sense a rear-guard action. Kepler's three laws of planetary motion and Galileo's law of free fall had already given a powerful boost to the conviction that behind the apparent complicatedness and imprecision of nature there lay laws that could be formulated with quantitative exactness. It was on that bedrock of precision that the vortices of Descartes met their demise after keeping under their spell for more than two generations all those who tried to come to grips with the principal scientific question of the times, the explanation of gravity both for terrestrial and celestial bodies. This rather enduring role of vortices for the explanation of planetary motions should be distinguished from their role in the evolution of planetary systems. In this latter respect Descartes' reliance on vortices prompted much less attention either from his critics or from his admirers.

The voice of critics and the silence of disciples

Among the first of the critics was Henry More in Cambridge who sharply questioned the formation of *striatae*. He saw in them exceedingly artificial figments of imagination and possibly because of this he cared not to comment on what they ultimately served, the formation of planets.[86] Far more could be expected on Descartes' cosmic system from More's compatriot, Thomas Hobbes, a personal acquaintance of Descartes, and one of the few whom Descartes asked for a criticism of his *Meditationes*. Their thinking seemed to have much in common. The elemental impact on Hobbes of his first encounter with the sweep and lucidity of Euclid's *Elements* had a close resemblance to young Descartes' famous dream, and Descartes found much to his liking in Hobbes' *De cive* (first printed privately in Paris in 1642), which amply showed its author's resolve to recast political, social, and ethical theory in a deterministic and postulational, if not mechanistic, framework. When Hobbes' natural philosophy and cosmology, *Elementa philosophiae*, saw print in 1655, Descartes was ignored, although the organization of the cosmological part of the work showed the influence of the *Principia*. To explain planetary motions Hobbes offered the idea, which he in all likelihood borrowed from Kepler, that the rotation of a body on its axis sets other bodies around it in an orbital motion.[87] Descartes had already been dead for eleven years when, in 1661, Hobbes referred to his explanation of the structure and properties of water and air as a construct that could hardly come 'from a sane man'.[88] A year later, in speaking of magnetism, Hobbes mentioned Descartes' theory of the origin of the earth but without any comment.[89]

Science, especially geometry, was the least convincing part of Hobbes' system of thought. His personal contact with the greatest scientists of his time did not seem to have much beneficial effect on him. Among them was Gassendi who rested his pioneering advocacy of atomism on Epicurus' philosophy. A chief opponent of Descartes among his contemporaries, Gassendi confined his interest in Descartes' cosmogony to showing that although Descartes rejected atomism, his ideas on the three basic forms of matter were a practical endorsement of it. No wonder that in Gassendi's one-column summary of the Cartesian genesis of the world only a single phrase was reserved to the capture by the sun's vortex of neighbouring vortices whose stars were thereby turning into its planets.[90] Gassendi did not refer to Descartes' cosmogony as he discussed the nature of sunspots[91] and of comets.[92] He did so briefly in connection with the distribution of stars in the universe.[93] Without mentioning the distribution of planets around the sun, Gassendi declared in general that the cause of the actual co-ordination of stars could only be known to the Creator. This was not to suggest, Gassendi added, that the actual position of stars was not without a 'natural necessity'. He, however, felt that the specific form of that 'necessity' could not be traced out by human ingenuity. At the same time he argued that assuming the ultimate causality of the Creator one could say with Epicurus that the sun, moon, and stars were not created separately and placed as such into the world, but rather evolved with the world in their vortices. About the Cartesian vortices Gassendi now stated that they were inferior replicas of those of Epicurus. The inferiority did not consist in the idea of vortex but in the alleged transformation of Descartes' first element into *striatae*. Descartes, Gassendi claimed, did not make it clear why certain particles of the first element should undergo that transformation and others not. That a stronger vortex could absorb a weaker one was a possibility that Gassendi admitted, though emphasising at the same time that 'whatever is said in the end, all this is mere hypothesis, and the difficulty remains; thus, one cannot do anything safer than to admit one's ignorance and refer the whole visible order of things to the decision of the Supreme Maker'.[94] His infinite power and glory warranted, so Gassendi declared in another context, that all the innumerable worlds should be inhabited.[95]

As to Descartes' disciples and champions, they provided unwitting evidence that his cosmogony was not to gain much credit. Henry de Roy (Regius) of Utrecht, who two years after the *Principia* composed a treatise which rightly appeared to Descartes as a rank plagiarism, offered only twenty lines on the origin of planets.[96] This cursory handling of the question, so contrary to Descartes' preferences, was carried over into the second and third editions, although they were considerably longer than the first.[97] Jacques

Rohault, the most important and articulate exponent of Cartesian physics, chose not to report on the Cartesian theory of planetary origins in 1671 in his *Traité de physique*, which served in uncounted editions and translations as the principal textbook on physics not only on the Continent but also in England for the next two generations. Rohault's procedure should appear all the more revealing as he described in detail Descartes' explanation of sunspots and novae through the formation of *striatae*.[98] Yet he said neither of the planets, nor of the comets that, according to Descartes, they had been former stars.[99] Again, while Rohault mentioned the vortex when speaking of the sun,[100] such was not the case in his short reference to the innumerable stars seen through the telescope,[101] the vortices of some of which were bound to collapse in the Cartesian theory. At the same time Rohault's *Traité* served ample evidence of his enthusiastic interest in elucidating gravity by vortex motion.[102]

While a fairly detailed account of Descartes' theory of the origin of planets was given in 1675 in a treatise of secondary importance by Claude Gadroys,[103] the topic was ignored in the far more important account of Cartesian physics and physiology, which Claude Perrault, a member of the Académie des Sciences and professor at the Sorbonne, published in 1680.[104] Ten years later the diffidence of Cartesians in their master's idea of planetary origins seemed to reach its high point in the massive, two-volume synthesis[105] of Cartesian thought by Pierre-Sylvain Régis, the foremost disciple and successor of Rohault. Régis defined planets as 'bodies composed of the third element, which being more solid than certain volumes of the second element similar to them . . . cannot escape from the vortex in which they were formed'.[106] This definition was as contrary to the views entertained by Descartes as was Régis' discussion of the formation of crust around stars. The process, which in Descartes' account was the triggering device of planetary genesis, produced, according to Régis, merely a decrease in the pressure originating in the star.[107] He assigned the formation of a planet to the agglomeration of the third matter into large bulks,[108] a patently non-Cartesian process, which he failed to analyse or to explain. His unqualified insistence that the vortices were 'indefectible' was as much at variance with Descartes' dicta as was the diagram, possibly the first of its kind, which showed the planets as massive spheres floating in the circular stream of the ethereal globules [*Illustration XII*].

The lonely enthusiast and the body scientific

Régis' deliberate departure from Descartes also represented a resistance to the enthusiasm by which the continual reprints and re-editions of Malebranche's *Recherches sur la vérité*, first published in 1674, endorsed the Cartesian concept of planetary origins. Malebranche, the last outstanding Cartesian philosopher, gave a spirited summary of Descartes' cosmogony insisting at almost every step that it was the inevitable consequence of relying on clear and simple ideas. Still, as Malebranche indignantly remarked, many rejected Descartes' cosmogony because it was so simple and easy to grasp that 'Women and Persons unskill'd in *Greek* and *Latine* are capable of learning it'.[109] The other reason was 'the chief thing that is found fault with in *des Cartes's* System', namely, its conflict with the biblical account of Creation.[110] In this connection Malebranche could quote Descartes' protestations of orthodoxy, but his cosmogonical thought was not given with proper nuances by Malebranche as he blithely reduced the circular motion of celestial vortices to simple motion, that is, motion in straight line: 'Supposing then that some part of Matter is moved in a Right Line, it will necessarily displace some other Portion of Matter, it shall find in its way, which latter shall Circularly move to take the room which the former has left: And if we conceive infinite Motions in a right Line in an infinite number of similar Parts of that immense Extension we consider; it will again necessarily follow, that all these Bodies mutually hindering each other, shall all conspire by their Reciprocal Action and Reaction, that is, by the Mutual Communication of all their particular Motions, to produce one that is Circular'.[111]

Such was the claim of one whose principal instinct was that of philosophers, not of physicists. The latter were soon to recognise that the orbiting of planets far transcended the potentialities of linear motion. No late-17th-century physicist of any standing gave his vote to Malebranche's insistence that the Cartesian vortices are a necessary outcome of linear motion: 'That first Consideration of the most simple Relations of our Ideas, already discovers to us the necessity of the *Vortexes* of *des Cartes*, that their number will be so much greater, as the Motions in a right Line of all the Parts of the Extension, having been more contrary to each other, shall with more difficulty have been reduc'd to the same Motion, and that amongst those *Vortexes* the greatest will be those in which most parts shall have concurr'd together to the same Motion, or whose parts shall have had more strength to continue their Motion in a right Line'.[112] The question of linear motion did not appear in Malebranche's claim that the minute globules themselves were small vortices.[113] The claim, however, certainly evidenced the hold which the notion of vortex had on his mind.

The contagiousness of the idea of cosmic vortices was very notable in the case of Guericke, who in spite of his antagonism for Descartes' science spoke of the earth's and of the planet's motion as taking place in the vortex of the sun.[114] Yet though Guericke showed much interest in cosmological topics, he seemed to eschew carefully any discussion of Descartes' theory of planetary origins, in all likelihood because of his low regard for it. Young Leibniz's first public plunge into science was even more expressive in that respect. His *Hypothesis physica nova*, published in 1671, was distinctly Cartesian in that he claimed to have achieved an explanation 'of all phenomena of nature from a single universal motion'.[115] Actually, Leibniz offered only an unconvincing verbalisation on Kepler's magnetic vortex theory of planetary motion while at the same time he dismissed Descartes' vortices, globules, fine particles, *striatae*, and fragments of all kinds, as too remote 'from nature's simplicity and from the evidence of experiments'.[116] The reprinting in London of the first part of the *Hypothesis* and the distribution of copies among members of the Royal Society during Leibniz's visit there in 1672 created no echo among leading British scientists, like Boyle, Hooke, and Newton, all of whom ignored the topic of Cartesian cosmogony.[117] Leibniz himself did not refer to cosmogony at all in the second part of his *Hypothesis*, an essay on 'abstract motion', which he dedicated to the

Académie des Sciences in Paris where he spent the year 1673. The essay signalled the start of his life-long campaign against the philosophy and theory of motion of that 'incomparable man, Descartes'. As it turned out, Leibniz was in basic agreement with Descartes' cosmogony, but of this the world of the learned received information only many years after Leibniz's death through the publication of his letter of March 22, 1714, to Bourguet. The same year saw the publication of Leibniz's *Monadology*, the principal ideas of which had already been circulating for some time. He now felt the need to state that his cosmic system of monads was evolutionary in character. In particular, Leibniz insisted to Bourguet that his system was compatible with the idea of an original chaos and he made it known that he was inclined to consider with Descartes the earth as a former fixed star. Indeed, he had ideas of his own on planetary evolution. He mentioned it as 'one of his own doctrines' that the earth might have had its origin in a huge sunspot 'which as a molten mass was ejected of the sun and was still trying to fall back into it'.[118]

Although Leibniz was practically alone among the prominent scientists of his age to endorse, privately though, what seemed to be the dearest to Descartes in his own system, the cosmogony of stars, planets, and comets, none of them could escape Descartes' influence. Following the publication of the *Principia*, it was immediately and generally sensed, what a century later d'Alembert stated, that its author was the first to base systematically and consistently the explanation of the physical world on some simple laws of motion.[119] Furthermore, to most readers of the *Principia* it must have been a great intellectual satisfaction to find there a physics both simple and plausible. Its development into a sweeping cosmic story, the novelty and originality of which were beyond dispute, could easily cast a magic spell on those reading it for the first time. The classic instance of this was recorded by Huygens who in 1645, at the age of 15, chanced upon a copy of the *Principia* in the library of his father whose home Descartes often visited. Almost half a century later Huygens' youthful experience still retained a touch of freshness: 'Descartes found the way to make others take his conjectures and fictions for truths. Much the same happened with the readers of the *Principia* as with those who read appealing novels that create the impression of being true stories. The novelty of the figures of his small particles and of his vortices was a source of great enjoyment. It appeared to me when I read for the first time this work of the *Principia* that all was well in the world, and I believed on finding some difficulties with the book that the fault was mine for not grasping well enough his thought. I was then but fifteen or sixteen. But having since discovered, as time went on, things obviously wrong, and others that are most improbable, I broke the spell that had seized me'.[120]

Huygens was the most outstanding among those who grew diffident of the cosmogonical superstructure that Descartes built while remaining basically Cartesian in their scientific methodology. The historic work of Huygens on centrifugal force was done within the framework of Descartes' vortices. After finding the correct form of that force he went on to fit it into the Cartesian explanation of gravity based on the vortex of globules around the earth. He believed that the rotational period of the ether should be equal to the period of a pendulum with the length of one earth-radius, or to one hour and twenty-four and a half minutes.[121] Like most friends and foes of Descartes' system of the world, Huygens was far more concerned about the finished form of the mechanism of the system of planets than of its genesis. In his last statement on vortices, made in his posthumously published *Cosmotheoros*, Huygens noted the mechanical difficulties inherent in the contiguous vortices of Descartes.[122] Huygens could not see how the original polyhedrical forms of vortices might start rotating. He instead proposed that star vortices should be pictured as 'so many little Whirlpools of Water, that one makes by the stirring of a stick in any large Pond or River, a great way distant from one another'.[123] This, of course, implied that one vortex could not envelop another and turn the latter's star into its own planet. Far from being ready to submit a substitute theory of planetary formation, Huygens took the view that one should not trouble one's head about matters that were, in his words, 'out of reach of human Knowledge or even Conjecture'.[124] Concerning Descartes' account of planetary and cosmic evolution as a whole, his erstwhile enthusiasm had in the long run turned into a jeer: 'All the whole story of Comets and Planets, and the Production of the World, is founded upon such poor and trifling grounds, that I have often wonder'd how an ingenious man could spend all that pains in making such fancies hang together'.[125]

Cavorting with planetarians

The foregoing sentence on which the *Cosmotheoros* came to a close showed that in a sense Huygens extricated himself from the spell of vortices while trying to the end of his life to clarify their dynamics. But his composing the *Cosmotheoros* proved that in a broader connection he remained fully under that spell. Although Descartes spoke only indirectly and with apparent diffidence about the possibility of planets being inhabited,[126] his cosmogony of planets implied around most if not all stars one or several planets, and it was all too tempting to picture them full of living beings in virtue of the 'Principle of Plenitude'.[127] By a strange twist of irony it remained for Huygens to expound in a 'scientific fashion' the existence of denizens on each and every planet in an immense if not infinite universe of countless stars. To crown the comedy he did this in the *Cosmotheoros* and in a manner for which he had just strictured Descartes, by basing a long story on 'poor and trifling grounds'. Cartesian was also the manner in which Huygens now boldly claimed the truth and logical necessity of his story as its successive phases unfolded. Once more the 'demonstration' of denizens of other planets, or planetarians in short, turned out to be a far less problematic enterprise than tracing out the formation of the system of planets, the very abode on which they were supposed to thrive.

Huygens' starting point set the tone of his whole discourse. If the planetarians were to be basically similar to us, they had to be based on dry land. That the surface of planets was solid followed for Huygens from the principle of similarity. Dissecting a dog would give a fair idea to any reasonable man, he argued, about the interiors of a 'Bullock, Hog, or any other Beast, tho he had never chanc'd to see the like opening of them'.[128] He coupled the principle of similarity with that of purposefulness to show that planetary surfaces could not be completely desert-like, for in that case the sun's heat would be radiated on them in vain. From this it 'naturally' followed the presence of water, of rivers, and of seas on any and all planets. In all this he did not feel that he reached conclusions too fast. As for the 'Growth and Nourishment' of planetarians of whatever domicile, ' 'tis no doubt the same with ours, seeing they have the same Sun to warm and enliven them as ours have'.[129] The 'spots' of Jupiter were for him watery clouds, although he added that Cassini might have been right in seeing in them the 'reflection from the Snow that covers the tops of the Hills in Jupiter'.[130] As to the presence of water on other planets the matter could not be more straightforward. If there was water on Jupiter, other planets could not be without it either.

From the occurrence of solid matter and moisture there followed in quick order the existence of plants and animals. The former had, of course, to be propagated by seeds on every planet; the latter could again multiply, so Huygens claimed, only by copulation of their male and female. Twice Huygens insisted in connection with the structure of planetary animals that they were like those on earth, because the forms of the latter exhausted all possibilities. The existence of rational beings on other planets rested on the consideration that it was not worthy of the Creator to leave the earthlings the sole contemplators of the richness of nature.[131] On the coat-tails of rationality came freedom of the will, so that the planetarians could be pictured as a mixture of good and bad traits. Their miseries, wars, and various afflictions served the same purpose as was the case with the inhabitants of the earth, namely, 'to exercise our Wits, and sharpen our Inventions'.[132] For, according to Huygens, if the planetarians were to lead their lives in an 'undisturbed continual Peace, in no fear of Poverty, in no danger of War', they would 'live little better than Brutes'.[133]

No less disputable was Huygens' claim that, since the nature of light was everywhere the same, so had to be the eye's structure. He applied the same argument to the hearing of planetarians, since according to him the air had to be the same everywhere in the universe. From the dependence of planetarians on a fauna and flora similar to that of the earth, it followed for him that their sense of touch, smell, and taste was closely patterned after ours. He even thought it to be clearly demonstrable that they, like the humans, must multiply by sexual copulation. Underlying this argument was Huygens' conviction that the earth, 'the smallest part almost of the Universe, was never design'd to monopolise so great a Blessing'.[134] It was on this rather arguable dictum that the first part of the first book of the *Cosmotheoros* came to a close.

Emboldened by the rapid progress of his 'intellectual' journey into the kingdoms of planetarians, Huygens, in the second part of the first book, went on to outline the various intellectual enterprises that necessarily had to develop among them. First to be mentioned was geometry, the most necessary and universal form of knowledge. The study of geometry inevitably brought about the cultivation of astronomy, and to the logic of this Huygens added the already overworked theme: 'For supposing the Earth, as we did, one of Planets of equal dignity and honour with the rest, who would venture to say, that no where else were to be found any that enjoy'd the glorious sight of Nature's Opera?'[135] This consideration in the hardly profound libretto, which Huygens was writing to that opera, was also his answer to a truly ominous objection: was he not constructing an edifice with the help of 'many Probabilities, one of which if it chance to be false, and contrary to our supposition, would, like a bad Foundation, ruin the whole Building, and make it fall to the ground'?[136] No wonder that such a degree of confidence in patently unsound reasoning could lead Huygens into drawing detailed pictures about the social living, architecture, navigation, music, agriculture, horticulture, and mining of the planetarians, emphasising at each step the close similarity of their achievements with ours. This was surely a breathtaking flight of fancy which he presented as verity. He was more sound in diagnosing his readers' condition when he gave them a 'breathing while'[137] as the first book of the *Cosmotheoros* concluded.

The second book consisted of an account of how the solar system would look to astronomers placed on various planets. This could be done without Huygens' consulting his planetarian colleagues. After all, two earthlings, Richer and Picard, had already measured in 1673 the parallax of Mars which yielded its true distance. This in turn gave through Kepler's Third Law the distances of the other planets and provided an apparently solid foundation for describing the true appearance of the solar system from either the closest or from the most distant planet. Here only the introductory pages of the second book need to be recalled. There Huygens took to task Athanasius Kircher for having 'cast off the only Foundation of Probability in such matters, which we have all the way made use of'.[138] The 'matters' concerned the planetarians for whom Kircher had no use in his *Iter exstaticum*, an imaginary journey from planet to planet.[139] Huygens might have been right in tracing Kircher's reluctance to populate the planets to some theological motivations, which Huygens clearly wanted to discredit by referring to Nicholas of Cusa, a Cardinal and a chief supporter of the view that there were rational beings everywhere in the universe.[140]

Worlds multiplied and two books ignored

At any rate, if Kircher turned out to be right, it was certainly for some wrong reasons. The physics of Kircher and Huygens and the scientific knowledge of those times hardly lent themselves to a serious discourse on physical conditions on planets, let alone on planetarians. Science then could not even set clear enough barriers for the flight of fancy that dominated speculation on the topic. Authors, who should and could have shown some rational restraint in the matter, fancied with free abandon the existence of beings on the moon and elsewhere. To achieve their aim they were willing to ignore or talk away glibly even the most obvious difficulties. Perhaps the most telling case in this respect was that of Bishop Wilkins who tried to show that the moon had an atmosphere in his *The Discovery of a World in the Moone, or a Discourse tending to Prove that 'tis Probable there may be another Habitable World in that Planet*.[141] To make matters worse he named Galileo as a chief witness who in his *Sidereus nuncius* had ascribed a real atmosphere to the moon.[142] Two decades later, in 1632, Galileo abandoned in his *Dialogue* this view by noting the total absence of water and clouds on the moon.[143] Unlike Wilkins, Galileo time and again emphasised that if there were living beings on the moon they were totally unimaginable to us.[144] This was for him the only conclusion justified on 'basic knowledge and natural reason'.[145]

That the dictates of both were honoured in the breach can be understood in the cases of Cyrano de Bergerac and of Francis Goodwin, whose accounts of a voyage to the moon and to the sun were plainly fictitious.[146] But Pierre Borel tried to write serious science in his *Discours nouveau prouvant la pluralité des mondes*.[147] Still one of his proofs depended on the mythical Paradise-bird. According to Borel its nest had to be on the moon, because the bird itself could only be seen dead, so he claimed, either on dry land or in the sea. Now, if there were birds on the moon, there were other animals as well, to say nothing of higher beings.[148] Borel, a great admirer of Descartes, turned even the Cartesian denial of the void to good advantage, for he claimed that it implied plenitude, that is, the same variety of things everywhere in the world as observed on earth.[149] Most of Borel's arguments rested on the principle of analogy. Any similarity, however generic, between the earth and the planets, such as eclipses there as on the earth,[150] or the planets' various colours,[151] or their motion through space,[152] was for him a conclusive evidence about the existence of planetarians.

The Cartesian who did most to spread the world view of Descartes was Bernard de Fontenelle, a novelist turned science populariser. He obtained the post of 'perpetual secretary' of the Académie des Sciences of Paris largely on the impact of his first effort in science popularisation which dealt, typically enough, with the plurality of worlds, or rather with the ubiquity of planetarians. When the book was first published,[153] it consisted of a dialogue distributed into five evenings. During the first evening Fontenelle explained to a Marchioness the Copernican ordering of planets. Here he made only a short reference to planets around other stars and to the earth being carried around the sun in a very subtle, rotating liquid. During the second evening he argued the existence of lunar inhabitants on the ground that the most surprising things can occur, and he recalled the astonishment of the aborigines of America on seeing huge ships with white men reach their shores. The third evening was devoted to the particularities of life on the moon and to the inhabitants of other planets. As to the former, Fontenelle's discourse was well exemplified by his readiness to declare the craters of the moon as convenient shelters against steady sunlight lasting two weeks at a time. As to the truth of the latter, he set great store by the observation that every chunk of soil was full of microscopic animals.

The topic of the fourth evening concerned the characteristics of the inhabitants of each planet. Fontenelle described the Mercurians as exceedingly vivacious people living in deserts divided by rivers of molten metal. For him Mercury was the 'Bedlam of the Universe'.[154] The Venusians fared better because they were less exposed to the sun's heat though they had enough of it to turn into vivacious little black people bent on perpetual celebrations. About the Martians Fontenelle found nothing much to note and similarly curt was his treatment of the Jovians and Saturnians. But the moons of Jupiter and the rings of Saturn forced him to delve into the topic of vortices. He was spellbound as he turned to the Marchioness with the words: 'Madame, if only you knew what the vortices of Descartes were!'[155] Needless to say, he saw no weak points in Descartes' account which he outlined only in generalities. One noteworthy part of the fifth evening was a vivid description of the perpetually blazing skies inside the Milky Way based on the assumed proximity of an immense number of small stars there. The other was the presentation of vortices as polyhedrons approaching the figure of spheres. It was during this evening that Fontenelle also described the Cartesian theory of the origin of comets and planets without, however, touching on the earlier phases of Descartes' cosmogony.

Eight years later, in 1694, the fourth edition of Fontenelle's work was augmented by a 'sixth evening', devoted to new discoveries in astronomy that allegedly brought further support to the proposition that all planets were inhabited. Rarely was a chapter 'on the latest' more behind its times. At any rate, no astronomical observation made during the previous eight years could be considered in any sense as being favourable to Descartes' system. As to astronomical writings, two books should have been mentioned by Fontenelle, if he really wished to be up to date. One was within his easy reach, the *Voyage du monde de Descartes* by Gabriel Daniel published in 1691.[156] In it both Descartes' dualism and his planetary theory were subjected to a virulent criticism. It was not difficult to find inconsistencies in Descartes' physics and cosmology. Daniel's best specific remark concerned the contradictory feature of a crucial phase of the Cartesian theory of planetary evolution, the capturing of one vortex by another. How could this, he asked, take place so that the captured star (planet) would still retain part of its original vortex around itself as postulated by the Cartesian theory?[157] Daniel was unquestionably right in noting the spell exercised on most minds by the reading of Descartes. He correctly traced this to one's willingness to accept uncritically the truth of some assumptions in the presence of a broad network of pleasing deductions.[158] He was also right in pointing out that the vortex had to be considered 'as the capital Point in the Cartesian System; and as the Foundation of that prodigious Edifice, which has been taken in our days by so many, for the compleatest Mastery of a Human Mind'.[159] His claim that 'the General Constructure of his World, and the Consequences he draws from it . . . have hitherto best escap'd the Censure',[160] implied unfamiliarity with Gassendi's dicta. But Daniel's really memorable misstatement came as he wrote that 'scarce any yet have given him [Descartes] disturbance upon the Hypothesis of his Vortices, which is notwithstanding the Foundation of all he says touching the motion of the Planets'.[161]

As a matter of fact, a momentous censure had already been passed on Descartes' vortices in Newton's *Principia*.[162] Most Cartesians were not to take immediate notice of a book which was forbidding reading except for a few, and which did not become widely available until twenty-six years later, in 1713, when it saw a second edition. In addition, there was the instinctive urge to seek physical reality behind Newton's mathematical formalism, and the vortices seemed to be a natural answer as a physical model. The discrepancy between the precision of Newton's formulae and the generality of the model was, however, too great. Vortices could not sustain elliptical planetary orbits, let alone cometary orbits of great eccentricity. One of the first to perceive this was Huygens, but in spite of this the vortices kept a strong hold on his thinking. Several years after he had studied the freshly printed *Principia* of Newton, he jotted down the baffling remark: 'The vortices destroyed by Newton. Replaced by vortices of spherical movement'.[163] The vortices cast indeed a powerful spell. They lived and lingered on though not as the framework of explanation of the evolution of planetary systems.

NOTES TO CHAPTER TWO

1. *Journal tenu par Isaac Beeckman de 1604 à 1634*, publié avec une introduction et des notes par C. de Waard (The Hague: M. Nijhoff, 1939–53), vol. 1, p. 244. Beeckman was one of the few to whom Descartes kept referring with unreserved admiration. On Descartes' relation to the scientists and ideas of his times and on Descartes' science the standard monographs are *Le développement de la physique cartésienne 1646–1712* by P. Mouy (Paris: J. Vrin, 1934), *The Scientific Work of René Descartes* by J. F. Scott (London: Taylor & Francis Ltd., 1952), and *The Vortex Theory of Planetary Motions* by E. J. Aiton (London: Macdonald, 1972).

2. See Part II of his *Discourse on the Method*, translated by E. S. Haldane and G. R. T. Ross (1911; New York: Dover, 1955), p. 87.

3. In his 'Olympica', a dozen or so manuscript pages, preserved only in the phrasing of A. Baillet; see *Oeuvres de Descartes*, edited by C. Adam and P. Tannery (Paris: L. Cerf, 1897–1913), vol. 10, pp. 186–87.

4. *Oeuvres*, vol. 11, pp. 1–118.

5. Descartes does not seem to have been directly influenced by the writings of Oresme, Buridan, and Albert of Saxony, some of which appeared in print during the first half-century of printing, but many of the salient ideas of these great figures of the 14th-century University of Paris had become widely shared and developed by Renaissance men of science whose ideas were familiar to Descartes. See on this, E. Gilson, *Études sur le rôle de la pensée médiévale dans la formation du système cartésien* (Paris: J. Vrin, 1930), p. 143.

6. *Le monde*, p. 34.

7. *Ibid.*, p. 49.

8. *Ibid.*, pp. 49–50 and 68.

9. *Ibid.*, pp. 58–60.

10. *Ibid.*, p. 72.

11. *Ibid.*, p. 71.

12. *Ibid.*, p. 110.

13. *Ibid.*, p. 48.

14. *Discourse on the Method*, p. 107.

15. *Ibid.*, p. 108.

16. *Le monde*, p. 47.

17. *Principia philosophiae* (Amsterdam: apud Ludovicum Elzevirium, 1644), in *Oeuvres*, vol. 8, pp. 1–348. References to the *Principia* will be given as III, 30, indicating the Part and Article, respectively.

18. *Les principes de la philosophie* escrites en Latin par René Des-Cartes et traduits en François par un de ses Amis (Paris: chez Henry le Gras, 1647), in *Oeuvres*, vol. 9. The translator,

identified on the title page as 'one of his friends', was the Abbé Picot, who inserted into the original text many explanatory phrases, mostly in italics, which reflected well the thinking of Descartes who wholeheartedly endorsed the translation with a special Preface.

19. *Principia*, III, 28–30. The French translation adds (III, 29) the picturesque detail that people sleeping in a boat going from Calais to Dover would not be aware of their being transported. The explanation rested on the recognition of the relativity of motion, but Descartes made no effort to unfold the consequences of this insight perceived and expressed before him with much greater emphasis by Nicholas of Cusa, Giordano Bruno, Galileo, and others. It is in this context that the word *vortex*, or *tourbillon* in the French translation, makes its first appearance in a cosmological connotation in the works which Descartes published. His previous use of the word *tourbillon* in *Les météores* (*Oeuvres*, vol. 6, p. 313) was in connection with small 'whirls' of dust produced by violent winds.

20. Letter of May 10, 1632; in *Oeuvres*, vol. 1, p. 250.

21. *Oeuvres*, vol. 2, pp. 484–85.

22. *Ibid.*, pp. 593–94.

23. Letter of July 30, 1640; in *Oeuvres*, vol. 3, pp. 134–35.

24. *Ibid.*, pp. 445–46.

25. *Principia*, III, 30.

26. *Ibid.*

27. This is the marginal title phrase of the last (64th) article of Part II.

28. See especially the concluding pages of Discours I of *Les météores*, in *Oeuvres*, vol. 6, pp. 233–39, where Descartes emphasizes that his explanation of atmospheric phenomena is not based on the substantial forms and qualities of the (Scholastic) philosophers, but on the interaction of very small (subtle) particles. The *Les météores*, *La dioptrique*, and *La géométrie* were first published as one volume with the *Discours de la méthode* (Leiden: de l'Imprimerie de Ian Maire, 1637) to illustrate the use of the latter.

29. Letter of July 27, 1638; in *Oeuvres*, vol. 2, p. 268.

30. *Strena seu de nive sexangula*, in *Johannes Kepler, Gesammelte Werke*, edited by W. von Dyck and M. Caspar (Munich: C. H. Beck, 1938–), vol. 4, pp. 261–80. For Descartes' discussion of snowflakes, see Discours VI of *Les météores*, in *Oeuvres*, vol. 6, pp. 291–302.

31. *Oeuvres*, vol. 1, p. 127.

32. *Ibid.*, p. 331.

33. In a letter of March 31, 1638, to Mersenne, in *Oeuvres*, vol. 2, pp. 85–86.

34. *Principia*, III, 153.

35. In a letter of April 20, 1646, to Mersenne, in *Oeuvres*, vol. 4, pp. 398–99.

36. See his letters of October 11, 1638, of November 15, 1638, and of January 29, 1640, to Mersenne, in *Oeuvres*, vol. 2, pp. 399, 443 and vol. 3, p. 11.

37. *Principia*, III, 37.

38. *Ibid.*, III, 6.

39. That this enlightening component of patristic tradition was sufficiently known in learned circles in Descartes' time can be seen from the theological information which Galileo received from friends in the clergy for the composition of his 'Letter to the Grand Duchess Christina' (1615). The first printing of the Italian original, together with a Latin translation, followed only in 1636, in Strasbourg, but the small edition saw free circulation only in non-Catholic countries, such as Holland, where Descartes lived.

40. *Principia*, III, 45. Here the French translation contains additional lines on the creation of Adam and Eve as adults.

41. See on this my *The Paradox of Olbers' Paradox* (New York: Herder & Herder, 1969), pp. 41–44.

42. *Principia*, III, 3.

43. In a letter of June 6, 1647, to Chanut; in *Oeuvres*, vol. 5, p. 56.

44. *Principia*, III, 46. Here Descartes said that the particles were 'almost equal'. In the next article he stated that they were 'equal with respect to one another'.

45. *Ibid.*, III, 47.

46. *Principia*, III, 47.
47. *Ibid.*, III, 46. The French translation here adds 'and comets' anticipating what Descartes was soon to spell out in detail.
48. *Ibid.*, III, 50.
49. *Ibid.*, III, 52.
50. *Ibid.*
51. *Ibid.*, III, 53.
52. See on this my *The Milky Way: An Elusive Road for Science* (New York: Science History Publications, 1972), p. 119.
53. *Principia*, III, 54.
54. *Ibid.*, III, 64.
55. *Ibid.*, III, 68.
56. *Ibid.*, III, 72.
57. *Ibid.*, III, 74, 75.
58. *Ibid.*, III, 83.
59. *Ibid.*, III, 90.
60. *Ibid.*, III, 91.
61. *Ibid.*, III, 94.
62. *Ibid.*, III, 96.
63. *Ibid.*, III, 98.
64. *Ibid.*, III, 99.
65. *Ibid.*, III, 104. As a case he mentioned the Pleiades in which formerly some astronomers counted seven and not six stars.
66. *Ibid.*
67. *Ibid.*, III, 115.
68. *Ibid.*
69. *Ibid.*, III, 115–16.
70. *Ibid.*, III, 121.
71. *Ibid.*, III, 127.
72. *Ibid.*, III, 119.
73. *Ibid.*, III, 122.
74. *Ibid.*, III, 123.
75. *Ibid.*, III, 139.
76. *Ibid.*
77. *Ibid.*
78. *Ibid.*, III, 141–45.
79. *Ibid.*, III, 146.
80. *Ibid.*, III, 147.
81. *Ibid.*, III, 148.
82. Bologna, ex typographia haeredis Victorij Benatij, 1651. See Lib. IV, cap. xxvii; Lib. VII, Sec. IV, cap. v; Lib. VII, Sec. VI, cap. iii.
83. *Principia*, III, 148.
84. *Ibid.*, III, 154.
85. *Ibid.*, III, 157.
86. In his letter of March 5, 1649, to Descartes; in *Oeuvres*, vol. 5, p. 313.
87. See the English version, *Elements of philosophy. The First Section, concerning Body*, in *The English Works of Thomas Hobbes*, edited by Sir William Molesworth (London: John Bohn, 1837), vol. 1, pp. 428–31. Being intent on mechanical 'purism', Hobbes insisted, obviously with an eye on Kepler's idea of a magnetic vortex, that magnetism was a purely mechanical phenomenon.
88. In his *Dialogus physicus de natura aeris*, which ended, in line with Hobbes' misguided obsession with geometry, on a section containing his latest 'solution' of doubling the cube. See *Thomae Hobbes . . . opera philosophica quae latine scripsit omnia*, edited by W. Molesworth (London: Longman, Brown, Green and Longman, 1845), vol. 4, p. 250.
89. In his *Problemata physica*, in *Opera philosophica . . . omnia*, vol. 4, p. 358.

90. *Syntagma philosophicum*, Pars Secunda, Sec. I, Lib. III, cap. v, in *Petri Gassendi opera omnia* (Lyons: sumptibus Laurentii Anisson & Ioan. Bapt. Devenet, 1658), vol. 1, p. 258.
91. *Ibid.*, Sec. II, Lib. II, cap. ii, p. 554.
92. *Ibid.*, Sec. II, Lib. V, cap. i, pp. 700–03.
93. *Ibid.*, Sec. II, Lib. II, cap. iii, p. 560.
94. *Ibid.* In the context Gassendi quoted Epicurus' dictum on vortices from his letter to Herodotus, but no reference to the role of vortices in Epicurus' cosmology was made by Gassendi in his *Philosophiae Epicuri syntagma*, in *Opera*, vol. 2, pp. 1–94.
95. *Syntagma philosophicum*, Sec. II, Lib. I, cap. vi, p. 529.
96. *Fundamenta physices* (Amsterdam: apud Ludovicum Elzevirium, 1646), . pp. 66–67. Curiously, those twenty lines did not occur in the section on 'the origin of planets and comets' (pp. 54–55), but in the section on 'the absorption of a celestial vortex'.
97. The second edition appeared under the title, *Philosophia naturalis* (Amsterdam: apud Ludovicum Elzevirium, 1654) and was about 150 pages longer than the first. The twenty lines in question were repeated almost verbatim on pp. 135–36. The third edition (Amsterdam: apud Ludovicum et Danielem Elzevirium, 1661) was about eighty pages longer than the second, and carried the same lines on pp. 154–55.
98. *Traité de physique* (Paris: chez la Veuve de Charles Savreux, 1671), Part [vol.] II, pp. 96–97. An uncanny evidence of the disrepute of the cosmogony of Descartes among his early disciples is the almost complete absence of any reference to it in P. Mouy's *Le développement de la physique cartésienne 1646–1712* (Paris: J. Vrin, 1934).
99. *Ibid.*, pp. 100–10.
100. *Ibid.*, p. 92.
101. *Ibid.*, p. 7.
102. *Ibid.*, pp. 118–28. Part of the discussion was taken up by the description of an experiment of Huygens. For further details, see Aiton, *op. cit.*, pp. 75–84.
103. *Le système du monde selon les trois hypothèses* . . . (Paris: chez Guillaume Desprez, 1675). See question iv in Chap. iii, sec. i, 'De la formation des Planètes', pp. 311–49.
104. *Essais de physique ou Recueil de plusieurs traitez touchant les choses naturelles* (Paris: chez Jean Baptiste Coignard, 1680). There is no reference to planets in Perrault's short discussion of the passing of a body from one vortex to another. See vol. 1, p. 119.
105. *Système de philosophie, contenant la logique, la métaphysique, la physique et la morale* (Paris: impr. de D. Thierry, 1690).
106. *Ibid.*, vol. 1, p. 422.
107. *Ibid.*, pp. 414–15.
108. *Ibid.*, pp. 421–24.
109. Quoted from the English translation, *Father Malebranche's Treatise concerning the Search after Truth, The Whole Work Compleat*, by T. Taylor (Oxford: printed by L. Lichfield, 1694), vol. 2, p. 61. The discussion of Descartes' cosmogony is in Book VI, chap. iv.
110. *Ibid.*, p. 66.
111. *Ibid.*, p. 62.
112. *Ibid.*
113. This idea was presented by Malebranche in full form in 1699, following his election as honorary member of the Académie des Sciences, in his essay, 'Reflexions sur la lumière et les couleurs et la generation du feu', in *Histoire de l'Académie Royale des Sciences, Année M.DC.XCIX. Avec les Mémoires de la Mathématique et de Physique pour la même Année*. Seconde edition, revûë, corrigée & augmentée (Paris: chez la Veuve de Jean Boudout, 1718), pp. 22–36.
114. See his chief work, *Experimenta nova (ut vocantur) Magdeburgica de vacuo spatio* (Amsterdam: apud Joannem Janssonium à Waesberge, 1672), pp. 159, 173 and 210.
115. The words quoted are part of the full title, *Hypothesis physica nova, quae phaenomenorum naturae plerumque causae ab unico quodam universali motu, in globo nostro supposito, neque Tychonicis, neque Copernicanis aspernando, repetuntur* (Mainz: typis Christophori Küchleri, 1671). Reprinted in *Die philosophischen Schriften von Gottfried Wilhelm Leibniz*, edited by

C. J. Gerhardt (1880; reprinted by Georg Olms Verlagsbuchhandlung: Hildesheim, 1960), vol. 4, pp.177–240.

116. *Ibid.*, p. 209. On Leibniz's admiration for Kepler, see Leibniz's *Theodicea*, §§ 360 and 380.

117. There is much interesting material in K. B. Collier's *Cosmogonies of our Fathers: Some Theories of the Seventeenth and Eighteenth Centuries* (New York: Columbia University Press, 1934) on authors who at that time discussed the past history of the earth without, however, going as far as to propose a theory of planetary evolution.

118. See *God. Guil. Leibnitii Opera philosophica quae exstant latina gallica germanica omnia*, edited by J. E. Erdmann (Berlin: G. Eichler, 1840), p. 722. In the same letter Leibniz referred to his 'Protogaea', a notice on the volcanic beginnings of the earth in the *Acta Eruditorum* (Leipzig, 1793), pp. 40–42. Although the notice contains no reference to the earth's origin as a planet, it evidences something of an evolutionary outlook in Leibniz. The same holds true of his many notes on geology, published for the first time in 1746 by C. L. Scheid and reprinted under the title, *Protogaea*, in *Gothofredi Guillelmi Leibnitii . . . opera omnia* (Geneva: apud Fratres de Tournes, 1768), vol. 2, pp. 181–240.

119. 'Discours préliminaire à l'Encyclopédie' (1751); see the English translation, *Preliminary Discourse to the Encyclopedia of Diderot*, by Richard N. Schwab, with the collaboration of Walter E. Rex, with an Introduction and Notes by Richard N. Schwab (Indianapolis: Bobbs-Merrill Company, Inc., 1963), p. 79.

120. Manuscript note Nr. 2791 in *Oeuvres complètes de Christiaan Huygens* (The Hague: Martinus Nijhoff, 1888–1950), vol. 10, p. 403.

121. 'Discours de la cause de la pesanteur', originally submitted in 1669 to the Académie des Sciences in Paris, but published only in 1690; *Oeuvres*, vol. 21, pp. 445–88. See especially p. 460.

122. References are to the English translation published in the same year, 1698, under the title, *The Celestial Worlds Discover'd, or Conjectures concerning the Inhabitants, Plants and Productions of the Worlds in the Planets* (London: printed for Timothy Childe, 1698), p. 157.

123. *Ibid.*, p. 159.

124. *Ibid.*, p. 160.

125. *Ibid.*

126. See note 43 above.

127. On the influence of this principle in 16th- and 17th-century cosmographical, philosophical, and literary works, see *The Great Chain of Being*, by Arthur O. Lovejoy (Cambridge, Mass.: Harvard University Press, 1936), chapters 4 and 5.

128. *The Celestial Worlds Discover'd*, p. 18.

129. *Ibid.*, p. 21.

130. *Ibid.*, p. 26.

131. *Ibid.*, p. 38.

132. *Ibid.*, p. 40.

133. *Ibid.*

134. *Ibid.*, p. 53.

135. *Ibid.*, p. 62.

136. *Ibid.*

137. *Ibid.*, p. 100.

138. *Ibid.*, p. 101.

139. Originally published in 1652. Less rare is the edition published by G. Schott, with Kircher's approval, in Würzburg (sumptibus Joh. Andr. & Wolffg. Jun. Endterorum haeredibus) in 1660.

140. *The Celestial Worlds Discover'd*, p. 3.

141. London: printed by E. G[riffin] for M. Sparke and E. Forrest, 1638. See Proposition X., 'That there is an Atmo-sphaera, or an Orb of gross, Vaporous Air, immediately encompassing the Body of the Moon'.

142. See the English translation in Stillman Drake, *Discoveries and Opinions of Galileo* (Garden City, N.Y.: Doubleday, 1957), pp. 39–40.

143. *Dialogue concerning the Two Chief World Systems*, translated with revised notes by Stillman Drake (Berkeley: University of California Press, 1962), p. 62.

144. *Ibid.*, pp. 61, 100 and 101.

145. *Ibid.*, p. 101.

146. *Histoire comique, par Monsieur de Cyrano de Bergerac, contenant les Estats et Empires de la Lune* (Paris: chez Charles de Sercy, 1657). See the modern English translation, *Voyages to the Moon and the Sun*, by Richard Aldington (New York: Orion Press, 1962). The text of Goodwin's *The Man in the Moone*, first printed in English in 1638, five years after its author's death, was re-edited by Grant McColley in *Smith College Studies in Modern Languages*, vol. XIX, Nr. 1, October 1937.

147. The remainder of the title states 'that the stars [planets] are inhabited lands, that the earth is a star [planet], that it is away from the centre [sun] in the third heaven [planetary orbit] and that it turns around the sun which is fixed and other very curious things'. Geneva [no printer], 1657.

148. *Ibid.*, p. 35.

149. *Ibid.*, p. 23.

150. *Ibid.*, pp. 36–37.

151. *Ibid.*, p. 22.

152. *Ibid.*, p. 17.

153. In 1686. References are to the very accurate English translation, enriched by explanatory appendices, *Conversations on the Plurality of Worlds*, A New Translation by a Gentleman of the Inner Temple (2nd ed.; London: printed for Thomas Caslon, 1767).

154. *Ibid.*, p. 189.

155. *Ibid.*, p. 197.

156. References are to the English translation by T. Taylor, *A Voyage to the World of Cartesius* (London: printed by Thomas Bennett, 1692).

157. *Ibid.*, p. 284.

158. *Ibid.*, [p. 4] of the introductory section, 'A General View of the Whole Work'.

159. *Ibid.*, p. 286.

160. *Ibid.*, [p. 4].

161. *Ibid.*, [p. 5].

162. In the Propositions on the 'circular motion of fluids' which constituted the concluding Section of the Second Book of the *Principia*.

163. In *Oeuvres*, vol. 21, p. 437. Huygens failed to note, as did before and after him all critics and students of Descartes, that the vortices were the source of an eventual 'disappearance' of the universe, a point which would have dismayed Descartes most of all. Because of the inequality of vortices and because of the inevitable incrustation of any star, the absorption of one vortex by another would have continued until all dead stars had come under the domination of the most powerful vortex. Since the star in its centre would have ultimately become incrusted, complete darkness had to be the final fate of that very world which Descartes had thought to be best discussed in terms of light!

CHAPTER THREE

GRAVITY AND GOD'S ARM

Chaos barren of planets

It has been guessed, and not without good reason, that Newton's choice of 'philosophiae naturalis principia mathematica' as the title of his immortal work was one of his many moves to indicate that Descartes had now been completely superseded. Other moves of Newton, such as his crossing out Descartes' name in his old manuscripts and his reticence about Descartes in print, were indications that Newton wanted to be remembered as a scientist being in a class wholly apart from Descartes. At any rate, the two *Principia* could not have been more different. One contained no mathematics at all, the other was in places forbiddingly difficult even for mathematicians. One consisted of lengthy verbalisations about the nature of basic physical entities and their interactions, the other offered about the same interactions mathematical formalism with no speculation on their nature. Most importantly, one had as its object the world as an evolving process, while in the other there were only brief hints that the world, or rather some of its principal parts, the stars to be specific, had existed at any time in a condition other than the actual one.[1]

The first edition of Newton's *Principia*, which saw review in only three journals,[2] did not cause any immediate and major disturbance in the ranks of the few Cartesians who read it. They could and did dismiss it as something which did not belong in the realm of physics but of mathematics.[3] None of them noticed that although the evolutionary aspect of the world was left wholly aside in Newton's work, it contained by implication a most serious threat to attempts aimed at an evolutionary portrayal of the cosmos and especially of the system of planets. The one to unfold this implicit message of Newton's *Principia* was Richard Bentley, chaplain to the Bishop of Worcester, who had professional training in classical philology and theology, not in science. But he lived in an age in which leading scientists firmly believed that the laws and discoveries of science were a direct pointer to a Lawgiver and Creator. Such a conviction was given a particular boost in England by the annual lecture-series founded by Robert Boyle, the greatest chemist of his time and a principal promoter of the Royal Society. As one of the highly literate divines intent on keeping science in God's service Bentley was chosen for the lectureship in 1692. He discharged his duty in eight lectures or sermons given in the church of St. Mary-le-Bow between March 7 and December 5. Of all the Boyle lectures Bentley's were the most successful and memorable. The collected text of his sermons went through five editions within the next five years following their original publication as separate booklets.[4]

Bentley grouped his sermons around three themes not counting the theme of his first sermon, the general folly of atheism. The first theme concerned the human mind as distinct from matter (Sermon II), the second theme related to the purposefulness of the human body (Sermons III–V). Although not cosmological in character, these sermons made clear the kind of atheism which constituted Bentley's target. It rested on the idea of an eternal, infinite world composed of atoms whose purely mechanical or chance combinations brought about every entity in the world. The only names mentioned by Bentley were those of the atomists of old, but his real antagonists were the latter-day supporters of Epicurean philosophy, including some Cartesians, some Gassendists, and certainly Hobbes and his admirers. It was, therefore, most natural that Bentley's third theme should be cosmological, namely, 'A Confutation of Atheism from the Origin and Frame of the World', which he set forth in the last three sermons.

Bentley's cosmological theme was elaborated in four points of which the first, the topic of his sixth sermon, claimed that it was 'impossible that the primary parts of our world, the sun and the planets, with their regular motions and revolutions, should have subsisted eternally in the present or a like frame and condition'.[5] For all its promise for clues on Bentley's thinking about planetary evolution, this claim of his bogged down in unconvincing argumentation about infinite quantities and periods. His fourth point, the topic of the ninth sermon, consisted in portraying the beauty and purposefulness of each and every part of the universe, especially the world of planets. This was preceded by the elaboration of the second and third points in the seventh sermon which certainly forms a major document in the history of theories of planetary origins. The second point concerned the impossibility of the eternity of matter and motion. The third point should be given in Bentley's own words: 'Thirdly, though universal matter should have endured from everlasting, divided into infinite particles in the Epicurean way; and though motion should have been coeval and coeternal with it; yet those particles or atoms could never of themselves, by omnifarious kinds of motion, whether fortuitous or mechanical, have fallen or been disposed into this or a like visible system'.[6] In modern phrasing this meant that mechanistic or purely scientific theories of the formation of planets and of their system could have no convincing measure of truth. Such was certainly a momentous and original statement. Its importance was enhanced by Bentley's resolve to justify it from Newton's *Principia*. Little did he guess that he combated well in advance important aspects of some future and famous theories on planetary origins, and that on one particular and decisive point he showed a scientific sensitivity which was not duplicated for almost two more centuries.

Bentley's systematic tackling of the idea of planetary origins strongly suggests that a particular form of it had already been given sufficient currency in some circles. Through Bentley's criticism of that form or theory its main features were brought together in a compact account which, in view of the great popularity of Bentley's sermons both in England and abroad, must have become widely known to philosophers, divines, and scientists alike during the next two generations. The starting point of that theory was the original chaos, a notion not too far from what within a century was spoken of as the nebular state of matter in the beginning. Bentley declared at the outset that his principal aim was to demonstrate that 'the atoms or particles which now constitute heaven and earth, being once separate and diffused in the mundane space, like the supposed *chaos*, could never, *without a God, by their mechanical affections*, have convened into this present frame of things, or any other like it'.[7] Bentley agreed wholeheartedly with his antagonists on the existence of atoms and the void in which they were supposed to move. He was quick to add that two important aspects of atomistic philosophy, or 'Epicurean physiology' as he called it, were raised from the status of precarious assertions to the rank of well demonstrated and uncontrovertible propositions. One was the universality of the gravitation of matter evidenced in the fact that all particles of matter in the vicinity of the earth gravitated toward its centre. With the experimental verification of this Bentley credited Boyle, an investigator of the behaviour of matter in vacuum pumps and the 'honourable founder' of the lecture series he was delivering. The other was the proportionality of the force of gravitation to the quantity of matter. The credit for this went, of course, to 'that very excellent and divine theorist, Mr. Isaac Newton'.[8] Bentley did not recite Newton's proof, but referred 'the curious in the auditory' to 'the book itself', that is, to the *Principia*. At the same time Bentley notified his listeners that to Newton's 'most admirable sagacity and industry we shall frequently be obliged in this and the following discourse'.[9]

Planets in need of God's arm

By noting that a strict notion of the void must imply the absence of a continuous ethereal fluid, Bentley was ready to construct the general framework in which only a recourse to God's mighty arm could account for the formation of planetary systems. The framework consisted in estimating the ratio of empty space to space occupied by matter within a sphere with a diameter corresponding to half the distance of the sun from the nearest stars. By assuming that the total mass within that sphere was 50,000 times the mass of the earth, he reached the conclusion that 'the empty space of our solar region . . . is 8,575 hundred thousand million million [8.5×10^{17}] times more ample than all the corporeal substance in it'. He added in the same breath that one 'may fairly suppose that the same proportion may hold through the whole extent of the universe'.[10] The really important point for Bentley was that the foregoing ratio permitted one to form a realistic picture of the chaos in which 'the atheists' saw the original status of things.

Since the protagonists of the original chaos described it as an even, or almost even, distribution of all available matter, Bentley could reasonably argue that such a distribution must have embodied the ratio already mentioned. In other words, each elementary particle or atom in that chaos had to be surrounded by a void 8.5×10^{17} times bigger than the dimensions of that particle. This also meant that every particle had to be 'above nine million times its own length from any other particle'.[11] Since in an even distribution 12 particles at most could be at that 'nearest' distance from a given particle, a chance meeting with neighbouring particles was exceedingly unlikely. Or in Bentley's words: if any of the particles 'should be moved mechanically (without direction or attraction) to the limit of that distance, 'tis above a hundred million millions odds to an unit that it would not strike upon any other atom, but glide through an empty interval without any contact'.[12] He noted that since the 'atheists' pictured all the atoms to be in constant motion, chances for a collision between them had to be considered as having a much smaller probability.

As an illustration of this he offered the case of two ships 'fitted with durable timber and rigging, but without pilot or mariners, to be placed in the vast Atlantic or Pacific Ocean, as far asunder as may be'. Thousands of years might expire, remarked Bentley, 'before those solitary vessels should happen to strike one another'.[13] In addition, there was the problem of the colliding atoms sticking together in very large quantities to form huge, stable bodies as the planets are. Even if this was expected to happen, its infinitesimally small probability notwithstanding, there remained the problem of putting that huge mass in a circular orbit. The 'Epicurean physiology' had at its disposal only two factors to achieve this, and even when combined they were hopelessly inadequate for the task. One was the chance co-ordination of the motion of an immense number of atoms toward a specific point so that a planet might be formed there. The other was the chance formation of vortices or whirlpools in the chaos. The former could only arise in wishful thinking, the latter, Bentley noted, contradicted the laws of motion. Purely linear motion, he wrote, never 'bends in a circle'.[14] As it turned out two centuries later, although it should have been recognised much earlier, this was as momentous a remark as was Bentley's claim about the necessity of a sufficiently dense, resisting medium for the purposes of planetary evolution. Whirlpools, he noted, which were compact enough to 'restrain' the planet in its orbit, could not be formed in a chaos that was basically a great emptiness. 'There is no such restraint in the supposed chaos, no want of empty room there; no possibility of effecting one single revolution in a way of a vortex, which necessarily requires (if attraction be not supposed) either an absolute fulness of matter, or a pretty close constipation and mutual contact of its particles'.[15]

There now remained the discussion of the mechanical formation of planetary systems, assuming that all particles of matter derived their motion from gravitating toward one another. According to Bentley, gravitation was even more damaging for the 'atheists' resting their cause with the initial chaos. If matter gravitated, then a chaos approximating an even distribution of particles in a practically empty space was hardly conceivable. One had, therefore, to assume that the present lumping of matter by gravitation existed since eternity, leaving no logical room for an original chaotic stage. Also, the fact of gravitation undermined their staple contention that the actual shape of the world could again dissolve into a chaos. Clearly, the 'atheists' seemed to be worse off with gravitation than without it, a point which Bentley put bluntly: 'This is absurder than the other; that only supposed innate gravity not to be exerted; this makes it to be defeated, and to act contrary to its own nature'.[16]

What was the nature of gravitation? In 1692 Bentley could not suspect that in a decade or so Newton was to reinstate in the Queries the continuous ethereal fluid as the mechanical vehicle of gravitational attraction. According to Bentley's reading of the *Principia*, and he was not off target, gravitation meant action-at-a-distance. Or as Bentley emphatically reminded his audience: 'Mutual gravitation or attraction, in our present acception of the words, is the same thing with this; 'tis an operation, or virtue, or influence of distant bodies upon each other through an empty interval, without any *effluvia*, or exhalations, or other corporeal medium to convey and transmit it'.[17] He did not postpone specifying that only 'an immaterial and divine power' could infuse into 'inanimate and brute matter' the ability to 'affect other matter without mutual contact'.[18]

Here was then what Bentley called 'a new and invincible argument for the being of God' and a 'direct and positive proof that an immaterial living mind doth inform and actuate the dead matter and support the frame of the world'.[19] That frame had some enticing aspects which seemed to be worth exploring and Bentley set about the task with undisguised enthusiasm. Some accurate information, which the author of the *Principia* disclosed about that frame, was now turning into a welcome grist for Bentley's mill. The first of these was the gravitational attraction exerted by all particles of matter in all directions. But could, for instance, the gravitational influence of a particle 'entombed and wedged in the very centre of the earth' be transmitted mechanically by the ether or any other effluvium to 'another particle in the centre of Saturn'?[20] If this already seemed to be unimaginable, what then of the fact that the same particle affected all other particles in the universe whatever their position and distance? If the transmission of the effect was to be carried along mechanically, then the particle, so Bentley believed, had to move 'all manners of ways and constantly in the same instant and moment' and this, as he put it, was 'flatly impossible'.[21]

Then there was the question of the formation of separate bodies by gravitation. This presented an insurmountable difficulty both in a finite and in an infinite universe. In the former, gravitation, which already presupposed a special act of God, would pull all matter into one huge body. Before broaching the case of the infinite universe, Bentley allowed for the sake of argument that distinct planetary bodies could nevertheless form out of chaos by gravitation. He could be generous, as he knew he had an ace up his sleeve. It was the orbiting of planets along nearly circular paths. He put forward the situation with due attention to details. First came the place of the formation of planets. Was it not preposterous to assume that all were formed at a point of their actual orbit? If, however, the formation took place at a point farther away from the sun, or closer to it, gravitation could not bring them to the correct distance. If they had been closer to the sun than necessary at the moment of their formation, gravitation could not have moved them away from the sun; if they had been farther than necessary, their gravitational approach toward the sun would have certainly carried them beyond their proper orbit, driving them inevitably into the very body of the sun. 'What natural agent', asked Bentley, 'could turn them aside, could impel them so strongly as falling?'[22]

Clearly, no natural solution was in sight. At best one could conceive a 'cross attraction' causing the planets to miss the sun narrowly. In that case, and here Bentley well displayed his familiarity with the *Principia*, the planets would pass far beyond the sun and achieve a very elongated, highly eccentric orbit, as comets do, around the sun. Since this was not the pattern in which the planets moved, Bentley could readily draw his memorable conclusion about the indispensable role of the divine arm in the formation of the planetary system: 'circular revolutions, in concentric orbs about the sun or other central body, could in nowise be attained without the power of the divine arm'.[23]

To make matters absolutely clear, Bentley called attention to the fact that the nearly circular motion of planets in their orbits could only be pictured from the viewpoint of dynamics as the combined effect of a 'gravitational energy' toward the great central mass of the sun and of 'a transverse impulse of a just quantity in each projecting them directly in tangents to those orbs'.[24] With an air of supreme confidence in his cause, Bentley once more conceded hypothetically that gravity might be an innate property of matter and not something due to a special provision of God. But the transverse impulse and its exact quantity were not ever to be derived from the ordinary workings of nature: 'certainly this projected, this transverse and violent motion can only be ascribed to the right hand of the *most high God, creator of heaven and earth*'.[25]

Such a spirited resolve to find in the divine arm the solution for a specific problem about nature's mechanism had its pitfalls. Gaps in scientific knowledge could most unexpectedly be filled. The best illustration of this was the *Principia* itself. More than in any other book either before or after, there were unveiled in it as if by a magic stroke solutions to a broad range of problems that until then appeared to be beyond man's acumen. Still one cannot help being impressed by Bentley's intuition which made him grasp in essence the magnitude of what was later to be known as the problem of angular momentum in the planetary system. It was only in that connection that he found it appropriate to utilise the expressive image of God's omnipotent arm. Such a religious imagery does not have today an appeal even remotely as strong as it did in Bentley's time. Among the many causes that contributed to that change of preferences one cannot find, however, a convincing scientific elucidation of the origin of the angular momentum of planets. There the gap in our knowledge has not narrowed significantly, a point which will have a crucial relevance for our story.

Bentley displayed an equally profound grasp of some fundamental questions of cosmology when he carried the discourse from the solar system to the system of stars and their planets. If the realm of stars was finite, it could be saved only in two manners from turning under the influence of gravitation into one huge lump of matter. Either all stars orbited around a huge central sun, or they were kept in fixed positions. In both cases God's direct action was necessary, a point which Bentley had already made abundantly clear. In the case of an infinite universe, the immobility of stars could be solved by assuming their distribution to be absolutely even. But the gain implied an even greater loss. First, it presupposed a perfectly homogeneous chaos, which could not give rise to the accretion of larger bodies by gravitation. Second, the perfectly homo-geneous distribution of an infinite number of stars eliminated gravitation itself without which planets could not remain in a circular orbit around the stars. Or as Bentley put it, 'an equal attraction on all sides of all matter is just equal to no attraction at all'.[26]

No matter what aspect of the frame of the universe was taken under scrutiny, Bentley soon noticed the inevitable need for a special, direct, and single act of God. Needless to say, such an approach could not fail to nip in the bud any inclination to entertain evolutionary ideas either about the solar system or about the universe at large. This was well illustrated in several of the ten facts, the topic of his eighth and last sermon, which in Bentley's eyes could only be explained by God's special providential concern for the inhabitants of the earth. The first of these was that the centre of our solar system was not a cold body. It did not occur to Bentley that the sun, let alone the planets, might have once had temperatures very different from what they now had. The second point was the nearly circular orbit of all planets and here he once more recalled the role of the divine arm.[27] Bentley was fully aware that this secured a measure of evenness of temperature without which human life could hardly be sustained. Had the earth passed very close to the sun and receded very far from it, the temperature changes would have extinguished all life on it. The third point concerned the sufficiently wide spacing of planets from one another. This eliminated undue mutual disturbances (here Bentley should have read the *Principia* more carefully), and as Bentley put it, 'no natural and neces-sary cause' could be responsible for this, but only 'divine art and conduct'.[28] The natural way of picturing the latter evoked a single act, not a protracted process calling for an evolutionary outlook. In the fourth point Bentley probed into the implications of the actual distance of the earth from the sun. Mindful of the decrease of 'the density of the sunbeams' with the 'square of the distance from the body of the sun', he pointed out that no life on earth could exist if it was at Mercury's distance or at Saturn's, a consideration which would have stood Fontenelle and Huygens alike in good stead. The earth was, therefore, put into its actual position 'by the wisdom of some voluntary agent, and not by the blind motions of fortune or fate'.[29]

Bentley invoked God's direct action to account for the daily rotation of the earth (fifth point) and for the actual speed of orbital motion (sixth point), inasmuch as life was basically affected by their absolute and relative values. Life, or more directly climate, was also greatly influenced by the plane of the ecliptic and by the constancy of the direction of the earth's axis (seventh point), which made Bentley once more adore 'the divine wisdom and goodness for this variety of seasons, for seed-time, and harvest, and cold and heat, and summer and winter'.[30] According to him, even the actual composition of the atmosphere (eighth point), the distribution of water and dry land (ninth point), and the relative proportion of flat land and mountains (tenth point) could not originate in mechanical causes but only in God's providential concern.

Bentley's awareness of the narrowly defined values of various physical parameters to make human life on earth possible is worth noting. Because of this he argued that if animals existed on Venus and Saturn, the hot and cold extremities of the solar system, they had to be of a shape and constitution entirely unimaginable to us. In the same context he also noted that the matter of each planet 'may have a different density, and texture, and form, which will dispose and qualify it to be acted on by greater or lesser degrees of heat, according to their several situations'.[31] Curiously enough, Bentley did not mix such scientific considerations into his discourse on the inhabitants of planets around each of the innumerable stars, which constituted the introductory part of the eighth sermon. He could have argued that it was hardly to be expected that God should produce innumerable exact replicas of the solar system. If, however, their variety was as wide as nature permitted, their inhabitants too must have been of inconceivably wide varieties. Bentley based the latter point solely on the plenitude of God. Rational beings, even if composed of body and soul, could be different in many ways from man, so argued Bentley who had but contempt for questions such as whether all planetarians shared in Adam's fall.[32]

Newton's stamp of approval

The question could not, of course, arise within the framework of the *Principia*, and the exchange of letters between Bentley and Newton did not touch upon such incidentals. In general, Newton was visibly pleased with Bentley's spelling out the broader implications of the *Principia*: 'When I wrote my treatise about our system, I had an eye upon such principles as might work with considering men for the belief of a Deity; and nothing can rejoice me more than to find it useful for that purpose'.[33] Newton's first specific remark seemed to undercut Bentley who claimed that in an infinite, homogeneous universe no large bodies could form by gravitational attraction. But Newton rose here to the defence of the workability of the infinite, homogeneous distribution of matter only to strengthen the point made by Bentley in connection with the heating and illuminating role of the sun. Natural causality could not explain, according to Newton, the difference between the hot, lucid bodies of stars and the cool, dark masses of planets. Without the intervention of a 'voluntary Agent'[34] Jupiter might as well be a hot star and the sun a dark body. For Newton it was inconceivable that there should be an evolutionary sequence of celestial bodies. He had only scorn for the great evolutionary picture of the heavens unfolded by Descartes: 'The Cartesian hypothesis of suns losing their light, and then turning into comets, and comets into planets, can have no place in my system, and is plainly erroneous'.[35]

Newton's conviction that the solar system came out ready-made from the hands of the Creator was also evident in the words by which he endorsed Bentley's interpretation of the orbital motion and arrangement of the planets. 'To make this system . . . with all its motions, required a cause which understood and compared together the quantities of matter in the several bodies of the sun and planets, and the gravitating powers resulting from thence'.[36] After naming the principal variables in the solar system Newton declared that 'to compare and adjust all these things together, in so great a variety of bodies, argues that cause to be, not blind and fortuitous, but very well skilled in mechanics and geometry'.[37] Apart from the naïve anthropomorphism of the simile underlying its scientific veneer, the inference has a double thrust. One brought out the role of God, the other implied that He acted once and for all and not by letting the forces of nature shape things in a truly gradual manner.

This reluctance to consider the merits of an evolutionary approach was rather strange on Newton's part as his next remark concerned the heating effect of the sun on the crust of planets. For Bentley the sun was the sole source of heat in the solar system. For Newton the bowels of the planets were fiery and he seemed to make the amount of fire proportional to the size of the planets. Thus Jupiter and Saturn did not have to be pictured as exceedingly cold bodies. Newton also called Bentley's attention to the fact that recent observations of Flamsteed indicated considerable perturbations by Jupiter and Saturn. Still Newton felt that it was, therefore, all the more justifiable to see in the great distances of the heaviest planets a providential act of the Creator. Newton was less keen on ascribing to a direct act of God the steadiness of the earth's axis or its inclination with respect to the sun's equatorial plane. On the whole, he was very enthusiastic about Bentley's forging from the characteristics of the solar system, as traced out in the *Principia*, proofs on behalf of the existence of God. In fact, Newton thought that he had perceived one such proof that appeared 'to be a very strong one', but he took the view that it was 'more advisable to let it sleep' until 'the principles on which it is grounded are better perceived'.[38]

Newton's second letter to Bentley earned a place in the history of science because of the rather questionable rebuttal which Newton administered to Bentley's advocacy to the finiteness of the universe.[39] The letter contained some telling points about the planets as well. Most importantly, Newton now made his own Bentley's phrase about the divine arm, and he did it precisely in connection with the transverse impulse needed for each planet to go into a nearly circular orbit around the sun. Or in Newton's emphatic words: 'I do not know any power in nature which would cause this transverse motion without the divine arm'.[40] Clearly then, a theory like the one allegedly offered by Plato, and repeated by Galileo and François Blondel, had to be considered basically deficient, as Newton noted in the same breath.[41] In his third letter to Bentley, Newton further pointed out that there was in nature no common place from which the planets might have started falling toward the sun, and that Galileo had no correct idea about the measure of acceleration by which the planets would have in that case rushed into the sun. Actually, as Newton noted, the planets would have gone into a parabolic orbit around the sun, that is, never to return to it again.[42]

Newton's detailed criticism of the 'Platonic theory' was only partly motivated by a desire to display the strength of the celestial mechanics he discovered. He was also visibly impatient with any evolutionary approach either to the planetary system or to the system of the world at large. Evolution was for him a bogey, which he saw even where it was not really present. Thus he read into Bentley's sixth sermon an implicit advocacy of the possibility that the present system of the world was preceded by others 'and so on to all past eternity'.[43] This, according to Newton, could easily imply the pernicious error that gravity was innate to matter. To avoid this pitfall one had to specify that 'old systems cannot gradually pass into new ones; or that this system had not its original [condition] from the exhaling matter of decaying systems, but from a chaos of matter evenly dispersed throughout all space'. To this he added a flat rejection of theories of repeated cosmic evolution: 'The growth of new systems out of old ones, without the mediation of a divine power, seems to me apparently absurd'.[44]

It seems that Newton's pen slipped when he introduced with a 'but' the foregoing remark containing the reference to 'a chaos'. He obviously meant to say 'or'! He had not only stated already several times the impossibility of deriving things from a chaos, that is, from matter evenly spread out in the beginning. He was now to repeat the same in his fourth letter to Bentley: 'I would now add, that the hypothesis of matter's being at first evenly spread through the heavens is, in my opinion, inconsistent with the hypothesis of innate gravity, without a supernatural power to reconcile them'.[45] The most telling part of his fourth letter lay elsewhere. In its first sentence Newton acknowledged that it was Bentley who directed his attention to some very important implications of the *Principia*: 'The hypothesis of deriving the frame of the world by mechanical principles from matter evenly spread through the heavens, being inconsistent with my system, I had considered it very little before your letter put me upon it'.[46]

Newton's endorsement of Bentley's utilisation of the *Principia* soon was an open secret among members of Newton's principate. Before any of them raised his voice against the evolutionary interpretation of the system of planets, the leader himself stated in print his opposition to it. He did so in 1706 in a Query added to the Latin edition of his *Opticks*. There he declared that it was 'unphilosophical . . . to pretend that it [the present order of the world] might arise out of a Chaos by the mere Laws of Nature'.[47] By the present order of the world he principally meant the world of planets. The uniformity of their motions, so Newton declared, 'must be allowed the Effect of Choice'.[48] He also strongly implied the role of a 'Choice', or direct intervention on the part of the Creator when the 'System wanted a Reformation'.[49] This reformation meant the restoration of the original arrangement of planets by God's mighty arm whenever their departure from their erstwhile orbits had become too large. As causes of that derangement Newton listed the mutual perturbation of planets and their gradual slowing down due to their passage through the ether.[50]

Newton's emphatic rejection of an evolutionary view of the system of planets was not without some covert inconsistencies. In the same context he also described God as having 'in the Beginning form'd Matter in solid, massy, hard, impenetrable, moveable Particles, of such Sizes and Figures, and with such other Properties, and in such Proportion to Space, as most conduced to the End for which he form'd them'.[51] He also ascribed all phenomena 'to the various Separation and new Associations and Motions of these permanent Particles'.[52] Such statements could readily suggest an evolutionary view. After all, while Newton rejected the notion of the original big Chaos, he compared particles of salt to a chaos, that is, a unit 'dense, hard, dry, and earthy in the Centre, and rare, soft, moist, and watery in the Circumference'.[53] More importantly, to offset his apparent advocacy in the *Principia* of gravitation by action-at-a-distance, he now endorsed as its carrier the ether, a material distinct from dense or ordinary fluid.[54] Was not he taking thereby an indirect step toward his claim made forty years earlier that all entities were formed out of the ether?[55] The relatively few readers of the *Principia* could also recall that there he spoke of the repeated 'rejuvenation of stars' by the falling of comets into them.[56] It could also be recalled that he had endorsed in 1696 the efforts of William Whiston, whom he later picked as his successor in the Lucasian chair, to give some role to natural causes in the formation of the system of planets. But Whiston soon ceased to be his friend and his work became disowned in Newton's circle. It was only half a century later and not in England that Whiston's speculations entered the mainstream of the history of theories on the formation of planetary systems.[57] It took even longer before it became known that privately at least Newton loved to speculate on the evolution of stars and worlds.[58]

Planets as bone of contention: Newtonians versus Leibniz

The official Newtonian record, which indicated a round rejection of evolutionary views, was amplified by Samuel Clarke through the publication in 1710 of the third edition of his annotated translation of Rohault's textbook.[59] With this edition Clarke's notes grew almost to one-third of the length of the book itself, which for another two or three decades served as the principal vehicle of instructing physics to undergraduates in Cambridge and elsewhere. Living under the auspices of Newton's rapidly growing glory they had to be protected with Clarke's notes from being contaminated by Cartesianism. Thus, for instance, Rohault's brief, innocuous reference to the motion of planets in the sun's vortex was given a lengthy commentary by Clarke in four points.[60] The fourth consisted in an emphatic rejection of Descartes' evolutionary view of the planetary system, a detail which Rohault did not even mention. Clarke's phrasing echoed Bentley and Newton of the *Opticks*. The orbital motion, Clarke stated, was to be conceived as the combined effect of the gravitational force and of 'Projectile Motion in straight lines', both of 'which were impressed upon them by God at the Beginning'.[61]

The next step in the Newtonian slighting of the evolutionary view of the origin of planets followed in 1713 with the publication of the second edition of the *Principia*. Its special fame was due to the 'General Scholium' written by Newton to vindicate the theistic foundation of his physics. Once more Newton emphasised the impossibility of the formation of the planetary system through 'mere mechanical causes'.[62] The various regular patterns characterising the motion and correlation of planets contrasted sharply, according to Newton, with the extreme variety of cometary orbits. Also, he saw a basic difference between the swarm of comets and the family of planets with respect to mutual perturbations. The former seemed to be largely free of them, whereas the latter did not. Consequently, he could not but make the memorable generalisation:

'This most beautiful system of the sun, planets, and comets, could only proceed from the counsel and domination of an intelligent and powerful Being'.[63] The same held true, he added, of planetary systems possibly existing around other stars. They too had to be 'formed by the like wise counsel', and 'be subject to the dominion of One'.[64]

In the Preface written by Roger Cotes to that second edition the same idea was stressed but on a less exalted level. By then Newton left debates with his opponents in the care of his lieutenants, of whose large number Cotes was one. His was now the assignment to further discredit Descartes whose evolutionary theory of planetary origins constituted the target of Cotes' acid remark: 'Those who assume hypotheses as first principles of their speculations, although they afterwards proceed with the greatest accuracy from those principles, may indeed form an ingenious romance, but romance it will still be'.[65] Plain sarcasm was thrown in for good measure into the fray as Cotes claimed with obvious reference to Descartes that even a 'miserable reptile' would have known that the actual order of nature was established not by fate and necessity but by the sovereign freedom of the Creator.[66]

The special symbol of free divine choice and of its powerful execution was the divine arm which began to loom large in the cosmological literature. The popularity of the symbol was no doubt helped by the fact that in one respect the divine arm could not be different from an ordinary one. The arm had as its necessary complement the hand on which craftsmen heavily depended. By Bentley's time it had become a commonplace to describe God as a craftsman, nay the incomparable clockmaker. The clock turned, in fact, into the chief paradigm of intelligibility in mechanical or classical physics. Once a process of nature could be approximated to the working of a clock, no further questions were asked. So it was with the world when described as the masterwork construed by God, the supreme clockmaker. Acquiescence into this outlook could only strengthen aversion to look at the system of planets, often taken for the world itself, as an evolutionary product. This is all too well illustrated in the exceptions Leibniz took to the explanation of the 'frame of the world' as given in the 'General Scholium' of the *Principia* and in the Queries of the *Opticks*. The criticism had its special weight not only because of Leibniz's stature but also because of the sharpness of some of his charges. Newton entrusted the job of rebuttal to Samuel Clarke who hardly guessed then that history would largely remember him for the replies he penned to the objections of Leibniz.[67] According to Leibniz, what Newton said about matter and motion helped only the materialists, while his definition of space as God's sensorium could delight only the atheists. Leibniz also charged that it was most unworthy of God to be represented as the maker of a clock which needed repairs: 'I maintain it [the world, or the system of planets] to be a Watch, that goes without wanting to be Mended by him: Otherwise we must say, that God bethinks himself again. No; God has foreseen every thing before-hand; There is in his Works a Harmony, a Beauty, already pre-established'.[68] Leibniz could not, of course, deny the numerous evidences of decay, or of apparent 'losses of motion', in nature. But he insisted that as far as the universe at large was concerned, the harmony constituted by the amount and proportion of forces at play never diminished: ''Tis true that every particular Machine of Nature is, in some measure, liable to be disordered; but not the whole Universe, which cannot diminish in Perfection'.[69]

Leibniz, who identified pre-established harmony in the best possible world with the actual succession of forms, could not bring himself to contemplate, openly at least, cosmic and planetary evolution to an extent comparable to the theory of Descartes. Leibniz seemed to be almost inhibited in this respect as can be seen from a short paragraph toward the end of his last letter to Clarke: 'As for the motions of the celestial bodies, and even the formation of plants and animals, there is nothing in them that looks like a miracle, except their beginning. The organism of animals is a mechanism, which supposes a divine preformation. What follows upon it, is purely natural, and entirely mechanical'.[70] It was easier for him to contemplate the growth, that is, the mechanical evolution of animals and plants than the evolution of planets, let alone of their system.

The manner in which Clarke tried to eschew the question of evolution was equally expressive. On the one hand, he had to defend the legitimacy of periodic repairs in the system of planets as this was Newton's position. On the other hand, he insisted that the cosmic amendments were real only with respect to us but not with respect to the plans and wisdom of the supreme Maker of the world clock.[71] At the same time the idea of a perfect world clock running without fail forever was for Clarke tantamount to 'the notion of materialism and fate', and he dismissed as invitation to scepticism the effort by which 'a philosopher can represent all things going on from the beginning of the creation, without any government or interpretation of providence'.[72]

Clarke's remark about 'a philosopher' was most likely aimed at Descartes, although by 1715 it was hardly necessary to censure him for his evolutionary picture of the universe. Even his most ardent followers would have by then been wholly satisfied if they could have saved at least the vortices. Philippe Villemot, who eight years earlier published a book-length defence of Descartes' vortex theory of planetary motions,[73] made only a brief evolutionary remark which suggested the repeated rise and dissolution of comets in the turbulent flow of the solar vortex.[74] About the origin of planets Villemot kept a meticulous silence even in contexts which logically invited a comment.[75] This procedure certainly reflected the general disrepute[76] of Descartes' theory, due as much to the intrinsic implausibility of the theory as to the tacit belief that somehow the paradigm of clockwork barred the perspective of evolution. Common reliance on that paradigm seemed to indicate precisely that. No one had followed up that curious speculation which Hooke offered half a century earlier in his *Micrographia*,[77] where he attributed an evolutionary wizardry to clockworks that was bordering on magic. If a watch stopped because its spring broke and because a few of its parts fell out, it was only necessary, he wrote, to shake the remaining parts long enough to effect their chance recombination into a new mechanism operating smoothly though rather differently from its previous pattern.[78]

Speculation of this type did not earn credit for Hooke. It was not until the late 19th century that such a conjecture, or a more sophisticated version of it, would have found sympathetic hearing among reputable men of science. In the early 18th century standard scientific opinion, represented by Newton's principate, endorsed at most the 'rejuvenation' of a star but not the idea of its gradual formation. There was a similar unanimity on the point that the planetary system was not only a clockwork but one with several features that could only be explained by a direct recourse to the 'omnipotent arm'. The theme in fact was made so popular by Bentley that twenty years after his lectures William Derham, rector of Upminster, devoted his own Boyle lectures to a 'Demonstration of the Being and Attributes of God from a Survey of the Heavens'.[79] Derham said nothing original by praising the 'infinite Hand' which must have put the planetary system into motion,[80] or by referring the doubtful to the *Principia* 'of the most sagacious Sir Is. Newton'.[81] The novelty of Derham's discourses lay in his tacit refusal to utilise God's arm for periodic repairs. He judged well the direction which intellectual preferences were soon to take.

No accent on Newton: Wolff and Swedenborg

Derham's book was duly quoted in Christian Wolff's lengthy account of the sundry evidences of the purposefulness of the created realm, first published in 1724.[82] Wolff, who was led from physics and mathematics to philosophy, did not entirely neglect Newton,[83] but there was no room for periodic repairs of the system of planets in Wolff's world view. First a protégé of Leibniz and later the foremost representative of Leibnizian thought, Wolff soon became a most influential spokesman for that rationalistic optimism which set the tone of thought in Germany during much of the 18th century. This optimism rested in no small measure on the conviction that everything was stable in the universal clockwork. Thus when Wolff discussed the wide separation of planets from one another he failed to note that this might account for the absence of great mutual perturbations. For Wolff those great distances prevented the eclipsing of one planet by another and also secured the development of their inhabitants without the influence from other planetarians.[84] Wolff firmly believed in the existence of denizens on the various planets, but even his verbosity could not add much new to the vivid portrayals available in the re-editions of Fontenelle's work.

Wolff's discussion of planetary distances had another feature which became the starting point of subsequent speculations about a possible arithmetic progression determining the initial separation of planets. The 'magic trick', as is often the case in science, consisted in an obvious over-simplification. According to Wolff, 'if one divided the distance of the earth from the sun into ten parts, then the distance of Mercury from it [sun] was 4, that of Venus 7, that of Mars 15, that of Jupiter 52, that of Saturn 95'. Then with his typical wordiness he rephrased the same by saying that if one placed all the planets on a line connecting the sun with Saturn, and divided it into 95 parts, the distances of each planet would be given by the foregoing numbers. To this he added that therefore 'Mercury and Venus are separated from one another by 3 such parts,

Venus and the earth also by 3, the earth and Mars by 5, Mars and Jupiter by 37, Jupiter and Saturn by 43'.[85] The jump from 5 to 37 had to appear enormous, and Wolff readily took up a speculation of Derham that Mars must have had a moon, or rather two, of its own to fill that place. This would have made a fine transition from the one moon of the earth to the four around Jupiter and Saturn,[86] but it hardly filled the gap between Mars and Jupiter.

Intent as Wolff was to see in every nook and cranny of the universe an evidence of a purposeful design by God, it was because of charges of advocating mechanistic determinism that he had to move from the University of Halle to Marburg in 1721 at the height of his career. This mechanistic determinism in Wolff's thought was more Leibnizian in origin than Newtonian. Leibniz in turn served for Wolff as the channel to Descartes' thriving at conceptual clarity and apriorism. This clarity was noticeably absent in most of the writings of Emanuel Swedenborg who published, in 1734, an ambitious and highly peculiar account of the evolution of the universe.[87] It was based on arcane speculations about the properties of the 'active' and 'passive', and of the first, second, third, and fourth 'finite', as Swedenborg called the successively higher formations of the elementary constituents of the material world. Fortunately, the obscurantism of his lengthy lucubrations on the 'finites' did not make wholly incomprehensible what he said about the formation of the planetary system.[88] Its interest for the history of the topic lies not so much in Swedenborg's thinly disguised Cartesianism but in the sequence of diagrams by which he illustrated his theory. The sequence was to remain for a century and a half the only one of its kind with respect to richness of details and had, of course, to come from the hands of one who believed he had unlocked every secret of the universe.

In a truly Cartesian fashion Swedenborg assumed that the 'fourth finite', or the grossest of his fundamental particles, formed a crust around the sun. The next step was a genuine product of the imagination of Swedenborg who did not espouse Descartes' idea of the collapse of the vortex of a star surrounded by a thick crust. He rather pictured the crust to recede slowly from the sun and develop fissures within itself [Illustration XIIIa]. From this shape of a fractured spherical shell the crust began to separate into an outer and an inner layer while collapsing toward the equatorial zone [Illustration XIIIb]. The inner layer consisting of 'actives' fell then back into the sun which itself was entirely constituted of 'actives', while the 'passive' outer layer turned into a belt only to break up at the same time into round masses [Illustration XIVa]. The smaller of these masses became the satellites around the larger ones, the planets [Illustration XIVb]. Unfortunately, Swedenborg himself obscured the relative clarity of this sequence by a diagram which showed a curious ring of satellites around the sun [Illustration XV]. In the text accompanying the diagrams the only worthy statement was that a 'thousand words will not convey the idea which may be derived from a single representation'.[89] Swedenborg did not explain why planets should go, as shown in his diagrams, 'toward the [respective] circle of their orbit'.[90] On the basis of his theory all planets should have moved in one belt, unless one assumed that the incrustation of the sun took place repeatedly.

This is what Swedenborg might have had in mind as he quoted at length the astronomical literature about variable stars.[91] The vague cosmogonical utterances in Ovid's *Metamorphoses* were quoted by him with equal zeal.[92]

Newton with French accent

While Swedenborg's work received some praise for its sections on minerals (he was a prominent mining engineer), his cosmogony earned him no credit. In spite of his Cartesianism, which was burdened with rank obscurantism, Swedenborg was not noticed in France. To be sure, by the mid-1730's France was on the verge of openly shifting her allegiance from Descartes to Newton. The opening act of that shift came with the publication in 1738 of Voltaire's *Elemens de la philosophie de Newton.* Voltaire's advocacy of Newton had some theological motivation which he put forward somewhat ostentatiously. According to him, Cartesianism led to Spinozean pantheism, whereas there was still to be seen a Newtonian 'who was not a theist in the most rigorous sense of the word'.[93] The theist of Voltaire, and he himself was an ardent one, was really a deist for whom it was inconceivable that God should do anything with the world once it had been created. Thus in presenting Clarke as one who vanquished Leibniz, Voltaire said not a word about their clash over the point whether the planetary clockwork needed repairs or not.[94] This was highly indicative of the deistic propaganda dear to Voltaire who possibly for the same reason did not seem to care that he had contradicted Newton with his claim that according to Newton the material universe was finite.[95]

Voltaire's universe, he said little about stars, was also immutable. In it there was room neither for the evolution of planets nor for the evolution of plants. He did not even tolerate the thought that various major features of the earth's surface could be the product of broad geological transformations. In defending his anti-evolutionist stance, he went as far as to claim – part in jest, part in despair – that shells found on some high Alpine slopes were placed there by pilgrims.[96] He wrote about the original chaos that it was 'an impossibility in the eyes of reason', but only to chastise biblical religion for its anthropomorphisms and myths.[97] For him the universe was conceivable only in terms of a machinery with neatly tooled components that stayed in their places once and for all. 'The universe', he wrote toward the end of his life, 'puzzles me and I cannot believe that this clockwork could exist and have no clockmaker'.[98] The puzzles had nothing to do with questions of physical origins. He remained to the end a worshipper of the supreme clockmaker whose mechanism ran unchanged and unimpeded once it had been started.

Voltaire was certainly an effective factor in swaying France behind Newton though not the most important factor in scholarly circles at least. The Abbé Pierre de Sigorgne, professor of physics at the Collège du Plessis, produced not only valuable treatises on Newtonian astronomy and philosophy but also a lengthy rebuttal of the last major exposition of the Cartesian system, Joseph Privat de Molières' *Leçons de physique,* the fourth or last volume of which appeared in 1739.[99] While Privat de Molières offered nothing new on Descartes' theory of planetary evolution,[100] de Sigorgne was emphatic in pointing out some of its basic weaknesses.[101] A chief among these concerned the plane of planetary orbits which, if planets had entered the vortex of the sun as Descartes imagined, must have shown a great variety of inclinations with respect to the sun's plane of rotation. The reason given for this by de Sigorgne was twofold. First, Descartes postulated that the plane of rotation of neighbouring vortices was at great angles with respect to one another. Second, it followed on the basis of Newton's physics that the planes of rotation and orbiting were invariable.

The reason why in the late 1730's Descartes' vortices still found ardent supporters in France was traced by d'Alembert, the youngest and scientifically most important figure among the early French supporters of Newton, to a mistaken sense of national pride. His remark might have been especially aimed at the eager crowning by the Académie des Sciences of an essay on vortical fluid motion, which Johann Bernoulli wrote in 1730 to vindicate both Descartes and Newton, though in different respects.[102] D'Alembert made his remark in the Preface to a long treatise on the equilibrium and motion of fluids; a work which established him, at the age of 23, as a leading scientist of his time.[103] It was also evidence of d'Alembert's Newtonianism, tainted with the rationalism of the Enlightenment, that he eschewed the origin of planetary systems. Newtonian as he was, he wanted no part of a Supreme Being immediately involved, as if through miracles, in the formation and maintenance of the system of planets.

The same motivations also determined the scattered dicta on the topic by Maupertuis, the most colourful advocate of Newton in France in the 1740's. The first part of his Essai de cosmologie, written in 1751, was devoted to the proofs, drawn from the contemplation of nature, of the existence of God. Except for his own 'law of least action' Maupertuis found none of the proofs sufficiently convincing.[104] Among those explicitly discarded by him were Newton's ideas on the topic. The most telling aspect of this was that Maupertuis referred to Query 31 of the Opticks, but reported only that Newton saw evidence of God's existence in the regular motion of planets in one plane. He failed to give his reader even an inkling about the emphasis which Newton put on the periodic restoration by God of the original balance among the planets.[105]

Such manhandling of the record had to be motivated by some extraneous aims and these were clearly discernible in Maupertuis' discourse. His overt distrust of almost all scientific proofs on behalf of the existence of God was one of them. It made him quote with favour Bernoulli's essay, crowned by the Académie des Sciences. If there was a slight probability to derive the common plane of planetary orbits from a physical process, such as vortical fluid motion, then, Maupertuis argued, there was no justification in tracing it directly back to God.[106] He should have exercised his memory with as much rigour as he did his reasoning. Years earlier, in 1732, Maupertuis himself pointed out a basic defect in Bernoulli's arguments. While Bernoulli seemed to be right in taking into account factors which Newton overlooked in estimating the orbital speed of the vortex fluid at various points from the centre, the velocity gradient could not be made such as to satisfy Kepler's Third Law for all planets at the same time.[107]

The other extraneous motivation which made Maupertuis pass over in silence the question of periodic repairs was his belief in the perfection of the cosmic mechanism. Once more consistency suffered. After all, he tried to enhance his law of least action by contrasting it with the fact that neither Descartes nor Leibniz could come up with a law of the conservation of motion which was not contradicted by experience.[108] In other words, Maupertuis readily recognised that in every physical process the quantity of motion or of energy (vis viva) diminished. Would it not have been reasonable to assume that something could steadily decrease in the case of planetary motions as well? His reluctance to deal with the matter was rather revealing as he admitted that the perturbations of Jupiter and Saturn were considerable.[109] Consideration of the 'decrease of motion' might have in turn led him to considering the reverse process, or evolution. But he was willing to speculate only on the accretion of rings around very large bodies, such as Saturn, and on the formation of systems of moons around large planets. As to the rings, he believed their material to come from the 'torrents that circulate around the planets'.[110] As to the moons, he considered them to be captured comets.

Beyond the moons he did not let evolution be at play. A good illustration of this was the long letter which he published following the appearance of a large comet in 1742.[111] He started out by debunking old superstitions about comets as harbingers of great calamities. This was not to be taken that comets, which Maupertuis believed to be solid bodies, could not be very destructive. His letter contained dramatic portrayals of near and head-on collisions of comets with the earth. He described how the earth's axis would as a result be deranged and how the earth's surface would be flooded by torrential waters and clouds. He even entertained the possibility of a collision turning the earth into hot lava or breaking it into thousands of pieces.

The lady to whom Maupertuis wrote the letter was not to be frightened after all. In Newton's system order ruled supreme. The comets obeyed the strictly defined elliptic orbits no less than the planets did. In addition, there were the immense spaces in which even the sun's huge body was a minute speck. Last but not least, none of the larger comets seemed to come really close to the central regions of the solar system. The small comets could hardly do much harm. In fact, a close encounter with a small comet could be imagined as highly beneficial. It could be captured by the earth as a second moon to make the night brighter. Another pleasant possibility was, according to Maupertuis, that in a particularly favourable encounter the inhabitants of the comet could readily be transferred to the earth. 'Who would be more surprised then', exclaimed Maupertuis, 'we or the inhabitants tossed from the comet upon the earth? What strange figures we would find each other!'[112] Once more obsession with planetarians turned a scientific mind oblivious to obviously unfavourable aspects of 'favourable' physical situations. A 'favourable' encounter could also reset the earth's axis in such a way as to secure perpetual spring for some areas of the earth. Maupertuis dismissed this prospect with the remark that men were not known to have been dissatisfied with the four seasons that had been around for the past five to six thousand years 'since our earth finds itself placed where she is in the heavens'.[113]

This was a remark fully in line with the picture of the 'system of the world' as a clockwork started by God. In it minor repairs could at best occur, such as the replenishing of the sun by comets and the rejuvenation of stars. In all these Maupertuis duly referred to Newton. Although when in speaking about major collisions Maupertuis mentioned Whiston, the latter's efforts to base, in part at least, planetary evolution on collisions with comets made no real dent on Maupertuis' mental world. It exuded order, lawfulness, and stability as had to be the case with any loyal Newtonian. In that world, in a much deeper sense than Maupertuis meant in his closing words on the harmlessness of the comet of 1742, there was 'nothing to fear, nothing to hope'.[114] The truth of this derived from the practical impossibility for much that was new to happen in a cosmic clockwork. Its main parts, the planets, were put into it ready-made and not to evolve into it. Maupertuis did not suspect that a major departure from that Newtonian outlook on the world of planets would soon be made in that France where he had so effectively helped to shift allegiance from Descartes to Newton.

NOTES TO CHAPTER THREE

1. One such hint came in connection with novae, pictured by Newton as the rejuvenation of aging stars by comets falling into them. See *Philosophiae naturalis principia mathematica* (2nd ed., Cambridge: 1713. p. 481, and 3rd ed., London: G. & Y. Innys, 1726, p. 525).
2. See E. J. Aiton, *The Vortex Theory of Planetary Motions* (London: Macdonald, 1972), p. 114.
3. According to P. S. Régis, probable author of the anonymous review of the *Principia* in the *Journal des sçavans* (June 1698, pp. 303–15), it contained not physics but 'pure mechanics'.
4. An easily accessible facsimile reprint of the original edition of the 7th and 8th Sermons is in *Isaac Newton's Papers and Letters on Natural Philosophy*, edited by I. B. Cohen (Cambridge: Harvard University Press, 1958), pp. 313–94. References will be to the text in *The Works of Richard Bentley*, edited by Alexander Dyce (London: Francis Macpherson, 1838; reprint by AMS Press, New York, 1966), vol. 3.
5. *Ibid.*, p. 132.
6. *Ibid.*
7. *Ibid.*, pp. 148–49.
8. *Ibid.*, p. 149.
9. *Ibid.*
10. *Ibid.*, p. 153.
11. *Ibid.*, p. 154.
12. *Ibid.*, p. 155.
13. *Ibid.*, p. 158.
14. *Ibid.*, p. 160.
15. *Ibid.*, pp. 160–61.
16. *Ibid.*, p. 162
17. *Ibid.*, p. 163.
18. *Ibid.*, pp. 162–63.
19. *Ibid.*, p. 163.

20. *The Works of Richard Bentley*, vol. 3, p. 164.

21. *Ibid.*, p. 165.

22. *Ibid.*, p. 166.

23. *Ibid.*, p. 167.

24. *Ibid.*

25. *Ibid.*

26. *Ibid.*, p. 171.

27. *Ibid.*, p. 178.

28. *Ibid.*, p. 180.

29. *Ibid.*, p. 181.

30. *Ibid.*, p. 190.

31. *Ibid.*, p. 182.

32. *Ibid.*, p. 175.

33. Of Bentley's four letters only one is extant. Newton's four replies will be quoted from Bentley's *Works*, vol. 3. For quotation, see p. 203.

34. *Ibid.*, p. 204.

35. *Ibid.*

36. *Ibid.*, p. 205.

37. *Ibid.*, p. 206.

38. *Ibid.*, p. 207.

39. See on this my *The Paradox of Olbers' Paradox* (New York: Herder & Herder, 1969), pp. 61–65.

40. Bentley's *Works*, vol. 3, p. 210.

41. *Ibid.*, p. 210. The context of Newton's reference to Blondel is as follows: 'I do not know any power in nature which would cause this transverse motion without the divine arm. Blondel tells us somewhere in his book of Bombs, that Plato affirms, that the motion of the planets is such, as if they had all of them been created by God in some region very remote from our system, and let fall from thence towards the sun, and so soon as they arrived at their several orbs, their motion of falling turned aside into a transverse one. And this is true, supposing the gravitating power of the sun was double at that moment of time in which they all arrive at their several orbs; but then the divine power is here required in a double respect, namely, to turn the descending motions of the falling planets into a side motion, and, at the same time, to double the attractive power of the sun. So, then, gravity may put the planets into motion, but, without the divine power, it could never put them into such a circulating motion as they have about the sun; and therefore, for this, as well as other reasons, I am compelled to ascribe the frame of this system to an intelligent Agent' *ibid.* The 'book of Bombs' of Francois Blondel (1618–1686), 'maréchal de camp' in the army of Louis XIV, was his *L'art de jetter les bombs* (Paris: chez l'autheur et N. Langlois, 1683) of which the first part comprised the pre-Galilean opinions on the trajectory of projectiles, the second part dealt with the practice of artillery, the third with the theory, and the fourth with the solution of some difficulties. In chapter viii of Part III, on 'some marvelous consequences of the properties of accelerated motion', Blondel wrote: 'Au reste comme il est difficile de comprendre qu'un mobile puisse d'abord acquerir un degré de vitesse determiné, sans avoir passé par tous les degrez precedens de moindre velocité; On peut ici juger pour quelle raison les Anciens ont été persuadez que les sentimens de Platon avoient quelque chose de divin. Car ce Philosophe dit sur ce sujet que Dieu ayant, peut-être, créé les Astres dans un même lieu de repos, les avoit laissé dans la liberté de se mouvoir en ligne droite vers un même point, à la manière des choses pesantes qui sont portées vers le centre de la terre, jusqu' à ce qu'ayant dans leur chûte passé par tous les degrez de vitesse, ils eussent acquis celui qui leur étoit destiné; après quoy il avoit converti ce mouvement droit et acceleré en mouvement circulaire pour le rendre égale et uniforme, afin qu'ils pussent le conserver éternellement' (p. 165). Blondel thought that one could, therefore, perhaps calculate the distance from the sun of that particular starting point of all planetary motions: 'Ce qu'il y a deplus admirable dans cette pensée, c'est que les pro-

portions qui se trouvent entre les distances des Astres et les différences de la vitesse de leurs mouvements, se trouvent assez conformes aux suites de ce raisonnement; et qu'il ne seroit, peut-être, pas absolument impossible de determiner le situation de ce premier lieu de repos, d'où ils auroient tous commencé de ce mouvoir' (p. 166). Blondel should have referred to Galileo's *Dialogue* as his obvious source. It is easier to excuse his silence on the calculations which M. Mersenne published in 1637 on the falsity of Galileo's claim. See on this W. R. Shea, *Galileo's Intellectual Revolution: Middle Period, 1610–1632* (New York: Science History Publications, 1972), p. 130.

42. Bentley's *Works*, vol. 3, p. 214.

43. *Ibid.*, p. 211.

44. *Ibid.*

45. *Ibid.*, p. 215.

46. *Ibid.*

47. See Query 31 of *Opticks, or a Treatise of the Reflections, Refractions, Inflections and Colours of Light* (reprint of the 4th edition of 1730; New York: Dover, 1952), p. 402.

48. *Ibid.*

49. *Ibid.*

50. *Ibid.*, p. 368 (Query 28, also added in 1706) and p. 402.

51. *Ibid.*, p. 400.

52. *Ibid.*

53. *Ibid.*, p. 386.

54. *Ibid.*, p. 376, and more explicitly pp. 350–51 (Query 21 added in 1717).

55. 'Perhaps the whole frame of Nature may be nothing but various Contextures of some certaine aethereall Spirits of vapours condens'd as it were by praecipitation, much after the manner that vapours are condensed into water or exhalations into grossed Substances', Newton wrote in an essay entitled, 'An Hypothesis explaining the Properties of Light', which he sent to Henry Oldenburg, secretary of the Royal Society, in 1675. See *The Correspondence of Isaac Newton*, edited by H. W. Turnbull (Cambridge: Cambridge University Press, 1959), vol. 1, p. 164.

56. See note 1 above.

57. As will be discussed in the next chapter.

58. In his 'Newton and the Cyclical Cosmos: Providence and the Mechanical Philosophy' (*Journal for the History of Ideas* 28 [1967]: 325–46), a careful collection of brief and private speculations of Newton on cosmic evolution, D. Kubrin seems to minimize the value of Newton's longer, more thematic, and public endorsements of a non-evolutionary world view.

59. On the development of Clarke's annotations, see M. A. Hoskin, ' "Mining All Within": Clarke's Notes to Rohault's *Traité de physique*', in *The Thomist* 24 (1961): 353–63.

60. See *Rohault's System of Natural Philosophy*, illustrated with Dr. Samuel Clarke's Notes taken mostly out of Sir Isaac Newton's Philosophy. Done into English by John Clarke (London: printed for James Knapton, 1723), vol. 2, pp. 73–76.

61. *Ibid.*, p. 74.

62. See Florian Cajori's revision of the English translation of the *Principia* by Motte (1729), *Sir Isaac Newton's Mathematical Principles of Natural Philosophy and his System of the World* (5th printing; Berkeley: University of California Press, 1962), p. 544.

63. *Ibid.*

64. *Ibid.*

65. *Ibid.*, p. xx.

66. *Ibid.*, p. xxxii.

67. *A Collection of Papers which passed between the late Learned Mr. Leibnitz and Dr. Clarke in the Years 1715 and 1716 relating to the Principles of Natural Philosophy and Religion* by Samuel Clarke, D.D. (London: printed for James Knapton, 1717).

68. *Ibid.*, p. 31.

69. *Ibid.*, pp. 111–12.

70. *Ibid.*, p. 267.

71. *A Collection of Papers*, p. 45.

72. *Ibid.*, p. 15.

73. Villemot, a 'doctor of theology and curé of the Guillotière', published his work, *Nouveau système ou nouvelle explication du mouvement des planètes* (Lyons: chez Louis Declustre, 1707), both in French and in Latin, the two versions facing one another. The work, now extremely rare, is divided into three parts. The first deals with the motion common to all planets, that is, with their elliptical orbiting in the sun's vortex; the second with the motions specific to each of them, namely, with their rotation and the inclination of their axis; the third with the 'interior motion of planets', or their gravity, apogee, and tides. Villemot ascribed the inclination of the earth's axis to the heterogeneous distribution of its mass (p. 128), and explained gravitation as the pushing of heavier objects toward the centre by the boiling of the earth's material (p. 182). In a note following the Preface, Fontenelle stated that the book deserved to be published because of its author's 'ingenious ideas'. Fontenelle's longer endorsement of the book's basic claims ('Sur les forces centrales des planètes', in *Histoire de l'Académie Royale des Sciences. Année MDCCVII* [Paris: chez Jean Boudot, 1708] pp. 97–103) evidenced both Fontenelle's amateurism in physics and celestial dynamics, and his addiction to the Cartesian vortices.

74. *Nouveau système*, p. 104.

75. Such contexts were the division of the sun's vortex into the planetary heavens (p. 40), the luminosity of sun as due to the rapid agitation of its particles (p. 66), the assigning of comets beyond the orbit of Saturn (p. 86), the irregular flow in the vortices (p. 94), and the allegedly steady pressure in the periphery of the sun's vortex to keep the planets' perihelia invariable (p. 220).

76. A good illustration of this is the *De vorticibus coelestibus dialogus, cui accedit quadratura circuli Archimedis et Hippocratis Chii analytice expressa* (Padua: ex typ. J.-B. Conzatti, 1712) by Giovanni Poleni (1683–1761), who at twenty-six became professor of astronomy at Padua and later distinguished himself in hydrology. Poleni's work was a sober and well-informed discussion of the various views offered on vortices as vehicles of the motion of planets. He referred to Kepler, Descartes, Bullialdus, Hevelius, David Gregory, Halley, Newton, Leibniz, Saurin, Varignon, and last but not least to Villemot. Toward the end of the book Poleni stated in a brief aside that the Cartesian theory of the origin of comets by the collapse of a stellar vortex cannot be admitted (p. 200). The main conclusion of his *dialogus* was a vindication of Newton's views on vortices.

77. *Micrographia or Some Physiological Description of Minute Bodies made by Magnifying Glasses with Observations and Inquiries thereupon* (London: printed for Jo. Martyn & Ja. Allestry, 1665).

78. *Ibid.*, p. 133.

79. This is the subtitle of his *Astro-theology*, first published in 1715. References are to the second, corrected edition (London: printed for W. Innys) published in the same year.

80. *Ibid.*, p. 67.

81. *Ibid.*, p. 147 note.

82. *Vernünftige Gedancken von den Absichten der natürlichen Dinge*. References are to the second edition (Frankfurt and Leipzig, 1726). On the allusion to Derham, see p. 58.

83. *Ibid.*, pp. 58, 131, 148.

84. *Ibid.*, pp. 141, 146.

85. *Ibid.*, p. 140. For further details and documentation, see my articles, 'The Early History of the Titius-Bode Law', *American Journal of Physics* 40 (1972): 1014–23, and 'Das Titius-Bodesche Gesetz im Licht der Originaltexte', *Nachrichten der Olbers Gesellschaft* (Bremen), Nr. 86 (October, 1972), pp. 1–18.

86. Derham already mentioned this in his *Astro-theology*, p. 185.

87. *Principia rerum naturalium sive novorum tentaminum phaenomena mundi elementaris philosophice explicandi* (Dresden and Leipzig: sumptibus Friderici Hekelii, 1734). References are to the English translation, *The Principia*, by James R. Rendell and Isaiah Tansley (London: Swedenborg Society, 1912).

88. *Ibid.*, see Part III, chapter iv, 'The Universal Solar and Planetary Nebular Matter and its

Separation into Planets and Satellites', pp. 172–94, and especially pp. 178–81. The pagination of the translation restarts with chapter xiv of Part II.

89. *Ibid.*, p. 185.

90. *Ibid.*, p. 189.

91. *Ibid.*, pp. 191–93. He referred to works by David Gregory, Tycho Brahe, and Hevelius.

92. *Ibid.*, pp. 176–77.

93. See the text of the *Elemens* in *Oeuvres complètes de Voltaire* (new ed.; Paris: Garnier Frères, 1877–83), vol. 22, p. 404.

94. *Ibid.*, p. 408.

95. *Ibid.*, p. 409. A happy claim, though in Voltaire's case it rested on sophism and not on the scientific insight displayed half a century earlier by Richard Bentley. When years later Voltaire took up again the question of an infinite universe in the article 'Infini' of his *Dictionnaire philosophique* (*Oeuvres*, vol. 19, p. 457), the thrust of his short series of questions aimed at the possibility of a privileged position were the stars infinite in number.

96. 'Dissertation sur les changements' (1746), in *Oeuvres*, vol. 23, p. 123. See also M. S. Libby, *The Attitude of Voltaire to Magic and the Sciences* (New York: Columbia University Press, 1935), p. 171.

97. See his *Le philosophe ignorant* (*Oeuvres*, vol. 26, p. 58) and 'Genèse' in his *Dictionnaire philosophique* (*Oeuvres*, vol. 19, p. 227).

98. In the satirical poem, 'Les Cabales' (1772), in *Oeuvres*, vol. 10, p. 182.

99. Paris, Vve Brocas, 1734–39.

100. *Ibid.*, vol. 4, pp. 16–26, Propositions iii and iv.

101. *Examen et réfutation des Leçons de physique expliquées par M. de Molières* (Paris: chez Jacques Clousier, 1741), pp. 318–32.

102. 'Nouvelles pensées sur le système de M. Descartes, et la manière d'en déduire les orbites et les aphélies des planètes', in *Johannis Bernoulli . . . opera omnia* (Lausanne and Geneva: M. M. Bousquet et Sociorum, 1742), vol. 3, pp. 133–73. Needless to say, there were no references in Bernoulli's essay to the cosmogonical parts of the system of Descartes. Bernoulli credited Descartes with formulating the first clear steps in reasoning in science (p. 141) and remarked that the 'vortices presented themselves most naturally to one's mind' (pp. 141–42).

103. See the end of the Preface of his *Traité de l'équilibre et du mouvement des fluides* (Paris: chez David, l'ainé, 1744), p. xxxii.

104. See the text of the *Essai de cosmologie* in *Oeuvres de Mr. de Maupertuis* (new ed.; Lyons, chez Jean-Marie Bruyset, 1756), vol. 1, pp. xv–xviii.

105. *Ibid.*, pp. 6–7.

106. *Ibid.*, p. 9.

107. *Discours sur les différentes figures des astres* (1732), in *Oeuvres*, vol. 1, p. 110.

108. *Essai de cosmologie*, p. xvi

109. *Discours . . .*, p. 130.

110. *Ibid.*, p. 157.

111. 'Lettre sur la Comète qui paroissoit en 1742', in *Oeuvres*, vol. 3, pp. 209–56.

112. *Ibid.*, p. 251.

113. *Ibid.*, p. 249.

114. *Ibid.*, p. 256.

CHAPTER FOUR

CONVENIENT COLLISIONS

William Whiston, Buffon's chief source

In 1749 there appeared the first three volumes of one of the largest and most widely read scientific works of the 18th century, the *Histoire naturelle* of Georges-Louis Leclerc, or Buffon, as he was to be remembered in history. Sold out in only six weeks the first of the three volumes dealt with the 'Theory of the Earth',[1] a necessary preamble for a proper description of the plant and animal kingdoms. The boldness and sweep of Buffon's approach was well indicated by the fact that his 'first discourse' on the proper method of studying natural history was followed by one on the formation of the system of planets. To be sure, Buffon started with the assertion, which he emphasised with a drawing [*Illustration XVI*], that the planetary system was a 'product of God's hand'.[2] This phrase, rather trite by then, was practically all that Buffon's exposition had in common with the views on planetary origins expressed by Newton. He was often quoted by Buffon but only with respect to details. Even there Buffon did not necessarily agree with the illustrious author of the *Principia*.

This is not to suggest that Buffon was eager to pick a fight with Newton or with others. He disagreed with grace and parried attacks with calm disdain. Most of his readers could hardly suspect that time and again he glossed over the serious debts he owed to others. Maupertuis was one of them. Buffon heaped praises[3] on his *Discours sur la figure des astres*, although he should have rather mentioned Maupertuis' long letter on the comet of 1742. Collision with a comet was the very heart of Buffon's theory of planetary evolution and in this respect he owed even more to William Whiston's work, *A New Theory of the Earth*, first published in 1696.[4] Buffon's lengthy account of it merely evidenced his ability to enhance his own glory by apparently putting others in the spotlight. The evidence was misleading in more than one respect. First, Buffon contradicted himself on an all-important point of Whiston's theory by stating that before the deluge the planetary orbits were 'perfect circles' and then claiming that during the same period the orbits were 'an almost circular ellipse'.[5] Second, Buffon represented Whiston's theory as being exclusively concerned with the geological changes implied in the six days' creation and in the deluge. He said not a word about the facts that the comet, which according to Whiston caused the deluge, put the earth into its present elliptic orbit, that Whiston tried to give scientific credibility to his ideas on the dynamical genesis of the earth as a planet, and that the story of the earth's turning into a planet was for Whiston the pattern for the formation of all planets around any and all stars.

Whiston spoke plainly of his indebtedness to Thomas Burnet's *Sacred Theory of the Earth*[6] in which biblical exegesis was bravely coupled with rudimentary geology, uncritical historiography, and naïve chronology to produce a 'scientific' account of the formation of a habitable earth from the chaos and of the mechanism and effects of the deluge. A tenet dear to Whiston that whatever happened to the earth constituted a pattern also for the other planets was clearly stated by Burnet,[7] who in turn learned it from Descartes. Burnet, an unabashed Cartesian, spoke also of turning stars into planets and of planets becoming stars again, another of Descartes' favourite ideas which also found its way to Whiston. In *The Sacred Theory* Burnet had almost exclusively the earth in view, and without going into any explanation or qualification he identified the biblical chaos with the original condition of the earth, already a planet of the sun. Burnet's avowed aim to write the history of cosmic evolution including all stars and planets never materialised.[8] Whiston, too, failed to discuss systematically the cosmic regions beyond the earth. He was certainly at one with Burnet in believing that a close correspondence could be established between the words of Genesis and the dicta of science. Whiston stated at the very outset that the biblical creation story was not to be viewed as a purely speculative or symbolic one: 'The Mosaick Creation is not a Nice and Philosophical account of the Origin of All Things; but an Historical and True Representation of the formation of our single Earth out of a confused Chaos, and of the successive and visible changes thereof each day, till it became the habitation of Mankind'.[9]

The vindication of such a programme, if it was feasible at all, demanded as much theology as science. Whiston seemed to be well grounded in both. While studying in Cambridge he endeared himself to Newton, received a fellowship at Clare College, and took holy orders. Then followed a chaplaincy with the Bishop of Norwich, a post which Whiston gave up only to return in 1701 to Cambridge as Newton's assistant. In Whiston, Newton picked not only a man of considerable mathematical talent but also an author of some fame. Newton himself praised Whiston's *New Theory of the Earth* which he consented to read in manuscript, but this was only indicative of Newton's own infatuation with chronology, exegesis, and ancient history. Actually, Whiston's ambitious programme was self-defeating from the very start. If Genesis gave indeed an accurate description of physical happenings, then not only the obvious meanings of some of its phrases had to be stretched beyond recognition, but also some conveniently new pages were to be added to what Whiston called the 'Book of External Nature'.[10] A habitable earth could hardly emerge from a chaotic globe in six days and Whiston went into lengthy details to show that a full day in the creation story of Genesis meant a whole year.[11] If, however, 'day and night' meant a year, then the earth could not be assigned its present rate of rotation from the very start. Moreover, if the earth rotated only once on its axis in one 'day-year', one side of it would have remained steadily turned toward the sun with the other side remaining in perpetual night. The daily rotation of the earth constituted, therefore, a major problem within Whiston's theory. It was through his efforts to provide a 'solution' for it that he earned a place in the history of speculations about planetary origins as a fore-runner of Buffon.

Whiston's sacred objective

The foundation which Whiston laid to the solution he proposed to the origin of the daily rotation of the earth was broad and meticulous. The need for that was probably justified in view of his ultimate aim, a unitary, mechanical explanation of both the six days' creation of the earth and of the deluge. The four books of his work were prefaced by a long 'Discourse concerning the Nature, Stile, and Extent of the Mosaick History of the Creation'.[12] There he tried to show that exegetically the six days could only mean six years and that the whole story concerned only the earth and not the incomparably larger realm of stars. According to Whiston, it was most inappropriate that of the six days four should be taken up by the shaping of the small earth and only two by the formation of the rest of the universe. This, in Whiston's eyes, implied a mis-planning unworthy of the divine Architect.

Whiston obviously realised that his reader's mind was to be prepared for a highly unexpected turn and this is why he gave in Book I a systematic account of Newtonian mechanics, both general and celestial. He did this in a chain of Lemmata, or short propositions, in which he emulated the style of the *Principia* without, however, becoming too technical. His reader was carefully instructed about the consequences of the inverse square law for planetary motions, about the characteristics of the planets, and about their orbits and periods. By informing his reader about the explanation by the new mechanics of such details as the libration of the moon and of the precession of the equi-noxes, Whiston obviously wanted to have him convinced of the credibility of what he was now to propose. Lemma XLII introducing a new topic, that of the comets, spoke for itself: 'Comets are a Species of Planets, or Bodies revolving about the *Sun* in *Elliptical* Orbits, whose periodical Times and Motions are as constant, certain, and regular as those of the Planets, tho' till very lately wholly unknown to the World'.[13] This meant that in spite of the great varieties of cometary orbits the motion of any comet could at long last be exactly calcu-lated and predicted. A careful realisation of this on the part of Whiston's readers was all the more necessary as their attention was now called to the fact that some comets deeply penetrated into the solar system. The all-important corollary of this should be given in Whiston's words: 'Hence we may observe a new possible Cause of vast Changes in the Planetary World, by the access and approach of these vast, and hitherto little known Bodies to any of the Planets'.[14]

With this Whiston came to the heart of the matter which he unfolded in quick strokes. First came the recognition of the possibility that close passing by comets might have very well been the cause of the fact that the orbits of the planets are not at all in the same plane. Then Whiston took up the case of a close encounter between a comet and a planet, both moving in the same plane. If the planet's orbit was perfectly circular before the encounter then the orbit would turn into an ellipse with an increase of the time of orbiting. The new dynamics held moreover the key to calculating such recondite points as the place and time of the encounter and even of the size (mass) of the comet itself. A more intriguing case was that of a planet with a satellite. Thus the encounter of the earth with a comet could either make the moon independent of the earth, or turn the moon's 'original' circular orbit into its present elliptical one. The latter case implied precious information not only about the parameters of the encounter but also about the original periods of the moon's and of the earth's rotation and orbiting.

The vistas of the process had to appear exciting, if this is to be judged by the exact diagram [*Illustration XVII*] which Whiston offered about it. Using the astronomical tables of the conjunctions of the sun and the moon he proceeded to determine the very day of the encounter.[15] He believed that he could also demonstrate that 'the Ancient [that is, pre-diluvian] Solar and Lunar years were exactly commensurate and equal; and 10 days, 1 hour, 30 minutes shorter than the present Solar year'.[16] Whiston's Lemmata leading to the encounter gave the impression of scientific exactness and rigour. The remainder of the Lemmata dealing with the various effects of the impact was a different matter. Now he needed some hypothetical details about the constitution of comets, and in particular about the comet which put the earth into its elliptical orbit, made it rotate on its axis once in every twenty-four hours, and produced the deluge by tidal effects. According to Whiston, comets were solid bodies with a vast atmosphere surrounding them. The encounter between the earth and a comet had to be such as to break up the earth's crust, release the subterranean or 'abysmal' waters, and turn them into gigantic tides with the help of precipitation coming from the atmosphere and tail of the comet. Since the deluge turned the outer layer of the now rotating earth into a muddy plasma, Whiston could present the flattening at the two poles as a post-diluvian feature of the earth, due in large part to the earth's passing through the comet's massive atmosphere [*Illustration XVIII*].

Book II of the work containing ten Hypotheses about the earth's pre-diluvian state, the encounter, and the deluge gave ample evidence of Whiston's readiness to build castles that only occasionally touched solid ground. Yet hardly anything in the Hypotheses was proposed by him as hypothetical. In speaking about the change of the terrestrial day from one year in length to 24 hours he brushed aside as 'very rash, not to say presumptuous' any doubt about the existence of people on the moon and Jupiter where the length of a day greatly differed from ours. From the difference one could only conclude, so Whiston argued, that 'as the State of external Nature appears to be in *Jupiter* and the *Moon*, very different from ours on *Earth* now; so most probably are the State and Circumstances, the Capacities and Operations of their several Inhabitants equally different from those of Mankind at present upon it; which is what I fully allow, and plead for, in the Case before us'.[17] Whiston's turning hypotheses into verities rested largely on what he called 'the universality of Correspondence'.[18] He meant by this the mutual corroboration which the various aspects of his scheme brought to each other. He felt that it 'sufficiently evinc'd the reality, and, in a proper Sense, certainty'[19] of his theory of the deluge. In particular, the calculations bearing on the date of the encounter and on the new orbit of the earth seemed to him to be above exception. It is well to recall that Whiston, to make more credible some of his assumptions about the pre-diluvian length of the year, did not refrain from imagining that Noah kept a 'journal of the Deluge'.[20]

This should help cast a proper light on Whiston's triumphant conclusions that the encounter with the comet took place on November 27th in the 2,349th year before the Christian era, and that the effective attraction of the comet lasted three and a half hours. As a result, the earth's velocity was increased by a ratio of 1,248/131,250, and its circular orbit changed into the present elliptical one.[21] Since, according to Whiston, a comet equal in size to the earth and crossing the orbit of the earth in front of it at a distance of 60,000 miles could produce the desired effects, he readily concluded that such had to be the case. He failed to see how deeply he was trapped in a *petitio principii*, to say nothing of the inadequacy of his mathematical techniques to calculate with any precision the mutual perturbations of two moving celestial bodies in a near approach. 'Which accuracy of correspondence,' he wrote, 'in the due quantity of Velocity, . . . cannot but be esteem'd a mighty Evidence for the reality of our *Hypotheses*: All whose consequents are so surprisingly true, and so fully bear Witness to one another'.[22]

The readiness by which Whiston found correspondence between his theory and the facts could be all too evident even to his non-mathematical readers when a few pages later Whiston submitted data on the tides of the deluge. These had to be, according to him, eight miles high. 'Which Elevation of the Abyss,' he declared, 'seems very agreeable to the Phaenomena afterwards to be observ'd, and so within a due Latitude establishes the foregoing Hypotheses of the nearness of the Comets [sic] approach, and the consequent bigness of the Comet it self before-mention'd'.[23] The comet turned out to be only half as big as the earth and its closest approach to the earth about 30,000 miles. This might have been in correspondence with the 'facts' but certainly not with the 'factual data' which Whiston gave shortly beforehand about the comet in question.

Just as the word 'Hypotheses' carried little of its original meaning in Book II of Whiston's work, Book III containing five Phaenomena should have rather been called 'Visions'. Such were in reality the pictures which Whiston painted about the conditions on the primitive earth, about the conditions prevailing immediately before the deluge, and about the deluge itself. The fifth Phaenomenon concerned the final conflagration of the earth to be effected by another collision with a comet. Before that, according to Whiston, there had to come a less violent encounter with a comet to secure conditions on the earth suited to the happy period of a millennium. The final conflagration meant also the end of the earth as a planet. Following the final judgment the earth, Whiston wrote, will 'desert its present Seat and Station in the World, and be no longer found among the Planetary Chorus'.[24] In this connection Whiston, who had found it important to denounce the vortices in the Lemmata,[25] once more disclosed his sympathies with a Cartesian view abhorred by Newton. It was the idea that a planet could turn into a comet. This possibility also conjured up before Whiston's eyes the cyclic succession of 'ages' in the universe.[26] He had already spoken admiringly of the Platonic fulness of years, or the Great Year.[27]

Whiston's mind was a receptive soil for odd speculations. Here no specifics can be given of the startling details he professed to know either about the millennium or about the conditions prevailing during the deluge and during the 1,600 or so years he believed preceded it. Suffice it to note that Whiston even calculated the total number of people who had lived before the deluge. He felt that his figure, 82,232 million, missed the true mark by being somewhat low.[28] Book IV concluding the work was entitled Solutions, a rather ironical feature in so many pages of carefree galloping across a terrain on which each step generated a multitude of problems that beg for solutions even today. Whiston could not conceal his amazement over seeing the pieces of a great puzzle fall into their places. Whatever could be learned about comets, planets, and the earth seemed in his eyes to correspond 'most nicely and wonderfully' and with 'the greatest accuracy' both to the facts of nature and to the words of Genesis. With studied modesty Whiston felt that it was not necessary to. note such a 'remarkable attestation' which brought a 'full confirmation of the most accurate Verity of the Mosaick History'. Such reflections 'when Just', he added, should be 'very Natural with every careful Reader'.[29]

A fragmentary cosmogony

Whiston was far less verbose on the general picture concerning his ideas on how the over-all frame of the world and other planetary systems might have originated. His scattered remarks permit, however, the reconstruction of the general cosmological background that lay behind his ideas of planetary origins. Whiston, a protégé and eventual successor of Newton in the Lucasian chair, dissented on several pivotal points of Newton's cosmology. First, Whiston sided with Bentley, an ardent advocate of the finiteness of the world.[30] There was also something un-Newtonian in Whiston's emphasis on the fiery status of the original chaos,[31] although he retained the basic Newtonian distinction between fiery and opaque matter. Like Newton, Whiston took the view that the fact of gravitation made it impossible to assume the primeval coalescence of chaotic matter into distinct lumps, or stars, by a purely mechanical process. It can only be stated by inference that Whiston assumed something more than a purely mechanical process for the formation of large units from the opaque component of the original chaos. As will be shown, Whiston did not follow Newton in assigning to God's specific action various features of the planetary mechanism.

Whiston's radical departure from Newtonian cosmology consisted in his claim that planets and planetary systems originated from comets. He emphatically stated that what happened to the earth was the pattern for all other planets both around the sun and around each star: 'Every unbyass'd Mind would easily allow, that like Effects had like Causes; and that Bodies of the same general Nature, Uses and Motions, were to be deriv'd from the same Originals'.[32] In the context Whiston argued only the different formation of stars and planets and the origination of each planet from its particular chaos. But the very first hypothesis of Book II of Whiston's work stated that 'the Ancient Chaos, the Origin of our Earth, was the Atmosphere of a Comet'.[33] This meant, in his words, that 'the Earth, in its Chaotic State, was a Comet'[34] and 'that a Planet is a Comet form'd into a regular and lasting constitution, and plac'd at proper distance from the Sun in a Circular Orbit, or one very little Eccentrical'.[35] Conversely, he stated that 'a Comet is a Chaos, i.e. a Planet unform'd, or in its primaeval state, plac'd in a very Eccentrical one'.[36] Clearly, Whiston must have pictured the initial phases of the shaping of the cosmos as a creation out of nothing[37] of a largely fiery chaos, followed by the formation of fiery lumps (stars) and of opaque lumps (comets) of matter, the latter being placed by God's hand[38] into very eccentric orbits around the former. Shortly afterwards he noted that some of the comets were placed in perfectly circular orbits around stars as their planets.

Distinctly un-Newtonian was Whiston's claim that systems with slight divergences from perfect uniformity were to be explained by secondary causes and not by a direct recourse to God's action. The actual shape of the solar system, the differences of the eccentricities of the planetary orbits, the varied positions of the aphelia represented such a system. Whereas Newton ascribed its chief features to God's direct action, for Whiston the solar system looked 'more like the result of Second Causes, in succeeding times, than the Primary Contrivance and Workmanship of the Creator himself'.[39] According to Whiston, only the 'regular, uniform, and harmonious' was to be viewed as a direct work of God.[40] On the other hand, it was equally 'philosophical and most pious' to ascribe the slightly irregular formations 'intirely to subsequent changes, which the mutual actions of Bodies one upon another, fore-ordain'd and adjusted by the Divine Providence, in various Periods, agreeably to the various exigencies of Creatures, might bring to pass'.[41]

Whiston's emphasis on the role of secondary causes was a far cry from according a genuine autonomy to nature. Behind that specific encounter with a comet, which shifted the planet from its circular orbit into a slightly elliptical one, there lay God's hand. The course of the comet had to be most carefully adjusted so that it might 'rub' the planet to the extent specified by the outcome to be obtained.[42] The need for that adjustment was voiced by Whiston time and again.[43] He made no secret either of the fact that his planetary theory served the moral and metaphysical perspective as specified by the timetable of the biblical plan of salvation. Since the Bible pictured the times immediately preceding the deluge as an age of 'unsufferable degree of Wickedness', how could one not see 'the Finger of God' in that 'remarkably and extraordinarily' timed encounter with a comet?[44] The situation spoke for itself: 'Had not God Almighty on purpose thus adjusted the Moments and Courses of each, 'twere infinite odds that such a Conjunction or Coincidence of a Comet and a Planet, wou'd never have happen'd during the whole space, between the Creation and Conflagration of this World; much more at such a critical Point of time when Mankind, by their unparellell'd Wickedness were deserving of, and only dis-pos'd for this unparallell'd Vengeance, no less than almost an utter Excision'.[45]

A cosmic Architect of divine perfection and power could, of course, play wizardry not only with one or a dozen, but with an uncounted number of balls of planetary size. He had to do precisely that as Whiston was convinced that other planets and planetary systems had also their denizens subject to moral trial and global judgments.[46] It should not be difficult to guess the weird details had he depicted the mores of those planetarians with the same 'expertise' with which he claimed to know the bowels of the earth and the murky depths of the original chaos. His portrayal of planetary origins in terms of God's hand hurling comets at countless planets neither served theology nor fostered the autonomy of scientific method. This latter point was passed over by Whiston's leading critic, John Keill of Balliol College and the future Savilian Professor of Astronomy at Oxford. To Keill's critique of Burnet's *Sacred Theory* there was appended a sharp rebuttal of Whiston's explanation of the deluge.[47] Keill fought Whiston on the amount of the waters of the deluge, on the porosity of the earth's crust and the like, points on which his information was as meagre as that of Whiston. He touched only indirectly on Whiston's planetary theory when he claimed that the earth could not originally have been a comet as no comet was known to have a moon of its own.[48] On the other hand, Keill found the collision of the earth with a comet very much within the realm of possibility and raised no questions about the biblical timetable which was rigidly adhered to by Newton himself.

Whiston's *Vindication*[49] of his theory, its *Examination* by Keill,[50] and Whiston's *Second Defence*,[51] were concerned with diluvial details, not with planetary theory. In 1707 Whiston served as Boyle lecturer with eight sermons entitled *The Accomplishment of Scripture Prophecies*.[52] There he treated the deluge as the fulfilment of a prophecy, but made only a short reference to the 'Astronomical Computations' which he gave 'elsewhere' as a striking confirmation of the truthfulness of the Bible.[53] He meant *The New Theory* of which he published a year later a second, considerably enlarged version.[54] It contained the phrase, new for its wording but not for its content, that the 'only assignable Cause' of the Deluge was 'the Impulse of a Comet with a little or no Atmosphere, or of a Central Solid hitting obliquely upon the Earth along some parts of its present Equator'.[55] Absence of any doubt in Whiston's mind about his theory was also evident from the 'improved' figures he offered about the change of the earth's orbital velocity and the height of the diluvial tides. He now tailored the 'time of the Comet's Attraction' to 35 minutes and from this he derived the fraction 468/48,993 as the measure of the increase of the velocity in question.[56] His readiness to find 'convenient' factors amply transpired also from his recalculating the height of diluvial tides. Since the comet, which he now assumed to be only one third of the earth, stayed only for a short while in the vicinity of the earth, the tides were not 82,944 times higher than the six-foot tides caused by the moon, but only half of the foregoing figure, that is, about 50 statute miles.[57] Whiston confidently set the start of the deluge at 2 o'clock in the morning, at the meridian of Peking, the habitat of Noah, whom he identified with Fohi, the legendary first monarch of the Chinese.[58]

Needless to say, Whiston again made much of the allegedly perfect agreement between the scientific and biblical timetables. The latter was a function of the unfolding of an ethical drama, and Whiston explicitly stated that major and minor happenings in the physical world were for the sake of its inhabitants: 'as properly suited to all their several conveniences, as was possible or reasonable to be expected'.[59] Convenience was, indeed, the key word in the planetary theory proposed by Whiston. Whatever seemed to suit the convenient explanation was readily espoused by him, fanciful as it could appear otherwise. The result was a tropical growth of wildly intertwined branches of reasoning which laid a forbidding cover on a perhaps valuable and certainly original idea. This fitted a conspicuous pattern in the history of science and reflected all too well Whiston's bent of mind. What he did with biblical chronology he did with ancient theological texts. As a result, there came his dismissal from the Lucasian chair on charges of supporting heretical views on the dogma of Trinity. One heresy trial easily led to another, but his judges were not particularly incensed by what appeared a minor offence, namely, Whiston's identification, in 1714, of the comet causing the deluge with the great comet of 1680.[60]

The paper was reprinted in 1717 at the end of Whiston's *Astronomical Principles of Religion, Natural and Reveal'd.*[61] There he gave a good sample of the facility by which a very capable man of science of his time could describe other planets of the solar system as 'most convenient and well contrived Habitations for all sorts of Sea and Land, visible and gross Animals; with such Plants as are useful for any of their Preservation and Sustenance, during their continuance thereon'.[62] With the imaginative powers typical of all believers in planetarians Whiston even claimed that the thinner parts of the atmospheres of planets 'appear to be the proper Places for the Habitation of not wholly Incorporeal, but Invisible Beings; or of such as have Bodies made of too subtle and aerial a Texture and Constitution to be ordinarily seen by our Eyes, or felt by our Hands'.[63] The one who wrote this was among the first at Cambridge to illustrate lectures in physics with experiments. He also co-authored with Hauksbee an able textbook on experimental physics. When it came to the planetarians, the sobriety of the new experimental philosophy rapidly yielded to intoxicating visions. Very typically, this matched the ease by which Whiston preferred to overlook the gaps in his theory of planetary origins.

The theory in question, a favourite brainchild of Whiston, must have had its fascination with others because in 1756, four years after Whiston's death, a sixth edition of his *New Theory* was printed. Together with the third, fourth, and fifth editions, which appeared in 1722, 1725, and 1737 respectively, it contained but minor changes with respect to the second edition. The repeated reappearance of the book prompted Keill to publish in 1734 a revised collection of his various criticisms of Whiston's *New Theory.*[64] Keill still failed to see that the real issue was not in such details as how Whiston could know anything about vegetables on Jupiter,[65] or why the bright atmosphere of the comet turned into a murky chaos when put into a circular orbit around the sun to start by God's word its evolution into a habitable earth. Keill's failure in this respect was compounded by his claim that the creation of the earth out of a chaos and the deluge were not to be explained by secondary causes. With an eye on Descartes and Whiston he noted that until 'this Age of World-makers, Christians have always thought them such works, as could never be produced by the Laws of Nature and Mechanism'.[66] While here he clearly overstated the case, he scored a point by the very last criticism which he offered on Whiston's theory. There Keill stressed that the best argument for considering both events as miracles was the patent inability of Whiston's theory to cope with them.[67] Keill never perceived that there was a far better way of turning back Whiston's charge that he was too fond of miracles. The truth of the matter was that compared with the two miracles postulated by Keill, Whiston's theory implied an infinitely large number of collisions between planets and comets all so carefully specified as to amount to as many miracles.

Buffon's originality

This fact escaped also Buffon who, while rejecting Whiston's theory, built his own explanation of planetary origins on equally miraculous, or at least, exceedingly convenient collisions. The superstructure erected by Buffon did not lack in striking originality. According to him the origin of the earth as a planet was a piece with the origin of the other members of the family of planets all of which turn in the same direction and in essentially the same plane around the sun.[68] Such a conformity left, in his judgment, no other choice than to assign the origin of all planets to one single cause. 'Certainly', he wrote, 'it was the hand of God that ultimately imparted to the stars (planets) the force of impulse' at a right angle to the direction of the gravitational force attracting them towards the centre 'when He set the universe in motion'.[69] Buffon's deference to God's hand was immediately balanced by his desire to save the autonomy of physical explanation. As he put it, 'in physics one should abstain as much as possible from resorting to causes which are outside the realm of nature', and he felt confident that 'one can explain that force of impulse in a manner likely enough and that one can find for it a cause the effect of which agrees with the rules of mechanics, and which is not far removed from the ideas one should have on the subject of the changes and revolutions that may and must occur in the universe'.[70]

Nothing could be more mechanical than a collision, although it remained to be seen whether the mechanism of a comet's collision with the sun could be imagined in such a way as to satisfy 'the ideas that one should have on the subject'. The subject was far more complicated than Buffon imagined and was beset with grave question marks which he hardly suspected. This was all the more ironical as Buffon was determined to keep his discussion strictly within the framework set by Newton's *Principia*. With Maupertuis and Voltaire, Buffon was a chief pioneer of Newtonianism in France. In 1740 he translated into French Newton's *Method of Fluxions and Infinite Series* which first saw print in 1734, seven years after Newton's death, in the English translation by John Colson from the original Latin manuscript. By 1740 Buffon, freshly appointed *intendant* of the Jardin Royal in Paris, or chief naturalist of the country, was already in the botanical section of the Académie des Sciences. He gained admittance there in 1734 on the strength of an essay on the calculus of probabilities.

Buffon did his best to show with actual calculations the high degree of improbability that all planets should move in the same direction and in essentially the same plane if the path of each planet was determined by a different event. The chances were 1/64 that all six planets should rotate in the same sense, and he gave the figure 1/7,692,624 as the probability that the planes or revolutions of the six planets should be confined by mere chance to within seven and half degree of an arc.[71] Since a chance ordering of six planets into their uniformity meant the chance co-ordination of six independent events, it seemed imperative to assume that this sharing by six planets of some fundamental properties was due to a single cause representing their common origin. The common properties demanded, in Buffon's words, a common 'impression of a movement of impulse, and this could only be had if the force and direction of the bodies' which communicated this impulse was also common, that is, one and the same. 'One may, therefore, conclude with a very high probability that the planets received their movement of impulsion by one single stroke'.[72] This probability was for Buffon 'almost equivalent to certainty' and he immediately stated that 'only comets were capable of communicating so great a motion to such vast bodies'.[73]

The comets seemed to be all the more the appropriate choice for the task as they were very numerous, approached the sun from all conceivable directions, and seemed to be wholly independent of one another. In view of this Buffon could confidently ask: 'May we not imagine with some sort of probability that a comet falling against the surface of the sun, displaced that star and separated from it some smallish parts to which then it communicated a movement of impulsion in the same sense and by one and the same shock, so that the planets which once belonged to the sun's body became detached from it by an impulsive force common to all of them, which they conserve even today'.[74] That the word 'conserve' could have an ominous ring never dawned on Buffon. Actually, a full century had to pass before it began to be heard, once an imaginative look had been taken at rather obvious data. Buffon certainly approached with vivid imagination the event he postulated, and fortunately so, because the event could appear a satisfactory solution only so long as it was handled with an imagination for which no difficulty was insurmountable.

A splashy but superficial theory

Buffon insisted that exact calculation of the speed and mass of the comet, capable of detaching a sufficient amount of mass from the sun, was out of question in the type of book he was writing. To be sure, the task was also beyond the measure of his familiarity with celestial dynamics. He felt reassured by the general consideration that a grazing collision of a comet with the sun seemed to be a more likely event than a direct hit, in which case the comet would have embedded itself into the sun's huge body. Buffon found another favourable circumstance for his theory in the relative smallness of the amount of matter to be detached from the sun, as the planets constituted only 1/650th of the sun's mass. A large comet seemed to be well suited for the purpose, since the speed of comets in the vicinity of the sun was at its maximum and 'it could hardly be doubted', Buffon claimed, that 'comets were composed of a very solid and very dense matter and that they contained in a small volume a large quantity of matter'.[75] In short, they were the ideal missiles for producing, if not planetary systems, at least new pages in planetary theories.

That comets had an extremely solid nucleus was a claim for which Buffon wanted to find some support in the *Principia*.[76] There Newton argued that the solidity and density of a body were proportional to the heat it had to withstand. Thus, according to Buffon, if a comet came as close to the sun as the comet of 1680 and if it equalled in size the earth, then its density had to be 112,000 times that of the sun and its mass as much as 1/9th of the solar mass. On such a generously conjectural basis Buffon could also be generous in conjecturing the size of the comet of such enormous density. If its size was only 1/100th of the earth's size, its mass would still have been 1/900th of the sun's mass. In view of the enormous velocities of comets at their nearest approach to the sun, a comet of very small size could therefore still easily dislodge 1/650th of the mass of the sun.[77] So it appeared within an analysis which, its numerical data notwithstanding, was only qualitative. It was of no substantial help that Buffon called attention to the close agreement between the density of the sun and the density of Jupiter and Saturn which between them had a mass 64 times larger than the four other planets. This certainly suggested that originally all planets belonged to the sun, but the real problem was to show that they could be separated from it by a comet and put into their respective orbits.

These crucial parts of the theory were given by Buffon a treatment that revealed his superficial grasp of some elementary truths of Newton's celestial dynamics, to say nothing of the impressive refinements added to it prior to 1745 when Buffon completed his manuscript. This was abundantly clear from Buffon's response to the first objection to his theory. It derived from the fact that according to the laws of dynamics the mass detached from the sun by the grazing impact could only go into an orbit of which the point of impact remained a necessary part. To perceive this one only needed to consult Newton's non-technical *De mundi systemate* where the very first diagram [*Illustration XIX*] helped visualise the problem of putting a projectile into orbit around the earth.[78] In the diagram the firing of the projectile took place from a mountain-top and at a right angle to the line connecting that point with the centre of the earth. The height of the projectile's circular orbit from the earth clearly depended on the height of the mountain from which the firing took place. From the context it was also clear that by increasing the power of the firearm one could only obtain a more and more elliptical orbit with the point of firing always to remain part of the orbit.

Buffon may have thought of that diagram in *De mundi systemate* when he acknowledged that a bullet shot from a sufficiently powerful firearm at the top of a high mountain would return to the same point in its rounds about the earth. He also glimpsed the road toward the correct solution when he proposed that the bullet of a firearm be replaced with a rocket which would keep accelerating itself for some time. Buffon pictured the material leaving the sun in the form of a long filament somewhat analogous to the flames propelling a rocket. But was there anything in that filament of hot plasma akin to a force that would propel it steadily for some time after the grazing impact of the comet with the sun, and at a right angle to the direction of that impact? On this latter and crucial point Buffon gave no information. That the height of the mountain was at a right angle to the direction of firing was a circumstance which did not appear to him of any consequence. At any rate, what he said of the hot filament implied that one could lift oneself by one's own bootstraps. According to Buffon, the leading part of the filament was pushed by the sections behind, while these in turn were speeded up by the attraction of the leading parts. The rest then seemed to follow naturally: 'Thus, provided that there was an acceleration in the impulse imparted to the filament of matter by the impact of the comet, it is very possible that the planets which were formed in that filament had thereby acquired the motion which we recognise in the circles or ellipses of which the sun is the centre of focus'.[79]

To provide for that acceleration was Buffon's crucial task but he sounded hardly convincing in his search for it. First, he offered an analogy from the sequence of eruptions in a volcano. In the case of Vesuvius it could be observed, he claimed, that the material of the first eruption received added impetus from the second, the second from the third, and so forth. As a result, the material of the first eruption was carried much farther than if no further eruptions had followed. Buffon believed that the filament was composed from eruptive segments that mutually reinforced each other's velocity. Second, to secure the sufficiently wide separation of the filament from the sun, Buffon suggested that the sun itself might have been pushed by the comet away from the filament. This implied that the sun was moving in a large circle around an imaginary centre of the solar system. This movement of the sun would appear, according to Buffon, as its motion toward other parts of the universe. Yet, if the sun recoiled by the impact, the same had to happen to the filament as well, thwarting the desired effect. Buffon's answer to that only pushed his game with the bootstrap one step further. He now assumed that the sun was already in a circular motion which the impact of the comet made only much larger. 'This would be sufficient to account for the actual motion of the planets'.[80]

The situation, which in fact was a complete impasse, was not remedied by the third mechanism which Buffon thought to be at play in the collision. 'The sun's material', he wrote, 'might be very elastic because the light, its only portion we know, seems through its effects to be perfectly elastic'.[81] To picture the separation of the filament from the sun as the rebounding of a very elastic material was not without originality, but it escaped Buffon that his interpretations amounted to getting more out of a mechanism than it contained or could deliver. This was precisely the point on which countless efforts to construct perpetual motion machines had run aground. Of those efforts no respectable scientist wanted to have any part long before Buffon's time. Nor was in line with reliable scientific reasoning the argument which brought to a close Buffon's theory of the mechanism of the origin of planets: 'I confess that I cannot say by which of the foregoing mechanisms was the direction of the first movement of impulsion of the planets changed; but these explanations are sufficient at least to show that this change is possible and even probable, and this is enough for my purpose'.[82]

It was on these hazy foundations that Buffon erected the consequences of his theory. First came the derivation of the fact that the most distant planets, Saturn and Jupiter, had a lower density than the other planets closer to the sun. The impact theory would readily account for this fact, so Buffon believed, because the lighter parts of the sun's mass necessarily received in the impact a much greater velocity than the denser parts. Buffon gave the respective densities of the planets as 67 for Saturn, $94\frac{1}{2}$ for Jupiter, 200 for Mars, 400 for the earth, 800 for Venus, and 2800 for Mercury after putting the sun's density at 100. Then he compared the ratios of the densities of Saturn and Jupiter with their orbital velocities (the all-important change of direction was no longer a question). The ratios, $67/94\frac{1}{2}$ and $88\frac{2}{3}/120\frac{1}{72}$, were indeed very close and Buffon exclaimed: 'It is very rare that from sheer conjectures one might derive so exact correlations!'[83] On the same basis the earth's density should have been $206\frac{7}{18}$ instead of 400,.a gross discrepancy, which Buffon brushed aside with the remark that 'from this one may conjecture that our globe was once less dense than it is today'.[84] The remark spoke for itself and so did the next: 'As to the other planets, Mars, Venus, and Mercury, since their densities are not known except by conjectures, we cannot know whether this would confirm or destroy our opinion on the relation between the velocity and density of planets in general'.[85] Only in the safe seclusion of Buffon's country estate, where these lines were written in the late summer of 1745, could one follow in matters of cosmology the zig-zagging line of greatest convenience.

The situation about the density–velocity relation looked all the more obscure as the foregoing data on the densities of planets were based, as Buffon now revealed,[86] on a convenient eclecticism. Unlike the figures for the densities of Mercury, Venus, Earth, and Mars, the figures for Jupiter and Saturn were not based on the principle advocated by Newton, namely, that the density of a planet was proportional to the heat radiated on it by the sun. On such a basis the densities of Jupiter and Saturn should have been $4\frac{7}{18}$ and $14\frac{17}{22}$ instead of 67 and $94\frac{1}{2}$. Buffon had, therefore, to indulge in convenient shuffling of considerations to show that the densities of planets had a closer connection with their velocities than with the quantities of heat they had to sustain. He offered no details on the condensation of the filament into six separate planets under the influence of mutual attraction,[87] though he was prolific on the rather secondary point of how the filament lost its fiery state. He brought in even the novae and the variable stars as proofs that quick movements could rekindle and extinguish the fire of entire stars. The quick impact of the comet did the latter with the fire of the filament.

It now remained to explain by the impact theory the rotation of planets on their axis and the formation of their satellites. This, Buffon noted with his customary confidence, 'far from adding difficulties and improbabilities to our hypotheses, seems on the contrary to confirm it'.[88] Undoubtedly, the grazing collision could readily appear as a satisfactory cause to give perhaps a rotational movement to the whole filament around its centre, but it was to be explained, instead of being assumed, that six different axes of rotation should arise in it. Of this Buffon seemed to be unaware. He believed he had clinched the whole argument and boldly made a statement which is the touchstone of the truth of all physical theories and of planetary theories in particular. It was 'the attention to all aspects of the phenomena'.[89] One such aspect was the rotation of the moon on its axis. Buffon ignored it as he did also the possibility that the five satellites of Jupiter and the four of Saturn might also rotate on their axes.

Buffon's neglect on these points should seem all the more curious as he paid much attention to the problem of the formation of satellites. He assigned it to an oblique impact of such a type (Buffon never ran short of convenient considerations) which 'could separate small particles of matter from the principal planet. These, while following the direction of the planet itself, would unite according to their densities at different distances from the planet by the force of their mutual attraction. At the same time they would necessarily accompany the planet on its course around the sun and keep turning around the planet pretty much in the same plane'.[90] The explanation patently turned the situation upside down as the oblique impact, convenient as it could be, had to precede the breaking down of the filament into six parts. It could certainly have nothing to do with the alleged separation of the small particles from the planets themselves.

Buffon saw a confirmation of his theory of the formation of satellites in the fact that they were observed around the faster rotating planets. His handling of an essentially quantitative consideration could hardly endear him to astronomers. While it was true that the velocity of a point on the equator of Mars, which then was not known to have two satellites, was only 15/24th of the velocity of a point of the earth's equator, he gave a patently wrong figure for the velocity of a point of the equator of Jupiter.[91] Most importantly, he conveniently overlooked the fact that his generalisation rested in part on the absence of any information about the rotation of Mercury and Venus. Also, he should have told his reader that much uncertainty surrounded then the corresponding value for Saturn and the question whether it rotated in the plane of its ring. Buffon saw in that ring (he failed to mention that G. D. Cassini had already in 1675 observed its double structure) the equatorial belt detached from the planet by its very fast rotation. This was a curious reasoning, because the available data indicated for Saturn an equatorial velocity which could not be higher than that of Jupiter.

In Buffon's own eyes the likelihood of the truth of his theory was very high. In accordance with his philosophy of scientific method he rested the measure of that likelihood 'on the power of mind to combine more or less disparate relationships'.[92] He certainly was able to see patterns of correlation but only at the price of being highhanded with data and with some basic laws of celestial dynamics. It was to his credit, however, that he came into the open with a speculation on a topic on which, as he put it, 'nothing has ever been written, although it was of great importance because the motion of impulsion of planets enter into at least one half of the composition of the whole universe, which the [gravitational] attraction alone cannot explain'.[93] The 'half' he had in mind were the planets which he believed to surround each and every star. In his reasoning, deeply steeped in the principle of analogy, the sun and its realm formed a pattern valid throughout the universe. Consequently, if it were true that 'it is almost necessary that a comet should at times fall into the sun'[94] and touch off the production of planets, this had to be true of other stars as well. Buffon left the unfolding of this inference to his reader. Speaking of other planetary systems would have inevitably involved conjectures about their denizens, and Buffon did not wish to put additional weapons against himself into the hands of the theologians of the Sorbonne. They were to be incensed enough by the many thousands of years which his theory implied for the formation of the earth.

Buffon's critics

It was largely details of geology that attracted the attention of critics in the first three volumes of Buffon's *Histoire naturelle*, although comments on his planetary theory were not completely missing. The anonymous review in the *Journal des Sçavans* contained such flattering remarks that Buffon was the first to attempt a synthesis of the whole natural history and that unlike Burnet and Whiston he avoided reliance on the miraculous. Yet, by emphasising that Buffon's theory of planetary origins was offered as strictly conjectural and by saying nothing more about it, the reviewer clearly conveyed his distrust.[95] In the simultaneous and much longer review of the work in the *Mémoires de Trévoux* it was, however, bluntly stated that Buffon's theory of planetary origins should be viewed as a dark cloud against which the factual reasoning of the rest of the work appears all the more brilliant.[96]

What really had to appear sombre for Buffon was the lack of reaction on the part of French astronomers. Nor was his theory to receive any mention in the massive volumes of the *Encyclopédie*. Its chief architects, Diderot and d'Alembert, were careful not to give credence to any view implying a miraculous intervention in nature by the Author of nature.[97] Curiously, Voltaire, who could hardly find to his liking Buffon's theory, a patent departure from Newton's views, kept aiming his barbs at Buffon's views on the evolution of plants and animals.[98] If Buffon's theory conflicted with Voltaire's belief in an inactive God, it did even more so with the conviction of d'Holbach whose *Système de la nature* represented the strictly materialistic sector of the *philosophe* movement. Thus once more Buffon was given the silent treatment as d'Holbach submitted the explanation of the origin of planets with the facile pen of those who recognised no major riddles in the universe. Taking his starting point from the claim that existence is a necessary property of matter d'Holbach opined that 'perhaps this earth is a mass detached, in the course of times, from another celestial body; perhaps it is the result of spots and crusts which astronomers sight on the disk of the sun, and which could from there penetrate into the planetary system; perhaps this earth is an extinct and displaced comet which previously occupied another place in the regions of outer space and which, consequently, was in the state of producing beings very different from those that are now found on it'.[99] As usual, superficial remarks about the problem of the origin of planets were duly followed by easy declarations on the features of their denizens.

D'Holbach's dictum deserves mention only because it offered opportunity to Georg Jonathan Holland to make incisive comments on theories previously offered on the origin of the system of planets. Holland, a graduate of the University of Tubingen and the tutor of the children of Prince Frederik-Eugen of Württemberg, earned fame by addressing himself in 1773 to the fallacies of d'Holbach's *Système de la nature* in a two-volume work which went through four editions in three years.[100] Well trained in mathematics, Holland could rightly claim that he touched on the topic of planetary origins in order 'to show the danger of taking one's imagination as guide in the realm of geometrical truths'.[101] The truths in question were those established by Newton about the necessary inclusion of the point of origin in a celestial body's orbit. Almost circular planetary orbits and concentric with reference to the sun could not, therefore, be assigned to the collision of a comet with the sun. Nor was it possible, Holland added in shifting his remarks from Buffon to Maupertuis, that a comet's highly eccentric orbit could be changed into a planetary orbit through a close encounter with the sun.[102]

It is not known whether Holland's criticism came to the attention of the author of the *Histoire naturelle*, the many volumes of which saw in the 1770's several re-editions. At any rate, in the most important of these, which carried for the first time supplementary volumes containing *Les Époques de la nature*,[103] the new material served ample evidence of Buffon's engrossment with his favourite brainchild. He now carried the origin of planetary system one step further by looking for the origin of comets which he assigned to the random explosion of stars.[104] About these he suggested that before their explosion they were similar to glass,[105] their heat being caused solely by the pressure of comets orbiting around them. The analogy which Buffon used in this connection could only appear crude even in the heyday of mechanistic philosophy: 'Each comet and each planet form a wheel, the spokes of which are the rays of the force of attraction; the sun is the common hub of all the different wheels; the comet and the planet are the moving rim and each contributes with all its weight and speed to the pressing of that general focus, the fire of which endures as long as do the motions and pressure of these vast bodies that produce it'.[106] Consequently, the exact number of comets in the solar system was for Buffon an important discovery to be made by posterity.[107]

While he seemed to be certain about a positive outcome of efforts in this connection, he voiced doubts whether man would ever observe the presence of planets around other stars. Their existence was for him a certainty. It never occurred to him that on the basis of his theory a planetary system had to be viewed as a very rare phenomenon. On the contrary, he held it 'not impossible' that our sun might give birth to another planetary system around it, were another comet to collide with it in a manner convenient for the purpose.[108] At any rate, he considered all planets around our sun and around any other star as convenient abodes for beings more or less similar to humans.[109] Buffon always extended his vistas far beyond the range of observational data and this explains his selective handling of new astronomical information relevant to his theory. On the one hand, he seized on the observation of the transit of Venus in 1769, on the basis of which previous estimates of the sun's distance and mass had to be revised. He reasoned that if the mass to be removed from the sun by the comet was not 1/650th but only 1/800th of the sun's mass, the weight of a possible objection against his theory was considerably alleviated.[110] On the other hand, he ignored the growing reluctance of astronomers to consider comets as solid bodies, a circumstance crucial for the success of his theory. The point on which he was immediately taken to task in connection with his theory of planetary origins as restated in the *Époques de la nature* concerned the shape of orbit resulting from a grazing collision between two celestial bodies. Jean-Baptiste G. A. Grossier, the founder and editor of the *Journal de littérature, des sciences et des arts*, put the matter bluntly as he referred to the great eccentricity of such orbits and to the necessary inclusion in them of the point of collision: 'Buffon knows all too well the laws of gravitation to be satisfied deep in his heart with his explanation'.[111]

A year later, in 1780, it became public knowledge that Buffon had been given in person the explanation of the reasons against his theory. The disclosure was made by the authors of the famed *Physique du monde*. Its first volume also contained the verdict that all the principal features of Buffon's theory were 'absolutely inadmissible in physics'.[112] Buffon's dicta on the glassy nature and explosion of stars were in turn lampooned in the very title, 'the universe of glass reduced to dust', of the book-length critique which Thomas-Marie Royou penned on the *Époques de la nature*. He kept enlivening his diction with remarks such as that in Buffon's theory comets were the twin children of a star, but stars were merely 'bastard sons', that is, products of unknown origin.[113]

More hurting for Buffon than such barbs must have been the icy silence by which French astronomers greeted the slightly revised version of his theory. Lalande had no comment on it as he praised in his short review of the *Époques de la nature* its brilliant style.[114] There also seemed to be a back-handed compliment in the remark of Bailly, author of the first modern history of astronomy, who commended Buffon for rejecting the basic difference between the opaque material composing the planets and the luminous substance of stars.[115] When finally the silence of professional astronomers yielded to specific comments, the result was devastating. Five years after the publication of the *Époques*, its section on planetary origins was reviewed in the classic monograph on comets by Alexander G. Pingré[116] who made no secret of the amateurism with which Buffon handled the physics of his theory. It amounted to an elementary mistake to leave aside, as Buffon did, the question of the periodic return of the comet if it had a grazing collision with the sun. Again, it followed from Buffon's presuppositions that the mass detached from the sun should move on as a single body of which all parts had the same velocity. The third, fourth, and fifth objections of Pingré concerned Buffon's inconsistency in correlating the densities, weights, and velocities of the planets. As a competent student of celestial mechanics, Pingré saw the weakest point of Buffon's theory in its inability to cope with the inevitable return of the detached mass to its point of origin. Furthermore, not even a comet with twice the total mass of all planets and with an almost parabolic orbit would have been powerful enough for the task envisioned by Buffon. At any rate, not only such a massive comet but even comets with considerably smaller mass were still to be observed. The fact that even the largest comets failed to produce noticeable perturbation in the orbits of planets was a clear enough proof of their exceedingly small mass. Ingenious and new as Buffon's theory was, Pingré concluded, 'it could not be reconciled with the laws of physics'.[117]

It did not escape Pingré that Buffon readily attributed mutually exclusive effects to one and the same mechanism. A case in point was Buffon's explanation of the formation of satellites. If the rotation of a planet was fast enough to have its equatorial belt detached, then it was clearly inconsistent to assume with Buffon that the same belt moved but to a given distance from the planet and coalesced there into one spherical body. Not surprisingly, Pingré wondered how a 'profound philosopher like Buffon could imagine a system so contrary to the most basic laws of physics'.[118] Events and processes, Pingré emphasised, which 'were belied by all the laws of sanest physics',[119] could not be endowed with likelihood by calculations of probability, however appealing. There was finally the question of the origin of the comets and of the stars which, as Pingré remarked with tongue in cheek, Buffon certainly did not wish to trace to the 'madness of the eternity of matter'.[120] If Buffon was firmly convinced of the truth of the creation of the universe by God, was it not then more reasonable to ascribe the planetary systems also to the creative action of God? Such a solution, Pingré wrote, 'will not, true enough, be physics, but will not at least collide head-on with the most solid foundations of true physics'.[121] As subsequent phases of speculations on the origin of planetary systems were to show, such a remark did not necessarily represent a facile abdication of a scientific problem.

When, following his death in 1789, Buffon was eulogised by Condorcet in the Académie des Sciences, the long-standing tacit scorn of the body scientific for Buffon's theory was translated into the severe sentence: 'In the Époques de la nature . . . Buffon seems to redouble his resolve . . . to counter with the grandeur of his ideas, with the magnificence of his style, with the weight of his name the unanimous authority of scientists and even that of the facts and of the calculus'.[122] It was only seven years later that Buffon received proper credit for his great insight that the various regularities of the solar system strongly suggested that it was the product of a single physical process. Coming as it did from Laplace, who by 1796 was the undisputed leader in celestial dynamics, the acknowledgment could not have been more authoritative. Yet the context was not without some biting irony. First, Laplace pointedly noted that the time had come for finding that process. Second, in 1796 it represented a not wholly pardonable ignorance to state as Laplace did: 'Buffon is the only one I know who since the discovery of the true system of the world [by Newton] tried to proceed to the question of the origin of planets and their satellites'.[123] In addition, the solution which Laplace now proposed for the evolution of the system of planets could appear at first sight as almost duplicating a theory developed half a century earlier by a philosopher whose name by 1796 turned into a household word in the intellectual world. He was none other than Immanuel Kant.

NOTES TO CHAPTER FOUR

1. *Histoire naturelle, générale et particulière, avec la description du Cabinet du Roi. Tome Premier* (Paris: de l'Imprimerie Royale, 1749). References are both to this first edition and in square brackets to the far more accessible text in the *Oeuvres complètes de Buffon* (Paris: Furne et Cie, 1858), vol. 1.

2. *Ibid.*, p. 131 [82].

3. *Ibid.*, pp. 148–49 [88].

4. *Ibid.*, pp. 168–80 [96–99]. Buffon used the second edition from 1708.

5. *Ibid.*, pp. 169 and 172 [96–97].

6. First published in Latin in 1680 and then, at the urging of Charles II, in English in 1684–89. A second English edition followed a year later, the text of which was reprinted with Basil Willey's Introduction in 1965 (London: Centaur Press).

7. *Ibid.*, p. 128, end of Book I.

8. See on this Katherine B. Collier, *Cosmogonies of our Fathers: Some Theories of the Seventeenth and Eighteenth Centuries* (New York: Columbia University Press, 1934), p. 70.

9. *A New Theory of the Earth, from its Original, to the Consummation of all Things. Wherein the Creation of the World in Six Days, the Universal Deluge, and the General Conflagration, as laid down in the Holy Scriptures, are shewn to be perfectly agreeable to Reason and Philosophy* (London; printed by R. Roberts for Benj. Tooke, 1696), p. (3) of the preliminary 'Discourse'. Since pagination restarts with Book I, pages of the preliminary 'Discourse' will be given in parentheses.

10. *Ibid.*, p. (39).

11. *Ibid.*, pp. 81–100.

12. *Ibid.*, pp. (1–94).

13. *Ibid.*, p. 36.

14. *Ibid.*, p. 37 (corollary to Lemma XLVI).

15. *Ibid.*, p. 41 (corollary to Lemma LII).

16. *Ibid.*, p. 46 (corollary to Lemma LVI).

17. *Ibid.*, p. 102.

18. *Ibid.*, p. 132.

19. *Ibid.*

20. *Ibid.*, p. 143.

21. *Ibid.*, p. 146.

22. *Ibid.*, p. 147.

23. *Ibid.*, pp. 153–54.

24. *Ibid.*, p. 215.

25. *Ibid.*, p. 36 (corollary to Lemma XLV).

26. *Ibid.*, p. 378 (solution C).

27. *Ibid.*, p. 212.

28. *Ibid.*, p. 179. Whiston acknowledged that the figure exceeded more than two hundred times the actual world population which he put at 350 million. He also noted that the actual earth could support only five times as many people, but claimed that physical conditions on the pre-diluvian earth were far more favourable (pp. 179–80).

29. *Ibid.*, p. 313.

30. *A New Theory of the Earth*, pp. 11–12.
31. *Ibid.*, pp. (35–36).
32. *Ibid.*, p. (40).
33. *Ibid.*, p. 69.
34. *Ibid.*, p. 73.
35. *Ibid.*, pp. 74–75.
36. *Ibid.*, p. 75.
37. *Ibid.*, pp. (4–5) and (36–41).
38. *Ibid.*, p. 113.
39. *Ibid.*, p. 114.
40. *Ibid.*, p. 116.
41. *Ibid.*
42. *Ibid.*, p. 378; see also p. 116.
43. *Ibid.*, pp. 43, 116, and 358–59.
44. *Ibid.*, p. 359.
45. *Ibid.*
46. *Ibid.*, p. 116.
47. *An Examination of Dr. Burnet's Theory of the Earth together with some Remarks on Mr. Whiston's New Theory of the Earth* (Oxford: printed at the Theatre, 1698), pp. 177–224.
48. *Ibid.*, p. 190.
49. *A Vindication of the New Theory of the Earth from the Exceptions of Mr. K.[eill] and Others. With a historical preface* [by William Whiston]. 1698.
50. *An Examination of the Reflections on the Theory of the Earth together with a Defence of the Remarks on Mr. Whiston's New Theory*, 1699.
51. *A Second Defence of the New Theory of the Earth from the Exceptions of Mr. John Keill* (London: Benj. Tooke, 1700).
52. The subtitle reads: *Being Eight Sermons preached at the Cathedral Church of St. Paul in the Year MDCCVII* . . . Its revised text printed in 1737 can also be found in *Defence of Natural and Revealed Religion: Being a Collection of the Sermons preached at the LECTURE founded by the Honourable Robert Boyle, Esq.; (from the Year 1691 to the Year 1732)* (London: printed for D. Midwinter *et al.*, 1739), vol. 2, pp. 259–348.
53. *Ibid.*, p. 292.
54. *A New Theory of the Earth* . . . The Second Edition with great Additions, Improvements and Correlations (Cambridge: printed at the University Press for Benj. Tooke . . ., 1708). The greatest change consisted in the addition of a new Hypothesis (pp. 144–81), in which Whiston claimed that 'The most Ancient Civil Year in most Parts of the World after the Deluge and also the Tropical Solar and Lunar Year before the Deluge, contain'd just 12 Months of 30 Days apiece, or 360 Days in the whole' (p. 144).
55. *Ibid.*, p. 111.
56. *Ibid.*, p. 203.
57. *Ibid.*, pp. 204–05.
58. *Ibid.*, pp. 140 and 202.
59. *Ibid.*, p. 130.
60. *The Cause of the Deluge Demonstrated . . . being an Appendix to the Second Edition of the New Theory of the Earth* (London, 1714). A modified version of it appeared under the title 'A New Theory of the Deluge' as an appendix to the fifth edition of *The New Theory* (London: printed for John Whiston, 1737), pp. 459–78.
61. London: printed for J. Senex, 1717.
62. *Ibid.*, p. 92.
63. *Ibid.*
64. *An Examination of Dr. Burnet's Theory of the Earth with some Remarks on Mr. Whiston's New Theory of the Earth . . . The Second Edition corrected . . . To the whole is annexed a Dissertation on the Different Figures of the Celestial Bodies . . . with a summary Exposition of the Cartesian and Newtonian Systems of Mons. de Maupertuis* (Oxford: printed for and sold by H. Clements, near the theatre in Oxford, 1734).

65. *An Examination of Dr. Burnet's Theory of the Earth*, p. 210.

66. *Ibid.*, p. 314.

67. *Ibid.*, p. 347.

68. 'De la formation des planètes', in *Histoire naturelle*, vol. 1, p. 133 [83].

69. *Ibid.*, p. 131 [82].

70. *Ibid.*, pp. 131–32 [82].

71. *Ibid.*, p. 134 [83]. Since $7\frac{1}{2}$ degrees are 1/24th of a half circle, the probability in question was $(1/24)^5$ or 1/7,962,624. The inversion of the second and third digits in the printing of Buffon's work remained unnoticed in the various editions I had access to.

72. *Ibid.*, pp. 134–35 [83].

73. *Ibid.*, p. 135 [83].

74. *Ibid.*, p. 133 [83].

75. *Ibid.*, p. 137 [84].

76. *Ibid.*

77. *Ibid.*, p. 138 [84].

78. Its publication in 1728 (London: impensis J. Tonson, J. Osborn & T. Longman) was accompanied by an English translation, *A Treatise of the System of the World* (London: printed for F. Fayram). Both the Latin and English texts came out in second edition in 1731.

79. 'De la formation des planètes', p. 140 [85].

80. *Ibid.*, p. 142 [86].

81. *Ibid.*, p. 143 [86].

82. *Ibid.*

83. *Ibid.*, p. 144 [86].

84. *Ibid.*, p. 145 [86].

85. *Ibid.*

86. *Ibid.*, pp. 145–46 [86].

87. *Ibid.*, p. 149 [87].

88. *Ibid.*, p. 150 [88].

89. *Ibid.*, p. 152 [89].

90. *Ibid.*, p. 151 [89].

91. *Ibid.*, p. 152 [89]. It was, according to Buffon, 500 to 600 times larger than speed of a point at the earth's equator. A value of about 350 should have followed from his own presuppositions and would have been in closer agreement with modern measurements.

92. *Ibid.*, p. 153 [89].

93. *Ibid.*

94. *Ibid.*, p. 135 [83].

95. In the issue of Octobre, 1749 (Volume CXIX,–CXX), pp. 648–57.

96. Article CXXV in the issue of Octobre, 1749 (Vol. I, pp. 2226–45). The *Mémoires de Trévoux* was popularly so called because its printing started in Trévoux in 1701. The official title was *Mémoires pour l'histoire des sciences et des beaux arts.* See especially p. 2243.

97. Failure to mention Buffon's theory could hardly be unintentional in such articles of the *Encyclopédie ou dictionnaire raisonné des sciences, des arts et des métiers* as 'Cosmogonie' (vol. 4; Paris: Briasson, 1754, pp. 292–93), 'Cosmologie' (*ibid.*, pp. 294–97), 'Mondes' (vol. 10; Neufchastel: chez Samuel Faulche, 1765, pp. 640–41), 'Planètes' (vol. 12, pp. 703–08), and 'Système du monde' (vol. 15, pp. 778–79).

98. See, especially, Voltaire's sarcastic *La défense de mon oncle* (1767), in *Oeuvres complètes* (Paris: Garnier Frères, 1877–83), vol. 26, pp. 405–10, and M. S. Libby, *The Attitude of Voltaire to Magic and the Sciences* (New York: Columbia University Press, 1935), pp. 168–204.

99. Paul-Henri Thiry (1723–1789), who is better known as Baron d'Holbach, published his *Système de la nature ou des lois du monde physique et du monde moral* in Amsterdam in 1770 under the name Mirabaud. Of the many re-editions of the work the most important was the one containing annotations and corrections by Diderot (Paris: E. Ledoux, 1821), reissued in facsimile reprint with an Introduction by Yvon Belaval in 1966 (Hildesheim:

Georg Olms Verlagsbuchhandlung). Diderot found nothing wrong with d'Holbach's dicta on the evolution of planetary systems which can be found in p. 99 (vol. 1) of the reprint edition.

100. *Réflexions philosophiques sur le Système de la nature* (Paris: chez Valade, 1773).

101. *Ibid.*, vol. 1, p. 53.

102. *Ibid.*, p. 54.

103. *Histoire naturelle, générale et particulière, contenant les Époques de la nature. Tome Douzième* (Paris: de l'Imprimerie Royale, 1778). 'Première Époque. Lorsque la Terre et les Planètes ont pris leur forme', (pp. 58–100) and 'Additions à l'article qui a pour titre: De la formation des Planètes', (pp. 365–74). There is also a modern critical edition by Jacques Roger, *Buffon, Les Époques de la nature* (Paris: Éditions de Muséum, 1962).

104. *Histoire naturelle . . . Tome Douzième*, p. 65.

105. *Ibid.*, p. 84.

106. *Ibid.*, p. 69.

107. *Ibid.*, p. 71.

108. *Ibid.*, p. 78.

109. It was to this topic that he devoted the second 'hypothetic' memoir (originally published in 1775), one of the numerous memoirs attached to the *Histoire naturelle*. See *Tome Neuvième*, pp. 301–370. It is worth noting that the *Encyclopédie* displayed in the article 'Mondes' studied reserve on this point: 'What should, therefore, one answer to those who ask whether the planets are inhabited? That one knows nothing of it' (see note 97 above, p. 640). In the article 'Planètes', in which Wolff was criticized for deriving a size of 14 feet for the denizens of Jupiter, praises were heaped on Fontenelle for his caution (!) in the matter. In the same article it was claimed that it was only very likely but not at all certain that the planets were inhabited (see note 97 above, p. 705).

110. *Histoire naturelle . . . Tome Douzième*, p. 366.

111. Numéro 13, 1779, pp. 409–57. For quotation, see pp. 438–39.

112. *Physique de monde* dédiée au Roi par M. le Baron de Marivetz et par M. [Louis Jacques] Goussier. Tome Premier (Paris: chez Quillau, 1780), p. 102. In this five-volume synthesis of geophysics and geology, the first hundred pages contained a summary of various theories of the geological history of the earth, together with a thirty-page-long critique of Buffon's theory.

113. *Le monde de verre réduit en poudre ou analyse et réfutation des Époques de la nature de M. le Comte de Buffon* (Paris: chez J. G. Merigot le jeune, 1780), p. 24. The criticism of the *Époques* by Giacinto Sigismondo Gerdil in his *Observations sur les Époques de la nature pour servir de suite à l'examen des systèmes relatifs à l'antiquité du monde* (Parma, 1789; reprinted in *Oeuvres du Cardinal Gerdil*, edited by l'abbé Migne [Paris: chez J. P. Migne, 1863], cols. 1305–26) was by far the most methodical of its kind. Gerdil was familiar with Holland's critique of d'Holbach but added several original observations on the various inconsistencies in Buffon's theory and with its inability to cope with the density of planets, and with the absence of atmosphere in the moon. According to Gerdil, the laws of celestial dynamics could explain the permanence of the planetary system but not its origin (col. 1313)!

114. *Journal des savants*, Juillet 1779, pp. 501–03. 'A novel in physics', 'a sublime novel', 'sublime poetry', 'cosmogonical novel' were the typical epithets accorded by some other critics to the *Époques*. See edition by Roger, pp. CXXIX, CXXXVIII, CXLIII, CXLIV.

115. *Histoire de l'astronomie moderne depuis la fondation de l'école d'Alexandrie jusqu' à l'époque de M.D.CCXXX. Tome Second* (Paris: chez les Frères de Bure, 1779), pp. 721–22. Curiously, Bailly did not enter into the difficulties which celestial dynamics posed for Buffon's theory.

116. *Cométographie ou Traité historique et théorique des comètes* (Paris: de l'Imprimerie Royale, 1784).

117. *Ibid.*, vol. 2, p. 175.

118. *Ibid.*, p. 176.

119. *Ibid.*, p. 177.

120. *Cométographie*, vol. 2, p. 177.

121. *Ibid.*

122. *Histoire de l'Académie Royale des Sciences. Année M.D.CCLXXXVIII* (Paris: de l'Imprimerie Royale, 1790), p. 57.

123. *Exposition du système du monde* (Paris: Imprimerie du Circle Social, An IV [1796]), vol. 2, p. 301.

CHAPTER FIVE

THE NEBULOUS ADVANCE

An amateur's self-confidence

The *Allgemeine Naturgeschichte und Theorie des Himmels* (or 'General Natural History and Theory of the Heavens') by the thirty-one-year-old Kant[1] was dedicated to the king of Prussia, known to posterity as Frederick the Great. The author, who kept his name from the title page, was a struggling member of the University of Königsberg looking for a secure academic status. Endearing himself to a king, who in a few years turned the Berlin Academy of Sciences into a major centre of learning, would have meant for Kant promotion and effective recognition of his theory. During the second half of the 18th century the Berlin Academy boasted of such presidents as Maupertuis, Euler, and Lagrange, but neither they nor Frederick the Great learned of Kant's work. J. F. Petersen, the publisher, went bankrupt and his holdings, which included the freshly printed copies of Kant's *Allgemeine Naturgeschichte*, were duly impounded.[2] Beyond that the fate of most copies of Kant's work is enveloped in mystery.[3] One copy reached the *Freye Urtheile*, a literary gazette published twice a week, where it was quickly given a review. It ended with the words: 'The style is lively and spirited; the author knows how to present with ease the most difficult astronomical propositions'.[4]

Ease was the tone on which the first part of the Preface of the *Allgemeine Naturgeschichte* came to a close. There Kant faced two objections of which the first had to do with the alleged materialism of 'naturalism', that is, of an explanation of the origin and evolution of the world in terms of mechanistic principles.[5] Kant's answer was that the successive differentiation of the physical forces and of material conditions could of themselves issue only in a chaos. That the product was nevertheless a harmonious, purposeful whole, Kant could ascribe only to the imparting by God of a specific direction to the mechanistic processes. The material universe, he wrote, 'has no freedom to deviate from this perfect plan'.[6] The other objection concerned the difficulty of the undertaking itself. 'I annihilate this objection', he declared, 'by clearly showing that of all inquiries which can be raised in connection with the study of nature, this is the one in which we can most easily and most certainly reach the ultimate'.[7]

Kant based his startling confidence in part on the exact explanation by Newtonian science of some of the large-scale features of the universe. His other source of confidence was the apparently perfect simplicity of the cosmic situation. The celestial bodies were round, their movements unimpeded, and their framework, the space, simplicity itself to comprehend. 'So I assert', Kant exclaimed, 'that among all the objects of nature whose first cause is investigated, the origin of the system of the world and the generation of the heavenly bodies, together with the causes of their motions, is that which we may hope to see first thoroughly understood'.[8] For Kant the unfolding of the origin of the actual structure of the universe was to yield to investigations much sooner than if 'the production of a single herb or a caterpillar' were to be understood 'distinctly and completely' by mechanical causes.[9]

This was no small claim, but Kant hinted clearly enough that the solution was at hand. His ardent vindication of the Cartesian motto, 'Give me matter and I will construct a world out of it', indicated precisely this. Kant pointedly recalled the endorsement of Descartes' evolutionary approach by the authors of *Universal History*,[10] a work which began with summaries of various theories on the geological prehistory of the earth, among them those by Burnet and Whiston. Kant did not mention these two by name, but he was eager to refer to Newton. The subtitle of Kant's work claimed that it was 'An Essay on the Constitution and Mechanical Origin of the Whole Universe Treated according to Newtonian Principles'. This was all the more surprising as Kant knew that Newton abhorred the idea of a mechanical explanation of the origin of the system of planets.[11] The non-mathematical 'General Scholium' in which Newton made this clear could readily be perused by Kant, not enough of a mathematician to wade through much of the *Principia*. An indirect recognition of this was his remark that whereas Newton cleared up the mathematical part of cosmology, he would do the same with the physical part. At any rate, he did not consider himself unqualified to reassure future investigators that the laws of determining the evolution of the physical universe were perhaps 'more capable of exact mathematical determination' than the laws of any other areas of the physical realm.[12]

In the second part of the Preface Kant gave a general outline of the three Parts of his work and insisted that from the interplay of the two Newtonian forces, attraction and repulsion of matter, one can derive the whole genesis of the universe. This tagging by Kant on Newton, and on his 'natural philosophy' in particular, a force of repulsion as being on equal footing with the force of attraction, had no more justification than had Kant's reference to hot vapour as the evidence of that repulsive force.[13] Kant was much more to the point when he spoke of Thomas Wright of Durham. The contents of Wright's *Original Theory and New Hypothesis of the Universe* came to Kant's attention through a lengthy review of that work published in 1751 in the *Freye Urtheile*.[14] Unfortunately, it was not pointed out in the review that after Wright had offered the hypothetical case of all stars being confined between two unlimited parallel planes, he saw this model realised in the segment of a huge ring or of a spherical shell containing all the stars.[15]

Kant certainly did not get from Wright the lentil model of the Milky Way, although he owed to Wright the idea that the stars moved around a centre in much the same way as did the planets around the sun. This was compatible with Wright's ring model of the Milky Way, and even more so with the lentil model which Kant proposed and which he saw realised in the often elliptical shapes of nebulae. With the analogy established between the planetary and stellar systems, Kant felt that he had fully justified the explanation of a topic in which, as he somewhat defensively stipulated, 'the greatest mathematical precision and mathematical infallibility can never be required'.[16] The analogous shapes of the solar system and of the galaxies were the key to his cosmogony and he took pains to emphasise that any such lentil-shaped configuration of planets, stars, and galaxies was to be denoted in his book as a 'systematic constitution'.[17]

The relatively short First Part of Kant's work was devoted to that analogy. He now spoke of the plane of the Milky Way as the zodiac of the starry realm and described the stars outside that plane as 'the comets among the suns'.[18] Kant's resolve to see the closest possible interconnectedness between all parts of the universe along the same pattern led him to conjure up the possibility of a good many planets beyond Saturn. He based this on the fact that with the exception of Mercury and Mars, the eccentricities of planets increased with the distance from the sun. He gave the figures, in terms of the semiaxis of the orbit, as 1/126 for Venus, 1/58 for the earth, 1/28 for Jupiter and 1/17 for Saturn. He then postulated that the ultra-Saturnian planets had even greater eccentricities to provide a gradual transition to the highly eccentric orbits of comets. As to the actual sighting of those planets he felt the need to warn: 'They would be perceptible only for a short time, namely, during the period of their perihelion, which circumstance, together with the slight degree of their approach to us and the feebleness of their light, has hitherto prevented the discovery of them, and will make their discovery difficult even in the future'.[19]

With this, Kant arrived at the crucial stage of his discussion and made no further secret of the fact that his and Newton's ideas on the origin of the planetary system were mutually exclusive, a situation which he likened to the absence of an intermediate situation between an empty space and one full of matter. Kant misrepresented Newton when he wrote that it was the fact of the emptiness of space which moved Newton to renounce a physical explanation of the origin of planets, as in an empty space there was no physical factor left to transform part of the linear attraction into a transverse impulse. Newton's speculation about the ether seemed to be as unfamiliar to Kant as were Newton's deeper preoccupations about the derivation of that transverse impulse without which the planets would rapidly fall into the sun. But even there, where his unfamiliarity with something very germane to the topic was conspicuous, namely, his lack of mastery of calculus, Kant was not reluctant to give a different impression. He made his reader believe that 'had he cared to do so' he could have shown conclusively the truth of his theory by advancing a 'series of necessarily deduced inferences according to the mathematical method . . . with all the parade that the method brings with it'.[20] The parading was done by Kant himself donning the garb of an accomplished mathematician. The truth was that not even the leading mathematical physicists of the time would have been up to the task to pour into the moulds of mathematics the processes which Kant's theory implied.

Planets out of Kantian a priori

According to Kant the material universe was created by the Divine Intelligence as a chaos in which the only distinctive feature was the presence of many kinds of elementary matter. The kinds of this elementary matter were, in his words, 'undoubtedly infinitely different, in accordance with the immensity which nature shows on all sides'.[21] The feature which differentiated the various kinds of matter was their specific densities. If matter was equally distributed into the various elements, it followed, that the elementary particles of the denser elements were much farther removed from one another than were the particles of elements of much lower density. Although in his model the gradation of specific density had to 'be thought to be as infinite as possible', he spoke of material particles denser than others only to the extent of the ratio of the radius of the planetary system to 'the thousandth part of a line'.[22] Accordingly, any two particles of that enormous density had to be located from one another by a distance equal to the foregoing radius. That Kant mentioned the radius of the planetary system and not that of a galaxy or of a supergalaxy as the implicit measure of the 'largest' density represents a point in his cosmology which will be discussed later. The full implications of his patently imprecise phrase, 'as infinite as possible', will also be better seen in the light of some of his other dicta on the infinity of the universe and on the grouping of galaxies into an endless chain of successively more inclusive supersystems.

Unaware of the stark apriorism of his postulates, Kant noted that if attraction was an inherent quality of matter, 'a universal repose could last only a moment in a region of space filled in the manner'[23] which he had specified. Particles of the densest elements would immediately act as foci of attraction for less dense elements with the final result being, so Kant reasoned, a distribution of all matter into motionless, huge, round bodies. Revealingly, he did not identify those bodies with stars. Nor did he face the crucial question of where the maximum cut-off point had to be on the density scale of elements which formed the very foundation of his cosmology. He offered but a minimum of statement on the equally pivotal point, the emergence of a rotational movement around the foci of attraction. He saw the cause of this in the various ways in which material particles repelled one another. This repulsive force was manifested, according to him, 'in the elasticity of vapours, the effluences of strong smelling bodies, and the diffusion of all spirituous matters'.[24] Since the existence of that force was 'incontestable', he readily declared: 'It is by it that the elements, which may be falling to the point attracting them, are turned sideways promiscuously from their movement in a straight line; and their perpendicular fall thereby issues in circular movements, which encompass the centre towards which they were falling'.[25]

This was hardly better than the chance 'swerving' of particles from their straight motion to which Epicurus assigned the origin of celestial bodies and of their motions, an explanation which Kant had already brushed aside as absurd.[26] Kant's own explanation was no more a solution to the problem than was the 'modern' theory of Buffon whose views Kant must have known, because a few pages later he quoted a small numerical detail from Buffon's work.[27] He was enough of an amateur in mathematical physics to rush ahead boldly on a terrain where the foremost experts could not spot a forward path. He merely advanced into a nebulous equivocation by simply asserting what was to be explained, namely, the emergence of a whirl of matter around the sun. Its first phase was as follows: 'When the mass of this central body has grown so great that the velocity with which it draws the particles to itself from great distances, is bent sideways by the feeble degrees of repulsion with which they impede each other, and when it issues in lateral movements which are capable by means of the centrifugal force of encompassing the central body in an orbit, then there are produced whirls or vortices of particles, each of which by itself describes a curved line by the composition of the attracting force and the force of revolution that has been bent sideways'.[28]

At this stage the solar system corresponded to a central mass around which particles orbited in every possible direction. Kant now had to explain the evolution of a unidirectional rotation in one plane. Once more he offered but words. According to him the one common direction arose 'naturally'[29] in every case where movements conflicted with one another in many different ways. The single plane of rotation was the result of the mutual limitation of the vertical component of movement of particles already put into orbit around the sun. This claim implied a *petitio principii*, because no geometrical and physical reason could be assigned to the differentiation along a 'vertical' direction in a distribution which was spherically symmetrical. The plane in question could not be determined by the rotation of the sun, since this latter point still was to be accounted for. Kant only noted that the sun was not yet in a fiery state.[30] He did not sense the pitfalls hidden in his apparently felicitous surmise that conflicting motions always evolved into a condition of what he called 'the state of the least reciprocal action'. Particles orbiting in a plane and in well defined orbits without colliding with one another was for him the ideal realisation of such a state. The foundation of that ideal had much less to do with reality than with Kant's *a priori* legislation about cosmic origins.

The stage of planetary evolution at this point was envisioned by Kant in the following words: 'We see a region of space extending from the centre of the sun to unknown distances, contained between two planes not far distant from each other, in the middle of which the general plane of reference is situated. And this elementary matter is diffused in this space within which all the contained particles – each according to the proportion of its distance and of the attraction which prevails there – perform regulated circular movements in free revolutions'.[31] The situation described by Kant seemed to imply that orbital velocities and the radii of their orbits varied on a practically continuous scale between some minimum and maximum values. Kant did not seem to be aware of the fact that he negated that distribution when he declared that 'planets are formed out of particles which at the distance at which they move, have exact movements in circular orbits'.[32] The question of the radial distribution of particles was even more acute when a little earlier Kant spoke of the 'seeds of planets',[33] that is, of some rudimentary agglomeration of particles which with respect to one another were motionless, because they orbited at the same rate and at the same radius from the sun. In that case, should not an immensely large number of 'seeds' have arisen instead of a few forming the nuclei of the present planets?

No speculation was offered by Kant on the equally crucial point of how the particles in one orbit gathered to one single point or 'seed' in that orbit. Instead he argued that chemical affinity between the various particles had to play as much a role as did gravitational attraction in the actual growth of the body of a planet.[34] In this respect his phrases were as vague as they were when he ascribed the greater eccentricities of the orbits of the more distant planets to the greater 'multiplicity of circumstances' prevailing in the more outlying areas of the solar system.[35] The nebulosity enveloping the origin of the solar system did not disappear as Kant offered his own appraisal of his theory: 'The view of the formation of the planets in this system has the advantage over every other possible theory in holding that the origin of the masses gives the origin of the movements, and the position of the orbits as arising at the same point of time; nay more, in showing that even the deviations from the greatest possible exactness in these determinations, as well as the accordances themselves, become clear at a glance'.[36] What all this proved was that Kant took but a glance and not a thorough look at the staggering problem of the origin of planetary systems. His explanation of it contained more nebulous statements at crucial junctures than there was nebulosity in the rudimentary form of the solar system. At any rate, contrary to the stereotype accounts of Kant's planetary theory, he did not compare the rudimentary form of the solar system to a nebulous agglomeration of matter.

Kantian theory versus the facts of the solar system

The inconsistencies, which plagued Kant's theory from the very start, kept showing up as he tried to derive the main features of the solar system from his initial assumptions. The first of these features concerned the densities of the sun and the planets, and the distribution of the total mass between the sun and the planets. As to the question of density, he ascribed the greater density of the inferior planets to the 'greater penetration' of denser particles toward the centre.[37] But then it remained to be explained why the density of the sun was considerably lower than that of the earth, Venus, and Mercury. As a solution Kant proposed that the sun collected its mass from the space well beyond Saturn, 'where the movements of the elementary matter have not been fitted to attain to regulated equilibrium with the central forces as in the region near the centre, but issue in an almost universal falling towards the centre and increase the sun with all the matter out of such widely extended ranges of space'.[38]

Here Kant seemed to forget that he had already pictured the distribution of elementary matter as that in which the particles were separated from one another in proportion to their respective densities. Consequently, if the sun's mass accrued from that very distant matter, the density of the sun should have become much lower than the densities of the planets. Of course, the question still remained unresolved about a crucial feature of the original density distribution. It concerned some well defined boundary lines in the three-dimensional density distribution to the analogy of large domains postulated by Descartes. Of this point Kant was wholly oblivious as he also seemed to forget that it was inconsistent on his part to make computations about the average density of the total mass of the solar system as if it had been originally confined between two conical surfaces forming an angle of $7\frac{1}{2}$ degrees and extending to a distance somewhat beyond Saturn. The density he obtained was as many times lower than that of the air as the latter was rarer than ordinary matter. Once more he called that original state of matter 'chaotic' and not 'nebulous'. More importantly, he was convinced that all was clear in the mechanical theory of the formation of the system of planets: 'Everything concurs as admirably as could be desired, in corroborating the sufficiency of a mechanical theory as an explanation of the origin of the construction of the world and of the heavenly bodies'.[39] It escaped him that he also failed to explain why any spherically distributed matter would shape itself into that flat disk which he described in the same context.

Kant's failure to see the affinity of some of his basic assumptions with those proposed by Descartes was of a piece with his omission to mention Buffon as he took up another feature of the solar system. It concerned 'the eccentricity of the orbits of the planets and the origin of the comets'.[40] There at the very outset Kant voiced the opinion, harking back to Buffon, that there was a gradual transition from planets to comets.[41] In other words, there had to exist planets beyond Saturn with larger and larger eccentricities and presumably with masses that increased with distance. Kant discounted the masses of comets, which belonged to still more distant regions, because of their very low specific density. Since they were formed, Kant noted, 'in the furthest regions of the universe, it is probable that the particles of which they are composed are of the lightest kind'.[42] By 'universe' he obviously meant the boundary surfaces separating neighbouring domains with particles of maximum density at their centres or nuclei. The halfway-distance between two such nuclei could then be described as an area of lowest density.

There was a distinct touch of inconsistency in the explanation which Kant offered for the large eccentricity of Mercury's orbit. It was caused, according to him, by the difference between the orbital speed of Mercury and the rotational speed of the sun. As the latter was much slower, it exerted a braking effect on Mercury and distorted its originally perfect circular motion. Kant imagined that the atmosphere of the sun extended well beyond Mercury, an idea already stressed by Descartes, and thus the braking effect could be mechanically transmitted. Apart from the highly arbitrary character of that assumption, Kant was still to explain that the sun (and the planets too) were rotating. The explanation which he finally gave in the fourth chapter of the Second Part was identical with the one he offered for the orbiting of planets. As particles condensed into a celestial body, they necessarily gave its nucleus many small, sidewise impulses which ultimately had to result, so Kant believed in patent disregard of the assumed symmetry of the original situation, in the rotation of the whole. He never considered the fact that on such a basis the rotation of all planets in the same sense was a most unlikely event.

Once gravitational attraction was duly recognised as the all-pervading factor in nature, one had, in Kant's words, a 'mode of explanation' of the formation of the planetary system which will 'authenticate its correctness by the natural claim of its fundamental conception and by such unforced inferences from it'.[43] The word 'unforced' has rarely been more out of place. It did not dawn on Kant that in his theory nature was 'forced' at almost every step to provide favourable circumstances for the formation of each and every feature of the planetary system. He failed to see that the long chain of these circumstances represented precisely that high degree of improbability the weight of which was forcefully set forth by Buffon. A good illustration of Kant's short-sightedness in this respect is his explanation of the ring of Saturn. (In 1755 it was still forgivable to speak of only one ring instead of two.) He traced its origin to the primordial and very eccentric orbiting of Saturn around the sun implied in his hypothesis. Such an orbiting had to take Saturn close to the sun with the result that due to the sun's heat heavy vapours rose from Saturn's originally liquid surface. Rotation around Saturn made that cloud turn into a ring which Kant, fortunately, did not let condense, let alone solidify. He pictured the ring as composed of an immense number of small particles. His application of Kepler's Third Law and of Huygens' formula for the centrifugal force as determining the period of rotation of the inner and outer edge of the ring bespoke the lucky amateur whose wrong method leads to the right result on rare occasions. He was far less fortunate with his claim that all planets were in their primitive stage surrounded by a ring. In the case of the primitive earth he saw evidence of that ring in the biblical references to the 'waters of the firmament' and to the 'coloured bow'[44] which appeared after the deluge. Thus, eagerness revenged itself on the one who wrote: 'I assume that Saturn had a rotation round its axis; and nothing more than this is necessary to unveil the whole mystery'.[45]

Basic problems and persistent illusions

Such dogmatism in Kant's dicta on planetary and cosmic evolution should not, however, distract attention from some basic problems and illusions of his theorising. The impossible consequences of the distribution of an infinite variety of elementary particles according to their densities came forcefully to the fore as Kant took up the question of the formation of the whole universe on the very same basis. Since for Kant the infinite universe was an infinite chain of hierarchically ordered systems, all patterned on the solar system, the question could not be avoided about enormously massive bodies forming the centres of galaxies and successively larger supergalaxies. In fact, Kant's infinite universe presupposed an infinitely large central body, but Kant carefully avoided considering such a destructive implication of his cosmogony. This was all the more revealing as Kant spoke explicitly of the need for a place where 'this primitive material had been most densely accumulated so as through the process of formation that was going on predominantly there, to have procured for the whole Universe a mass which might serve as its fulcrum'.[46] He seemed to recognise something of the difficulty as he tried to justify the assumption of an absolute centre in an infinite universe. That place could be designated, according to him, 'by means of a certain ratio, which is founded upon the essential degrees of the density of the primitive matter'. Such a point, he added, 'may have the privilege of being called the centre; and it really becomes this through the formation of the central mass by the strongest attraction prevailing in it'.[47]

Kant did not consider the possible conflict of the emergence of that absolute central body with his original ideas of the distribution of particles in the primordial chaos according to a linearly repetitive function of density. His attention was now focused on the start of all evolution in the absolute centre. Planetary systems and galaxies in their full formation represented the crests of an evolutionary wave which spread from the centre toward the infinitely distant peripheries. The troughs of that wave were the areas where such systems had reached their complete dissolution, ready to rise again from the cosmic ashes like the 'Phoenix of nature'.[48] Kant's universe was the paragon of a cyclic transformation including a perennial life-and-death cycle for planetary civilisations everywhere. He suggested this latter point by inference from what the situation on other planets suggested to him. About the existence of denizens on other planets of the solar system Kant had no doubt. It was to that topic that he devoted the third and concluding Part of his work,[49] following faithfully an already hallowed pattern in speculations on planetary origins.

Kant's case was not without originality. For the first time wishful thinking about planetarians closely matched the illusoriness of the mechanism producing the planetary system itself. Equally telling should seem the fact that Kant's assumption about the original density distribution of matter played its tragicomic part to the end. First, he belaboured the point of the mind's dependence on the structure and composition of the body and of the general physical habitat. From this he blandly concluded that the more refined was the material substratum on which the mind depended, the more refined were the corresponding mental and emotional operations. With these 'verities' firmly laid down, what could have been more natural than to populate each planet with denizens befitting its density? The densest planet, Mercury, with its allegedly crudest matter, hosted the rudest denizens, whereas the much less dense Jupiter and Saturn were the homes of an altogether superior race. In Kant's own simile, a Hottentot of the earth would pass for a Newton among the Mercurians, but our own Newton would not outrank a monkey on Saturn.[50]

The measure of Kant's self-deception could best be gauged by the eagerness with which he exploited the farcical implications of the foundations he had laid. The ten-hour rotation of Jupiter on its axis was for him a matching circumstance for the superior agility of the Jovians who in five hours of daylight did as much as the sluggish earthlings achieved in twelve hours.[51] Again, because of the refined composition of their bodies, the Saturnians were depicted by Kant as practically immortal.[52] Such was a heedless foray into a no-man's-land the general features of which Kant professed to know with fair certainty. He claimed that it was 'more than a probable opinion that the respective perfections of the planetarians . . . followed the specific rule according to which their habitats also were more perfect in proportion to their distance from the sun'.[53] The extreme heat on Mercury and the extreme cold on Jupiter and Saturn presented no problem for Kant. The gross matter of Mercury readily withstood the sun's heat, whereas the very refined matter of Jupiter and Saturn readily absorbed it. No wonder that Kant fondly recalled that wit from The Hague who compared disbelief in planetarians to the reluctance of fleas living in a beggar's hair that heard with incredulity the report of one of their numbers which had the good luck of spending a day in the wig of a nobleman.[54] According to Kant the incredulity was typical of anthropocentric thinking. He was not the last to misplace warning about the fact that effort to overcome anthropocentrism also had traps of its own.

This Third Part of the *Allgemeine Naturgeschichte*, so contrary to the rigour in reasoning on which Kant later prided himself, was in his eyes an integral part of his theory of planetary origins. He never suspected that the loss of almost all copies of his book was not an unqualified misfortune. Apart from its few pages on the correct explanation of the visual appearance of the Milky Way, the *Allgemeine Naturgeschichte* offered no reliable science and certainly no sound ideas on the evolution of planetary systems. Kant's own evaluation of the merits of his theory was very different as can be seen from his inclusion, in 1763, of a lengthy summary of it in his *Der einzig mögliche Beweisgrund zu einer Demonstration des Daseins Gottes*, a work dealing with the only possible proof of the existence of God.[55] The inclusion was rather baffling because it served to illustrate the argument from design and not another speculative argument, the only proof which Kant still accepted at that time. The work certainly gave a foretaste of the post-critical Kant, but it also suggested with its four editions between 1763 and 1794[56] that Kant was not to form a critical appraisal of his theory as time went on. This lack of criticism was also the hallmark of the reference to Kant's cosmogony by Herder in the opening section of his famed philosophy of history first published in 1785: 'The human intellect perhaps never attempted a longer flight and completed it successfully in part at least, when through Copernicus, Kepler, Newton, Huygens, and Kant it perceived and laid down firmly the simple, eternal, and perfect laws of the formation and motion of planets'.[57] The passage merely proved that Herder remained basically a poet in writing intellectual history. Unfortunately, not only poets, like Herder, presented the respective contributions of prominent men of science in such a confusing manner.

Steady neglect and isolated critique

Herder's widely read encomium of Kant, the scientist, made no dent on the indifference of scientists in Germany, let alone abroad, toward Kant's theory. Part of this indifference may have been due to reluctance to consider an evolutionary cosmology, but the principal cause must have been the 'science' of Kant's *Allgemeine Naturgeschichte*, a work which Kant specifically mentioned in 1765 to Johann Heinrich Lambert, freshly appointed member of the Berlin Academy, together with the summary of its contents in *Der einzig mögliche*

Beweisgrund.[58] To be sure, in the exchange of letters between Kant and Lambert only the Milky Way was discussed, possibly because Lambert advocated a strictly non-evolutionary cosmology in his *Cosmologische Briefe* published in 1761.[59] But even Johann Elert Bode, whom Lambert brought to Berlin in 1770 to be his eventual successor as editor of the *Astronomisches Jahrbuch* and who after Lambert's death in 1777 became the head of German astronomy, made no mention whatever in his many and influential publications of Kant's planetary theory. In 1777, when Bode published the third edition of his popular astronomy, he merely added to Lambert's name that of Kant as he briefly discussed the structure of the universe.[60] A year later, when Bode's more technical textbook of astronomy saw print, the same generous though brief praise was accorded to both Lambert and Kant, but once more no mention whatever was made of Kant's ideas on the origin of the system of planets.[61] In this respect Bode felt no need to make any change, although he devoted much care to improving and enlarging the numerous re-editions of both works which dominated German astronomical literature for the next half-century.[62]

The silence of Bode meant also the silence of that highly active enterprise into which German astronomy developed toward the end of the 18th century. Those who began to see Kant as one who had anticipated Herschel's ideas on the universe consisting of countless Milky Ways were some of Kant's admirers in Königsberg. One of them, Johann Friedrich Gensichen, published in 1791 lengthy extracts from the *Allgemeine Naturgeschichte* as a sequel to the German translation of Herschel's great memoirs on the construction of the heavens.[63] The summary gave a good idea of Kant's theory of planetary origins, but German astronomers paid no attention. Its sole discussion came in a short-lived periodical whose editor, J. A. Eberhard, tried but in vain to stem the tide of the sudden adulation accorded to Kant after the publication in 1787 of the second edition of the *Critique der reinen Vernunft*. The sundry non-sequiturs of Kant's theory provided plenty of ammunition for J. C. Schwab, one of the contributors to Eberhard's periodical, who aimed both at the inconsistencies of Kant's theory and at its conflict with Newton's ideas on the system of planets.[64] Schwab emphasised that the densities of planets and of the sun showed a sequence incompatible with Kant's assumption about the original distribution of material particles according to their density in the still 'chaotic' matter. Kant's claim that Saturn's rings were due to the rise of vapours was also attacked by Schwab for its departure from the basic mechanism, namely, the formation of large masses through attraction, on which Kant rested his theory.[65] According to Schwab, Kant contradicted Newton by defining attraction, to say nothing of repulsion, as an innate property of matter.[66] Schwab, who cited in this connection Maclaurin's *Account of Sir Isaac Newton's Philosophical Discoveries*, found there more than one emphatic denial of the claim that the orbiting and rotation of planets might be derived from attraction and collision.[67] Concerning this last point Schwab chided Kant for not being clear enough about the measure of the role which chance collisions were supposed to play in his theory. 'Lucretius was at least candid on this score, but you, Mr. Kant,' Schwab remarked, 'claim that your system is purely mechanical'.[68]

Actually, 'the Kantian hypothesis', went Schwab's verdict, 'leads us into a chaos of whirls out of which we are no longer able to escape'.[69]

In Schwab's criticism of Kant's theory two more details are worth mentioning. One of them was Schwab's remark that he was unable to secure a copy of the original edition of the *Allgemeine Naturgeschichte*. The other was his reference to 'one very strong in the mathematics of physical astronomy', who wanted to remain anonymous while volunteering to Schwab the conclusion that 'it did not follow from Kant's theory that in planetary systems all planets must move in much the same plane'.[70] While the full text of the *Allgemeine Naturgeschichte* saw three editions between 1797 and 1807,[71] the refusal of German professors of astronomy and mathematics to discuss Kant's theory of planetary origins remained unchanged, although ideas of cosmic evolution were discussed with distinct sympathy by some of the most prominent among them.[72]

Laplace's theory: the first form

Silence about Kant may have been due to distrust in an evolutionary outlook when in 1797 a German periodical on the latest in science carried an enthusiastic review[73] of a German translation of Laplace's *Exposition du système du monde*, a work published only a year earlier. In the review no mention was made of the fact that Laplace's book came to a close with a chapter entitled 'Considérations sur le système du monde, et sur le future progrès de l'astronomie'.[74] It was Chapter vi of Book V. Its future renown derived not from Laplace's guessing the future of astronomy, but from his 'considerations' which stood for a novel theory of the origin of planetary system.

Unlike Kant, Laplace merely touched on the question of denizens on other planets. According to Laplace 'analogy impels us to believe' that 'the beneficial action of the sun which brings forth the animals and plants on earth . . . would produce similar effects on the planets'.[75] Laplace did not think it proper to assume that the same matter, so fertile on earth, should be sterile on a large planet like Jupiter 'which, like the terrestrial globe, has its days, nights, and years, and on which the observations point to changes which presuppose very active forces'.[76] Laplace's use of the word 'active' was somewhat euphemic, as being 'very active' did not exclude being very devastating. Laplace was, of course, enough of a sober scientist to refrain from speaking of men-like beings on Jupiter. He realised that man, conditioned for a specific range of temperature, could not survive on any of the other planets. Did this, however, mean that there should not be 'an infinity of organisms suited for the different temperatures of the globes of that universe?' It was in the same vein that he asked again: 'If the mere difference of materials and climates is matched with so great a variety in the living formations on earth, how much more should differ those on the planets and on their satellites?' Misleading as one's imagination could be in giving an idea of that variety of organisms, their existence was 'most likely'.[77] But in that famed Chapter vi of Book V Laplace's last remark was reserved for man in the form of a pointed contrast between his physical puniness in a vast universe and his excellence as a thinking being.[78]

Needless to say, Laplace did not have to be familiar with Kant's work to argue the likelihood of extraterrestrial life. Neither on this particular point nor on the theory of the origin of planets could Laplace's presentation have been more different from Kant's affectation of expertise. As a starting point Laplace listed five characteristics of the various motions in the solar system which an acceptable theory of its origin had to explain satisfactorily. 'One is astonished', he wrote, 'to see all the planets move around the sun from west to east and in almost the same plane; to see all the satellites move around their planets in the same sense and almost in the same plane as the planets; finally, to see the sun, the planets and the satellites, whose rotation had been observed, turn on their axes in the sense and also almost in the plane of their orbital motion'. Laplace's verdict was that 'such an extraordinary phenomenon is not at all the effect of chance'.[79] To prove this quantitatively he took stock of the co-ordination of thirty motions in the planetary system, namely, the orbiting of the seven planets and of fourteen satellites, the rotation of the sun, of five planets, of the moon, of the ring of Saturn, and of one of its satellites. That at least one of these motions should considerably deviate from the general plane or from the direction of motion in the solar system was then equivalent to $1 - \frac{1}{2^{29}}$, or $\frac{536,870,911}{536,870,912}$, a fraction[80] practically equal to 1. This meant that, if the thirty planetary motions were the effect of chance, it was a practical certainty, that is, a probability almost equal to 1, that one such deviation should have indeed been realised. In addition to the orbiting of planets in the same direction and in the same plane, to the motion of satellites in the direction and plane of the planets, and to the rotation of both planets and satellites in the same sense, Laplace singled out the very small eccentricity of planetary orbits and the very large eccentricity of cometary orbits as the fourth and fifth features of the solar system that could hardly be explained on the basis of chance.

Laplace now recalled Buffon as the only one who tried to derive from a single physical process the arrangement of planets.[81] While Laplace admitted that the first feature of the solar system, the orbiting of all planets in the same sense and in the same plane, could be explained by Buffon's theory, it could not cope with the four other features because of basic laws of celestial mechanics. Laplace did not mention the well known fact that these laws had been made known some thirty years earlier to Buffon, who refused to concede defeat.[82] What Laplace said of Buffon implied that Laplace did not know of Kant, which is also borne out by a comparison free of chauvinistic motivations of their cosmogonies.[83]

At this point Laplace was already past three-fourths of his famed theory which in its first form ran to slightly less than three thousand words. All he said up to that point amounted to a general conclusion that the planetary system had to derive from a single physical process. On the face of it this conclusion could seem unobjectionable, but on closer look the situation did not look so perfect. Laplace's count of fourteen satellites could only be obtained by taking into consideration the first two satellites of Uranus. In the second part of the chapter dealing with stars, nebulae, and the future prospects of astronomy, Laplace mentioned Uranus and its 'recently discovered satellites',[84] but only as supports of the hope of finding even more distant planets in the future. He mentioned neither Herschel nor the fact that according to Herschel the orbits of those satellites made a 'considerable angle with the ecliptic'. Herschel added with his customary caution that 'to assign the real quantity of the inclination . . . will require a great deal of attention and much contrivance'.[85] In fact, within two years of the first publication of the *Exposition* Herschel had to declare those motions to be retrograde.[86] Was Laplace suspecting such a development, and therefore trying to forestall a heavy blow to his argument from probability, as he hastened to specify that only a deviation of 100 degrees[87] from the ecliptic would constitute a retrograde motion? Whatever the help of that proviso for his theory concerning the planets and their satellites, it was of no help with respect to the widely differing orbital planes of comets. No wonder that Laplace considered the comets to be bodies alien to the solar system.

Laplace now had to specify the common cause of the common features of planetary motions. Since Buffon's theory did not work, 'let us see', Laplace wrote, 'whether we can ascend to the true cause'. He found it in a quasi-liquid medium which surrounded the sun as if it were its atmosphere, extending beyond the limits of the actual planetary system. To provide that extension, the atmosphere of the sun had to be 'excessively hot'.[88] Because of its excessive heat it had to be very bright and Laplace recalled the nova of 1572 as an analogy. The brightness of that solar atmosphere could perhaps be exemplified by that nova, but hardly the successive shrinking of that atmosphere resulting from its gradual cooling, a point which he had just mentioned. This was not the only point where Laplace's cosmogony exuded a studied vagueness. Thus, for instance, while he had said that the great extension of the sun's atmosphere was needed to account for the orbital motion of even the outermost planet, he did not say that the solar atmosphere was rotating. He did not spell out the fact that a rotating primitive solar atmosphere had to be flattened considerably, nor did he refer, in the context at least, to the cause of that rotation. Instead, he spent on comets more than half of the remaining three hundred words of his cosmogony. Whereas Buffon claimed a gradual transition to exist between planets and comets, Laplace viewed comets as not forming part of the primitive solar atmosphere. His reason was that comets, if born in that primitive atmosphere, would have before long fallen into the sun due to retardation of their motion. Comets, therefore, formed outside the primitive solar atmosphere, and only those survived which entered the sphere of influence of the sun after its original atmosphere had shrunk considerably.[89]

Since the same retardation also threatened the planets, Laplace specified that they formed at the outer edge of that atmosphere. This was the essence of the last paragraph of his cosmogony which deserves to be quoted in full: 'But how did that atmosphere determine the orbital and rotational motions of the planets? If these bodies had penetrated that fluid, its resistance would have made them fall into the sun; one may, therefore, conjecture that they were formed on the successive borders of that atmosphere by the condensation of zones, which that fluid had to leave behind in the plane of its equator as it was cooling and condensing toward that star [sun], as this was shown in the preceding book.

One can also conjecture that the satellites were formed in a similar fashion from the atmosphere of the planets. The five features listed above follow naturally from these hypotheses to which the rings of Saturn add a further measure of likelihood'.[90]

Laplace's reference to an earlier part of the *Exposition* was unavoidable. The really decisive part of his cosmogony, which he merely stated here, was given in some detail in a brief chapter 'on the atmosphere of celestial bodies'.[91] There he started with the following definition: 'A thin, transparent, compressible and elastic fluid, which surrounds a body, this is what one calls its atmosphere'. His next point was that owing to gravity the density of such atmosphere is decreasing with distance from the centre. As a result the atmosphere will consist of layers, the mutual friction of which will equalise the orbital motion of molecules constituting the atmosphere. Such was the physics underlying Laplace's declaration that 'all atmospheric layers must take in the long run a rotational motion common with that of the body which they surround'. Laplace had, of course, to give the impression that the physics in question obeyed the laws of mechanics, and especially the law of the conservation of angular momentum: 'In these changes, and in general in all changes experienced by the atmosphere, the sum of the product resulting from multiplying the molecules of the body and of its atmosphere with the areas which their radius vectors projected on the plane of the equator describe around their common centre of gravity always remains the same in equal times'.[92] From this it followed that if these areas were continually diminishing owing to the contraction by gravity of the atmosphere, the orbital velocity of its molecules had to increase. What Laplace added here on the zodiacal light, on Saturn's rings, and on the moon merely served as illustrations of the foregoing reasoning.

Those illustrations could easily distract from some grave problems raised by Laplace's theory. Even worse, he could easily have given special attention to them. It was certainly not beyond Laplace's mathematical techniques to investigate the question whether those zones could coalesce into single planets. More importantly, since his extensive investigations of the mutual perturbations of Jupiter and Saturn[93] largely depended on the application of the conservation of angular momentum, he could have most naturally asked a question or two about the possibility of the formation of those zones in a rotating atmosphere. Was the angular momentum of that rotating atmosphere sufficient for the separation of those zones from the main body? Consideration of this question would have immediately drawn his attention to the actual distribution of angular momentum in the solar system which he could easily have calculated. Had he done this he might have discovered right there and then that his theory could not cope with the distribution in question.

He looked at his theory for the rest of his life as one of his chief glories, although its questionable status might have been revealed to him by his true glory, namely, his study of the mutual perturbation of Jupiter and Saturn. A pivotal part of that study, strongly suggesting the stability of the solar system, found the clue to that stability in the conservation of angular momentum in the interaction of those planets. The clue to that stability meant, however, less to

him than the stability itself, as can be seen from the fact that he went on dis-
coursing about the permanence of life on earth.[94] What he said was a rehash of
Condorcet's favourite theme about the unlimited progress of mankind in virtue
of the stable position of the earth in the planetary system.[95] The theme must
have been all the more a favourite with Laplace as Condorcet had made much
of Laplace's studies of Jupiter and Saturn. It was already noted that what
Laplace said about life elsewhere in the universe had brevity but not originality
as its chief merit.

In extending his vistas beyond the solar system, Laplace spoke of the
Milky Way and of the host of nebulous stars or nebulae, but he did so in terms
that were very brief and generic in comparison with what had already been
disclosed on the construction of the heavens a decade or so earlier by Herschel,
whom Laplace did not mention. What he specifically said about nebulous stars
was that the extent of the refraction of light passing through their 'atmospheres'
had to be enormous.[96] The possible contraction of nebulous stars was not,
however, the mechanism which he suggested as an explanation of variable
stars, some of which, as he noted, disappeared completely for a longer or shorter
period of time. Rather, he spoke of changes operating on stellar surfaces,
changes that were far more powerful than the ones producing the sunspots. It
was in that connection that he conjured up stars whose density was equal to
that of the earth but whose diameter surpassed by a factor of 250 the diameter
of the sun. Since the enormous gravitational attraction of such stars would not
permit their light rays to escape from them and thereby reach us, these stars,
Laplace remarked, would be opaque, that is, invisible to us.[97] In the universe,
according to Laplace, there could be as many opaque stars as there were visible
stars.[98] Although he did not in any sense speak of a gravitational collapse of
stars, there is some similarity between his opaque stars and the presently much
discussed black holes.[99] Since Laplace's opaque stars were intrinsically luminous,
it is doubtful that he was influenced by the enormously large, but intrinsically
dark regents or central bodies, by which Lambert had set so great a store in his
cosmology fairly well known in France from 1770 on.[100]

The second and third forms

Whatever the silent treatment given to Laplace's theory of the origin of planets
in that German review to which reference has already been made, the review
certainly was indicative of the success of the *Exposition* itself. Another sign
of that success was the second edition of the *Exposition* within three years
(1799).[101] The further elaborations, which Laplace offered on his cosmogony,
showed something of the hollowness of a remark of his in the first edition.
There he spoke of his cosmogony as deserving the same mistrust which ought
to be shown toward any theory which is not 'the result of observations and
calculus'.[102] Still the principal novelty in the second version of his cosmogony
was an updating of his calculation of the enormous improbability of a chance
production of the solar system.[103] The vanishingly small probability of a
chance causation of the solar system hinged once more on the occurrence of a
major deviation from uniformity in planetary motions. Laplace's figure was

now $1 - \frac{1}{2^{37}}$, the number 37 being the result of counting four more satellites, which Herschel discovered around Uranus in 1797,[104] a year after the first edition of the *Exposition*. Clearly, Laplace followed closely the work of Herschel, but in a peculiar way. While Laplace reported the four new satellites, he did not report Herschel's further observations which made him conclude that the orbiting of the first two satellites of Uranus was retrograde. About this fact, so much at variance with the initial presupposition of his cosmogony, Laplace was to observe a complete silence throughout the subsequent editions of the *Exposition*! There was indeed something peculiar about that 'mistrust' with which he claimed to submit his cosmogony.

Equally noteworthy was the positivist touch which Laplace injected into his cosmogony in its second form: 'If the conjectures which I have just proposed about the origin of the solar system are well founded, the stability of the system is a consequence of the general laws of motion. These phenomena, and some others apparently explained, entitle us to think that all depend on those laws through relations more or less hidden which should be the chief aim of our investigations; it is, however, wiser to admit ignorance about these relations than to substitute to them imaginary causes'.[105] But this sound precept was preceded by another, directed at 'the philosopher'. Reflection on the stability of the system and on the permanence of species 'would alone explain the arrangement of the planetary system, were not the philosopher to extend farther his view and search in the primordial laws of nature the causes best indicated by the order of nature'.[106] The remark certainly was not meant to endorse metaphysics and natural theology. In 1799 the tide of official sentiments in France was still running heavily against God and religion, and Laplace, the political chameleon (he served with equal safety the *ancien régime*, Robespierre's Terror, Napoleon's Empire, and the Bourbon Restoration), was not to advocate unpopular philosophy. His reference to the 'philosopher' could have meant Kant to whose cosmogony he may have been alerted in the meantime. This conjecture has a support in the fact that Laplace could quickly recognise Kant's amateurism in celestial mechanics, his recourse to imaginary correlation of causes, and his hollow anchoring of scientific cosmogony in the act of the Creator.

There was, however, one 'philosopher' Laplace could not easily dispose of. He was none other than Herschel who early in his career took to task the debunkers of metaphysics, possibly some Humeans,[107] and who at the height of his career declared that of the two errors of speculating too much or too little he preferred to be guilty of the former.[108] Yet, whatever his love for speculations and his evolutionary view of galaxies, he never seemed to respond to Laplace's speculations on the origin of the solar system. Herschel's account of his well-known meeting with Laplace in Paris in 1802 is very telling in this respect because it recalls not so much a discussion of Laplace's theory as Herschel's concern about the ultimate cause of the construction of the heavens. In this concern of his he found an ally in the First Consul, with Laplace being their antagonist. The summary of the conversation of a certainly illustrious threesome touching on this particular point is to be read in Herschel's own words:

'The first Consul then asked a few questions relating to Astronomy and the construction of the heavens to which I made such answers as seemed to give him great satisfaction. He also addressed himself to Mr. Laplace on the same subject, and he held a considerable argument with him in which he differed from that eminent mathematician. The difference was occasioned by an exclamation of the first Consul, who asked in a tone of exclamation or admiration (when we were speaking of the extent of the sidereal heavens): "And who is the author of all this"! Mons. De la Place wished to shew that a chain of natural causes would account for the construction and preservation of the wonderful system. This the first Consul rather opposed. Much may be said on the subject; by joining the arguments of both we shall be led to Nature and to nature's God'.[109]

The new details which Laplace added in 1808 in the third edition of his *Exposition*[110] to his theory revealed even more of his partiality for it. The first decade of the 19th century was memorable in astronomy because of the discovery of the first four asteroids. Laplace seized eagerly on the newcomers to the solar family and added them to the planets. He now could count a total of forty-two movements and concluded in a tone of full assurance: 'It is more than four thousand milliards [billions] against one that this arrangement [of the planetary system] is not the effect of chance; this constitutes a probability far superior to the one enjoyed by the most certain events of history about which we admit no doubt whatever. We have, therefore, to believe with at least the same confidence that one primordial cause arranged the planetary motions; especially when we consider that the inclination of the greatest number of these motions with respect to the solar equator is hardly considerable and well below one-fourth of the circumference'.[111] The 'one-fourth of the circumference', or the 100 degrees mentioned in the first edition, was now needed also in view of the asteroids whose orbits heavily departed from the ecliptic, a circumstance not mentioned by Laplace. That the theory of Laplace had not so much a support as a serious difficulty in the discovery of the asteroids was argued to Laplace by none other than Olbers, the discoverer of Pallas (1802) and Vesta (1807), the second and fourth asteroids. This remarkable historical detail is recorded in a letter which Olbers sent to Bessel on July 10, 1812, from Paris where he spent some time as a member of the official delegation from Bremen. In the letter Olbers mentioned Laplace's reluctance to consider in their true weight the large deviations of the planes of Ceres and Vesta from the ecliptic and the fact that the density of Uranus was not less but greater than that of Saturn. 'Before the imposing greatness of Laplace all bow',[112] Olbers summarised the science-politics prevailing at the discussions held at the Bureau des Longitudes. One can, therefore, easily understand the unwillingness of Lagrange, Delambre, Poisson, Prony, Rossel, Arago, and Biot (scientists mentioned by Olbers) to take issue with Laplace's favourite brainchild. The general mistrust about Laplace's theory of most French scientists during the remainder of the 19th century should reveal something of its unconvincing character when viewed with eyes no longer conditioned by the presence of its powerful author.

Laplace, the cosmogonist, was handling the facts by making the most of the apparently favourable ones and glossing over the unfavourable findings. One of these was the recognition of the curious regularity of the spacing of planets, originally noted by J. D. Titius of Wittenberg in 1766 and popularised by Bode during the rest of the century.[113] By 1808 the law, now known as the Titius–Bode law, was enjoying a very high reputation. The discovery of Uranus was seen as an indisputable corroboration of it, as was the discovery of asteroids, if the theory of Olbers, who considered them as fragments of a planet between Mars and Jupiter, was correct. In fact, the search for Neptune was already receiving its first impetus precisely with a view on the Titius–Bode law. Laplace at long last mentioned that 'double progression' of planetary distances, but made no effort to derive it from his contracting solar atmosphere. He skirted the issue by merely voicing an already familiar theme that the law could serve as a basis for the discovery of further planets.[114] If the law had such a predictive value was it not precisely because it was an integral aspect of the physical process producing the planetary system? It was not too scientific to admit the benefits of a regularity but not to face the burden of its physical explanation.

The only noteworthy detail which Laplace added to the physics of his cosmogony was very brief. This detail concerned the explanation of the sense of the rotation of planets. It could indeed be argued with some plausibility that once the zones or rings formed, the 'true [orbital] velocities of the parts [molecules] of the vaporous ring were increasing with distance from the sun; the globes produced by their aggregation had to turn on themselves in the direction of their orbiting'.[115] This remark, in which there was some exact physics, was preceded by remarks on the ring and the formation of globes in it. The remarks were utterly void of anything exact. Laplace merely stated the formation of several globes within each ring, and their being gradually attracted to the biggest one among them. Thus Laplace believed to have secured the formation of a single planet from a given ring. Long before F. R. Moulton showed in 1900 the impossibility of this process,[116] Laplace could have easily caught a glimpse of the lack of soundness in his speculation. The mathematical techniques used by Moulton were not beyond those at Laplace's command. While Laplace was not far from the truth that the formation of permanent rings was very rare, very far from the truth was his long-standing claim that Saturn's rings, or any similar rings, were liquid or solid.[117] In 1859 Maxwell's analysis of the dynamical stability of Saturn's rings showed that they could only be a swarm of very small particles,[118] a point which had become increasingly plausible to assume already in Laplace's time on the basis of telescopic observations. Perceptive contemporaries of Laplace needed neither foresight nor hindsight to sense the inherent weaknesses of his theory and perhaps out of deference to Laplace they preferred not to refer to it in print. No other reason seems to explain the absence of any mention of the nebular hypothesis in the lengthy reviews which Biot and Gauss wrote of the third edition of the *Exposition* immediately after its publication. Was the nebular hypothesis so much less important than Laplace's incompetent discourse on the history of astronomy, to

which both Biot and Gauss paid special attention?[119] If it did appear unimportant to both it was precisely because its weaknesses revealed a crack in Laplace's
vaunted respect for the facts and laws of exact science.

That the most important of the weaknesses of Laplace's theory had been
a topic of conversation at that time can be surmised from that peculiar review
of Laplace's *Théorie analytique des probabilités* which appeared in the French
official daily, *Le Moniteur universel*, on July 7, 1812. The lengthy review, covering three folio columns, consisted of two parts. The first was a set of excerpts
from Laplace's preface to the *Théorie*; the second, which abruptly followed the
first, was an account and defence of Laplace's nebular hypothesis! The gist of
that second part was not so much the claim of the reviewer, who identified himself as X, that Herschel's observations of nebulae were an evidence of Laplace's
derivation of the solar system from a nebula. The gist was rather the reviewer's
emphatic claim that in a nebula several points of condensation were most
likely to arise and in such a way as to make each rotate on its axis but not in
the same sense. Consequently, the reviewer concluded, 'it is not true what has
been advanced by several famous philosophers that the universal attraction cannot produce in a system of originally motionless bodies any permanent
movement and that it has to unite them in the long run to their common centre
of gravity'.[120] A nebula giving rise to a number of stars whose different rotations
amounted to no net rotation for the entire nebula: was not this a covert defence
of Laplace's theory against the charge that it did not do justice to the conservation of angular momentum?

The fourth and fifth forms

In fact, the defence might have been Laplace's self-defence. Three days after the
appearance of the review, Olbers, writing his already mentioned letter to
Bessel, obviously reported the latest Parisian gossip which attributed to
Laplace's inordinate fondness for his theory that curious page on it in the
Moniteur. Less than a year later, when the fourth edition of the *Exposition* saw
print,[121] Laplace provided for that gossip an authoritative support. It consisted
in his professed intent to carry the origin of the solar atmosphere 'as far back as
possible'. He pictured its primordial state as being as 'diluted as to be hardly
observable' and spoke in the same breath of stellar groups such as the Pleiades
as 'condensation of nebulous matter with several nuclei' and of double stars
as cases of nebulous matter with two nuclei.[122] The original picture which he
offered about the primitive solar atmosphere could now be arrived at from
two directions. One was that of the formation of planets postulating a primordial quasi-liquid atmosphere; the other was the evolution of nebulae into
primitive solar atmospheres. Laplace hastened to add that such a convergence
of two opposite approaches gave 'the existence of that previous state of the sun
a probability closely approaching certainty'.[123]

In the fourth edition Laplace took up for the first time in the context of
his cosmogony the question of the origin of rotation in the planetary system. He
now seemed to perceive that what he had previously offered in connection
with the atmosphere of celestial bodies needed a more rigorous formulation so

as to be immune against a specific objection: 'If all the molecules of an agglo-meration of luminous material ultimately unite as a result of their condensation into one single liquid or solid mass, this mass will have a movement of rotation whose equator will be the primitive [original] place of the maximum of [orbital] areas passing through the common center of gravity, and the rotation shall be such that the sum of areas projected on this plane will remain the same which it was at the start [of condensation]; whence it follows that, if all mole-cules started from a state of rest, the body which they will form in the end will be immobile. But such will not be the case if these molecules formed several nuclei which could then have a movement of rotation provided that these movements directed in different directions were such that the sum of areas described by the radius vector of all these bodies [nuclei of condensation] around any point is constantly zero'.[124]

This statement was Laplace's answer to 'philosophers', who claimed that bodies originally at rest can only form a body at rest. As a further illustration of his argument against them Laplace referred to the case of three bodies at rest, of which two were much larger than the third. There were an infinite number of possibilities, Laplace argued, that following the fusion by gravita-tion of the two larger bodies the third could rotate around their common centre of gravity. The case of a final state with complete rest was 'infinitely improbable'.[125] Clearly, what Laplace had in mind was that an asymmetrical original distribution of molecules and of their condensations (nuclei) was most probable, and therefore the ultimate rotation of the system was inevitable. Without saying this explicitly Laplace concluded this crucial paragraph with the following words: 'The living force [kinetic energy] of the system, zero at first, grows by the mutual approach of the molecules toward one another and becomes very large if the motions of the system do not experience abrupt changes. The only factors that must always remain zero are the motion of the centre of gravity and the sum of areas described around this point by all the molecules projected at any plane'. With this recognition of the need to satisfy the principle of the conservation of angular momentum Laplace felt free to make the grand conclusion: 'Thus, attraction alone is sufficient to explain all the motions of this universe'.[126]

The explanation was hardly as free of problems as Laplace would have his readers believe. Problems were rather obvious in his two other references to the rotation of the solar atmosphere. In one of them Laplace used that rotation as a proof that the solar system had its limits, which coincided with the distance where the centrifugal force was equal to the gravitational attraction.[127] The other reference concerned the increase in the rotation of the sun due to the contraction of its atmosphere.[128] The solar atmosphere as envisaged in the former statement was still to condense and therefore to rotate if it indeed could rotate by condensation. The latter statement had to be reconciled with the very slow rotation of the sun, a crucial point which Laplace left unexplored.

According to Laplace, the primitive atmosphere of the sun was also sub-ject to cooling. The latter evidenced itself in condensation that led to the detachment of ring-like zones in the outermost regions of the equatorial place, where the centrifugal force was larger than elsewhere.[129] Laplace offered no proofs in support of his claim that the cooling was taking place in distinct phases rather than continuously. His theory needed the distinct zones not so much to account for the rings of Saturn as to provide the mass of each planet for its specific orbit and period. Here Laplace once more considered the formation of a single planet from the fragments of a ring and referred to the asteroids as a case where the fragments remained separate.[130]

In the fourth edition Laplace made more explicit his views on comets with respect to his 'nebular' theory. 'The comets are foreign to our hypothesis',[131] he wrote, and ought to be 'considered small nebulosities passing from one solar system to another and formed by the condensation of the nebular material spread out in great abundance in the universe'.[132] Underlying this conception, there seemed to be a breaking-up of the original cosmic nebulosity into at least two distinct classes with respect to size. One was leading to the formation of suns (and their planets) and another, much smaller, represented the original matter of comets. Laplace failed to make this point, while insisting that comets were altogether different from planets.[133] He was also silent on the closely related question of the average distance between stars. Instead, he claimed that his hypothesis 'explains in a happy manner the great size of the comets' heads and tails as they approach the sun, and the extreme thinness of their tails'.[134]

Laplace's remark that the flattening of the planets at their poles proved their previous fluid state[135] had its merits but his nebular hypothesis could still be wrong. He certainly indulged in rhetorics as he stated that 'the examination in depth of all circumstances of that system [of planets] increases still the probability of our hypothesis'.[136] Of all those circumstances he or any other astronomer knew but a few, and Laplace patently made an unsubstantiated claim when he wrote that 'one of the most singular phenomena of the solar system is the rigorous equality between the angular movements [speeds] of rotation and revolution of each satellite'.[137] This equality, Laplace stated, was 'infinitely improbable' on the basis of chance, but it followed 'naturally' on the basis of the inverse square law. The trouble with this was that Laplace had observational evidence of this equality only in the case of the earth's satellite, but none for the satellites of Jupiter, Saturn, and Uranus. Still Laplace unhesitatingly ascribed to that equality the fact that no ring developed around any of the satellites and that none of them was accompanied by a secondary satellite.[138]

One wonders why Laplace, after he repeated almost verbatim what he had said in the third edition on the stability of the solar system, replaced the word 'philosopher' with 'geometer' in the fourth edition.[139] Did he want to disclaim even an indirect connection with Kant? After all, he himself waxed philosophical as he took up the critique of Newton's explanation of the solar system by a recourse to God's arm. First, Laplace quoted the respective words of the Scholium of the *Principia*, then he added that Newton 'reproduced' the same statement in the *Optics*. This hardly did justice to the historical sequence that anyhow began with Newton's letters to Bentley, of which Laplace seemed to be unaware. 'Could this arrangement of the planets', Laplace asked, 'not be an effect of the laws of motion and could the supreme intelligence, which Newton lets intervene, not make it [this system] dependent on a more general phenomenon?'[140] This more general phenomenon was, according to Laplace, 'a nebulous material scattered in various quantities in the immensity of the heavens'.[141] Laplace, who as a philosopher did not seem to care about the problem of infinite regress in the chain of causation, was as a scientist not concerned about the properties of that nebulous material. Being exceedingly vague in its properties it could be manipulated with ease to provide the factors needed by the theory.

There was more merit in Laplace's next question: 'Is it still possible to affirm that the conservation of the planetary system is envisaged by the author of nature?'[142] The question and the answer he gave to it represented a major departure from his unqualified assertion of the stability of the solar system in the previous editions. He still claimed, as one would have expected, that the 'mutual attraction of the bodies of that system cannot alter its stability, unlike Newton supposed it'.[143] The studies on that stability were Laplace's chief glory which was not to be tarnished by doubts. Another factor which Newton himself described as a cause of the eventual collapse of the solar system was a different matter. 'Should not there be in the celestial spaces', Laplace wrote, 'any other fluid but light, its resistance and the decrease of the sun's mass due to its emission [of light], must in the long run, destroy the arrangement of the planets'.[144] Even more so was the case if such fluid existed, and here Laplace should have remembered that only a few pages earlier he spoke of 'an ethereal milieux' offering resistance to the motion of comets.[145] He certainly did not wish to entertain the prospect of an eventual 'reform' of the planetary system into its original status.

Much more congenial for Laplace were the vistas of a universal transformation, and he interpreted in that sense the description by Cuvier of an immense array of fossil species. Laplace now spoke of the eventual dissolution of the solar system which, for all its grandeur, was but an infinitesimal part of the universe. He also seemed to be intent on discrediting Newton's metaphysical cosmology by emphasising the apparent inability of man to formulate a definitive explanation of nature in its entirety. Ultimate explanations were for Laplace, and for his ideal philosopher, 'the expression of the ignorance in which we are concerning the true causes'[146] Such a remark might have been the sign of a healthy scepticism had Laplace carried it over into his discourse on planetary origins. There he was wont to throw caution to the wind. His last consideration in this respect was particularly telling. It touched the very heart of his theory, the origin of angular momentum in the solar system. In voicing the practical certainty of the existence of planets around other stars, he not only set great store by the principle of analogy, but simply declared: 'These stars are endowed as the sun is with a rotational movement, since they were, as also was the sun, surrounded by a vast atmosphere. It is only natural to attribute to their condensations the same effects which were produced by the condensation of the solar atmosphere'.[147]

That the same process, rotation by condensation, was taking place in the stars, because such was the case with the sun, was an argument that could easily turn into a vicious circle. In the case of the sun Laplace offered only generalities. Rotation in the solar system was not, however, a vague entity. The distribution of rotational momentum in the solar system had startling features easy to evaluate quantitatively. If departure from dissymmetry in the original solar atmosphere was the cause of a most specific and well-ordered rotation, then at least some model of the dissymmetry could be constructed. Laplace, the meticulous investigator of the stability of the solar system, could easily have offered something specific about a dissymmetry that could act as its source of rotation. He certainly had time during those eleven years that separated the fourth edition from the fifth.[148] Curiously, the principal change in his theory was the elimination of much of what he had said in the fourth edition about the derivation of rotation from condensations (nuclei) of molecules in a nebulous mass. What he kept was transferred to Chapter x of Book IV on the atmosphere of planets. In its fifth edition the *Exposition*, which from the third edition on had an Appendix of six Notes dealing with ancient Chinese, Hindu, Babylonian, Greek and Arabic astronomical observations, had a seventh Note. It contained that part of the last chapter which dealt not with the future progress of astronomy but with the evolution of the primordial solar atmosphere into the planetary system. The separation of the nebular hypothesis from the main body of a most informative science popularisation was an uncanny symbol of what was not of lasting value in a great classic of science. Another symbolic expression of this was the absence of any mention of the nebular hypothesis in the lengthy review of the fifth edition in the *Moniteur*.[149]

German myth-making

While it can safely be assumed that Laplace had no knowledge of Kant's *Allgemeine Naturgeschichte* in 1796, it is difficult to believe that he remained entirely unaware of its contents during almost three decades in which he saw through press five editions of the *Exposition*. Being in frequent contact during the Napoleonic era with German astronomers, he could have easily heard from them of Kant, the cosmologist. Of course, even if Laplace had been philosophically interested, he would have hardly learned of Kant who remained for a long time a notoriously unknown quantity in France. The metaphysical taste of Kant had little in common with the positivism fostered by the French Encyclopedists, and even wider was the gulf between the spirit of the French Enlightenment and the aspirations of those who in Germany carried Kant's thought to its logical extreme and whom the ageing Kant tried in vain to discredit and disown. References to Kant in the works of Fichte, Schelling, and Hegel were at best lip service to the greatness of the master. These leaders of 19th-century German idealism had certainly no use for Kant's admiration for exact Newtonian science. Their scientific ideals were embodied in the vagueness of the subjective intuitionism of Naturphilosophie. In that sadly mistaken approach to nature, even the plainest scientific truths and theories were not safe from astonishing 'reinterpretations'. Thus Schelling derived in 1802 the elliptical orbit of planets with two foci from the 'necessarily' twofold polarisation of each and every physical force.[150] Equally blatant was the case of Hegel who claimed in 1801 that there could be no planetary body between Mars and Jupiter because the Platonic (Pythagorean) sequence of planetary distances provided for no empty room there,[151] and who long after the discovery of the first asteroids praised the grandiose insight of the Pythagoreans into the structure of the planetary world.[152] This stubborn assertion of one's preconceived ideas in the teeth of the facts of nature was a wholly natural product of a way of thinking in which the external world was the evolution of thought and in which planets were defined as the 'most concrete' actualisation of externality.[153]

The cosmic evolution that interested astronomers was a very different kind, and partly because of Kant's role in preparing the way of Hegelian Idealism his theory of planetary origins continued to be passed over in silence in German scientific circles. A good illustration of this is the article 'Planeten', which appeared in 1833 in the famed revised edition of Gehler's *Physikalisches Wörterbuch*.[154] That Kant was not mentioned there can readily be understood by taking a look at the article, 'Physik-Geschichte', in which he was described, with tongue in cheek to be sure, as the 'great reformator of philosophy' opening the road to the scientific obscurantism of Naturphilosophie.[155] The article 'Planeten', written by H. W. Brandes, the best known professor of physics in Germany in the 1830's, is also worth mentioning for a very pertinent reason. In discussing Laplace's theory Brandes insisted on the extremely meagre knowledge that could be had about the initial conditions of the primitive nebula and consequently on the very tentative character of any hypothesis on the origin of planets.[156]

This badly needed warning was absent in the lengthy section which J. J. Littrow, Director of the Vienna Observatory, reserved on the origin of planetary systems in his *Populäre Astronomie* in 1825.[157] In addition to his wholly approving presentation of Laplace's theory Littrow also offered some details, though unreliable, on the history of the question prior to Buffon, but not a word on Kant. The same was true in every respect in the second edition of the work which appeared in 1836 under the title, *Die Wunder des Himmels*,[158] and served in subsequent editions as the standard German popular exposition of astronomy for the rest of the century. It was only in the next edition of *Die Wunder des Himmels*, published in 1842, that a brief reference was made to Kant as the one of whom the 'Fatherland can rightly be proud' for having anticipated Laplace by half a century.[159]

German national awareness was by then rapidly rising and eager to put its cultural heritage in the best possible light. Part of the effort aimed at discrediting Naturphilosophie and the strategy consisted in rebutting the Hegelians' claim to Kant. Schopenhauer, one of the first to oppose Hegelianism, became something of a pioneer when in 1851 he directed attention to Kant's ideas on the evolution of the system of planets.[160] He was perhaps the first to use the expression 'Kant-Laplace'sche Kosmogonie' which was to give strong support to the belief that Laplace merely rediscovered what Kant had already formulated with 'extraordinary acumen'.[161] The fundamental weaknesses of Kant's theory escaped Schopenhauer as much as did the fact that the two theories were in several basic respects very different. Both the differences and the weaknesses in question were wholly ignored by Helmholtz. who in his famous address 'On the Interaction of Natural Forces', given in 1854, officially anointed Kant as the full-fledged originator of Laplace's nebular hypothesis.[162] As if by fate, the place had to be Königsberg and the year almost the exact centenary of the publication of the *Allgemeine Naturgeschichte.*

It should be noted that the principal topic of Helmholtz's address was the conservation of force (energy) and its eventual dissipation (entropy). The planetary system came in only as a good illustration of the potentialities of the foregoing principles. If the various forces of nature were convertible on the basis of the conservation of energy, the enunciation of which in 1847 catapulted Helmholtz to international fame, the heat of the sun could readily be explained as the natural consequence of the nebular theory of planetary origins. 'When, through condensation of the masses, their particles came into collision and clung to each other, the *vis viva* [kinetic energy] of their motion would thereby be annihilated and must reappear as heat'.[163] Helmholtz took pains to note that before the enunciation of the principle of the conservation of energy, the nebular theory 'gave no information with regard to the origin of heat and light'.[164]

Helmholtz saw the essence of the theory in 'the notion that the same attractive force of all ponderable matter which now supports the motion of the planets must also aforetime have been able to form from matter loosely scattered in space the planetary system'.[165] Kant received high praise from Helmholtz for undertaking 'the labour of studying the works of Newton' and for the depth to which 'he had penetrated into the fundamental ideas of Newton'.[166] All this hardly reflected well on the care by which Helmholtz read Kant and Newton. Contrary to Helmholtz's presentation, the idea of a solar nebula spreading far beyond the orbit of the most distant planet was not proposed by Kant but by Laplace. Again, it was proposed by Laplace, not by Kant, that 'by the centrifugal force, which must act most energetically in the neighbourhood of the equator of the nebulous sphere, masses could from time to time be torn away, which afterwards would continue their courses separate from the main mass, forming themselves into single planets with satellites and rings, until finally the principal mass condensed itself into the sun'.[167] Helmholtz failed to do justice to the historical record as he pointed out that an originally slow rotational motion of the nebulous mass 'must be assumed'.[168] This was certainly true as far as the laws of dynamics were concerned, but Kant wholly ignored this crucial point, whereas Laplace asserted it implicitly, if he did so at all during the first stages of his speculations.

While Helmholtz helped create a mythical 'Kant-Laplace theory', his address focused interest, primarily in Germany, on the theories of both Kant and Laplace, and on cosmogony in general. Neglect of the theories of both Kant and Laplace had, indeed, been almost flagrant in the largest German scientific publication of the mid-19th century, Humboldt's *Kosmos.* There Kant was commended only for his ideas on the Milky Way, while Laplace, 'the immortal author of the *Mécanique céleste*', was taken to task for holding that comets were passing from one solar system to another.[169] Humboldt mentioned neither Kant nor Laplace as he took, in two short passages, a dim view of efforts aimed at deriving the planetary system from the condensation of a nebular mass.[170] It was not lack of information but plain mistrust that motivated Humboldt in this connection. The best-informed German of the times, Humboldt was perfectly familiar with Kant, and he had close ties with Arago, if that was needed for an awareness of Laplace's theory.

French disdain and faux-pas

Arago, the undisputed head of the French scientific establishment in the 1830's and 1840's, was an exception among French scientists of the time with his discussion of Laplace's theory. The discussion formed a small part of Arago's long essay on Laplace's life and achievements, which he wrote in 1843 to further the financing from public funds of a new edition of Laplace's works. There he emphasised in a positivist vein Laplace's lifelong reluctance to broach speculative topics. 'Once, and once only', wrote Arago, 'did Laplace launch forward into the regions of conjectures'.[171] The occasion was the nebular hypothesis. Its summary by Arago had more than one telling point. He first claimed that Laplace assumed the primitive solar atmosphere to be in rotation. Later he praised Laplace for departing from the general consensus according to which a condensing nebular mass should have ultimately turned into an immobile spherical solid. It was with the same ambivalence that Arago endorsed Laplace's claim that it was most unlikely that the mutual attraction of three unequal bodies should issue in a rotating system. For Arago also wrote that 'it is perhaps especially to be regretted that Laplace should have only briefly alluded to what he considered the obvious possibility of movements of revolution having their origin in the action of simple attractive forces, and to other questions of similar nature'.[172] This was a revealing reservation on a decisive point. Less important were Arago's two explicit criticisms of Laplace's theory. One touched on the lack of details concerning 'the division of the matter into distinct rings'.[173] The other related to the absence of particulars on the 'primitive physical condition, the molecular condition of the nebula'.[174]

In spite of these reservations, the ideas of the author of the *Mécanique céleste* (as Arago stylishly identified Laplace whose *Exposition* he did not mention in the context) were for Arago 'still the only speculations of the kind which by their magnitude, coherence, and mathematical character may be justly considered as forming a physical cosmogony'.[175] Such a conclusion was already unjustified because of Arago's disparaging remarks on conjectures, and was made contradictory both by Arago's silence on Kant's theory and by his attributing a 'mathematical character' to Laplace's hypothesis. It had no mathematical character at all and the first attempt to garb it (with Arago's backing) in mathematical formulae had very dubious merits. The author of the attempt was the Father of positivism, August Comte, for whom Arago made it possible to read a memoir on the subject before the Académie des Sciences on January 21 and 26, 1835. Comte, who spoke of the 'ingenious conception of Herschel and Laplace', singled out the absence of mathematical formalism in Laplace's hypothesis as the cause of 'the reluctance which almost all geometers and astronomers' have to discuss it, although they consider it 'in general very plausible'.[176]

According to Comte one could verify the hypothesis mathematically in two ways. One consisted in calculating the moment of inertia of the solar nebula at successive stages of its contraction, and deriving from it the orbital period of each planet as all owed their origin to rings detached from the rotating nebula. This otherwise sound method was not feasible because nothing was known about the distribution of mass within the nebula while it contracted. Yet the actual distribution of masses in the solar system was well known and the actual distribution of the moments of inertia, or of angular momentum, could have easily been calculated to a fair degree of accuracy. The result would have immediately revealed to Comte and to Arago the real weakness in Laplace's hypothesis. Failing to follow up this track, Comte became trapped in another approach which amounted to the simple statement in a simple mathematical formula[177] that at any actual planetary distance the gravitational attraction of the rotating solar nebula was equal to the centrifugal force of particles on its equatorial rim. This merely showed that rings could detach themselves from the nebula upon its subsequent contraction, but it was still to be shown that the rings separated from the nebula at the actual distances of planets from the sun.

Behind Comte's failure to see the *petitio principii* in his mathematisation of the nebular hypothesis there lay strong psychological motivations. He had already sent to the printer the second volume of his *Cours de philosophie positive*, in which he presented the regularity of the system of planets not only as a feature perfectly explained by the nebular hypothesis, but also as the perfect evidence that nature obeyed exact laws and that man was capable of finding at least those laws of nature that regulated his immediate cosmic surroundings.[178] Comte based on this his conviction that definitive and exact laws could also be established in sociology and that consequently a society structured on his positivist creed was feasible. Any search that could weaken the definitiveness of the laws of the solar system was therefore an anathema for Comte. This is why he frowned upon efforts to open up such new frontiers as stellar astronomy and molecular physics. Part of the price he had to pay can be seen in his list of the basic assumptions of Laplace's theory. There he referred to gravitational force and to contraction by cooling, but not to the indispensable original rotation. He brought in rotation in the same ambiguous and furtive manner in which Laplace did. Comte's position in this respect was all the more logical as he emphatically claimed that scientific cosmology must dispense with questions about ultimate conditions and especially with the question of creation. Cosmology had to restrict itself to the analysis of the interaction of material particles 'that have no beginning and end'.[179]

By the mid-1840's Comte, once an instructor of mathematical physics at the famed Polytechnique, had almost completely lost his credibility in French scientific circles for turning positivism into a brand of mysticism. As to Arago, the other French scientist to go on record on Laplace's theory, he did not sound so enthusiastic as to spark interest in it among French men of science. Natural choice as these should have been for the task, they did not seize, perhaps for good reasons, on the memoir which J. Plateau read on January 15, 1842, before the Belgian Academy of Sciences.[180] There Plateau described an experiment which, according to him, 'offers in small an image of the formation of planets, according to the hypothesis of Laplace, through the rupture of cosmic rings due to the condensation of the solar atmosphere'.[181] The experiment consisted in putting into rotation a spherical mass of olive oil suspended in a mixture of water and alcohol by turning a disk in the centre of the globe of oil. By increasing the rate of rotation of the disk the rotating globe of oil transformed itself into a ring which Plateau boldly compared to that of Saturn [*Illustration XX*]. The ring could through some further manipulation be also broken up into several parts which formed spheres revolving on their vertical axes in the sense of the original motion of the ring. He saw the process analogous to the formation of planets and described small drops of oil floating between larger ones as 'satellites'.[182] The basic weakness in Plateau's interpretation of his experiment came to the fore as he described the formation of a ring of oil with a globe of oil inside it. This was the exact replica of Saturn's system (curiously, Plateau did not recall that the solar system too contained a huge central body), but the special device of two pulleys needed to produce the effect made Plateau confess that one should not see in the procedure more than a 'scientific game'.[183] The circumstances, he added, 'which give rise to the result, have evidently no analogy whatever with those that might have produced the configuration of Saturn's system'.[184]

A decade or so later, when enthusiasm for Laplace's hypothesis became a fashion, Plateau's fascinating game in experimentation was taken uncritically as tangible evidence that the hypothesis was truth. In the absence of such enthusiasm even important discoveries made in the 1830's and 1840's about the system of planets failed to draw attention to the explanation of its evolution proposed by Laplace. The most memorable of these discoveries was the detection of Neptune in 1846. Although the lion's share of glory went to the French astronomer, Urbaine J. J. Leverrier, British and German astronomers were prominently involved in the efforts to find the long-suspected planet. Another major discovery consisted in spotting around Uranus further satellites with retrograde motion. That work too involved astronomers from France as well as from England and Russia. None of them cared to elaborate on the bearing of their findings on Laplace's theory.

Anglo-Saxon exploits

A rather revealing instance in this connection was provided by Sir John F. W. Herschel. In his famous *Outlines of Astronomy*, first published in 1849, he spoke of 'nebular hypothesis' with no connection whatever with Laplace's theory, which he did not mention at all. Herschel used the expression with reference to truly 'nebulous' nebulae, that is, large celestial bodies, perhaps galaxies, com-

posed of diffused 'self-luminous' material and not of distinct stars.[185] Indicative
of the lack of interest among professionals in Laplace's theory was also the
absence of any reference to it in the classic work on the history of astronomy
which Robert Grant, future Director of the Glasgow Observatory, published
in 1852.[186] Grant's case deserves to be noted all the more as he spent the years
1845–47 in Paris studying from original sources the tremendous contributions
which French astronomers, especially Laplace, made during the previous
hundred years, and because he discussed in detail the retrograde motion of the
satellites of Uranus.[187]

    Though the famed Master of Trinity, William Whewell, was not an
astronomer by profession, he had the qualifications to discourse on matters
astronomical, which he did both in his classic compendia on the history and
philosophy of the inductive sciences, and, especially, in two monographs.
Although in those compendia he had excellent opportunity to give a detailed
criticism of Laplace's hypothesis, he merely referred to it as illustrating a
specific way of reasoning in science.[188] In his *Astronomy and General Physics
considered with Reference to Natural Theology*, published in 1833 as a Bridgewater
Treatise, Whewell devoted a whole chapter to the nebular hypothesis taken in
a broader sense which related to the origin of nebulae, but all he wanted to
show was that nebulae and their workings still implied a higher cause. He left
'to other persons and to future ages to decide upon the scientific merits' of the
theory of Laplace.[189] Exactly two decades later, in his monograph on *The
Plurality of Worlds*,[190] Whewell spiritedly fought the proof based on analogy, or
the claim that our sun was one of the countless stars surrounded by planetary
systems like ours. It was in this connection that he declared that the 'nebular
hypothesis is too imperfectly worked out, as yet, to enable us to know, what it
will or will not account for'.[191] He offered no specifics when shortly afterwards
he remarked that the coalescence of rings into spherical, planetary masses 'seems
most likely', adding in the same breath that it 'does not appear that it has been
very clearly shown'.[192] Actually, the only objection he specified against the
nebular hypothesis was that it did not account for Bode's law.[193] The ironic
touch on Whewell's cryptic claims about the weaknesses of the nebular
hypothesis came when he declared that rotation was a universal phenomenon
in nature securing stability to its various systems.[194] He did not guess that in
saying this he took for a firm foundation what actually contained the seeds of
a fatal weakness.

    The silence of Herschel and Grant and the vagueness of Whewell should
seem all the more curious because by 1852 tens of thousands in the English-
speaking world had become familiar with Laplace's hypothesis through John P.
Nichol's popularisations of astronomy. Nichol, professor of astronomy at the
University of Glasgow from 1836 until his death in 1859, saw through press
eleven editions of his *Views of the Architecture of the Heavens*, a 'Series of Letters
to a Lady', first published in 1837. The next-to-last or Seventh Letter was
devoted to the nebular hypothesis, or 'Laplace's bold and brilliant *induction*' as
Nichol put it.[195] For him there were incontrovertible evidences of the once
nebular condition of the sun, and he listed as such the zodiacal light and the

retardation of Encke's comet in a 'RESISTING MEDIUM OR ETHER, occupying the planetary spaces'.[196] With the same devotion to the grandeur and truth of Laplace's ideas Nichol listed as obvious verities the inevitable emergence of whirlpools in the sundry streams of the contracting nebula and their ultimate co-ordination into one single rotation. As an additional support Nichol mentioned Herschel's 'fine induction'[197] on the rotation of nebulae (galaxies) and of globular clusters. He was more original with the two diagrams in his book that illustrated the development and detachment of rings in the rotating solar nebula [*Illustration XXI*]. He argued that each of these rings most likely condensed into a single planet. True enough, it was most unlikely that the rings should solidify, but Nichol made a wrong guess in claiming that it was also most unlikely that the rings should remain composed of minute particles of matter. Nichol's fondness for the nebular hypothesis must have played a part in his facile acceptance of the 'mathematical demonstration' by the 'young French geometer', Comte,[198] and in his eager reference to ring-like nebulae observed by Herschel as evidences of Laplace's ideas on the phases which a rotating nebula should undergo. It should not be surprising that the retrograde motion of the satellites of Uranus could not, in Nichol's eyes, invalidate 'a theory so beautiful and perfect',[199] especially when there was a convenient recourse to the influence of external factors on the outermost edge of the solar system. The crowning touch on Nichol's panegyrics on the nebular hypothesis was his claim that the hypothesis even brought solution to Newton's concern about the stability of planetary configurations. Since the hypothesis implied for all planets a basically circular motion in the same direction, their perturbations could easily be periodical within certain limits, so Nichol interpreted the more remote grounds of Lagrange's and Laplace's work on the stability of the solar system.[200]

Almost exactly ten years after the first publication of Nichol's *Views*, he journeyed for a lecture tour to the United States. The fifth of his seven lectures given in January and February 1848 before the Mercantile Library Association of New York was on the nebular hypothesis and showed heavy traces of the 'Seventh Letter' of his *Views*. While Nichol now described the theory of Laplace as a 'hypothetical cosmogony',[201] the tone of his diction made no secret of his faith in its truth. The same ambivalence characterised his dicta on the nebular hypothesis in his *Thoughts on Some Important Points relating to the System of the World*, published in the same year.[202] The real value of his *The Planetary System: Its Origin and Physical Structure*, which appeared in 1850, consisted in incidentals. One was a diagram [*Illustration XXII*] to help explain the rotation of planets in the direction of their orbiting. The other was a note added in press.[203] It contained the first occurrence in a printed discussion of the nebular hypothesis of the 'Law of the Spheres of Attraction', a brainchild of Daniel Kirkwood, an amateur student of astronomy and teacher in the local school in Spottiswood, Pennsylvania. After many fruitless efforts to discover a new correlation among the various data relating to the planets, he found on August 12, 1848, his Law which states that the squares of the number of times any two planets rotate during one revolution in their orbits are in the same proportion as are the cubes of the widths of their respective spheres of attraction.[204] In the late summer of the next year the Law received wide publicity at the meeting of the American Association for the Advancement of Science in Cambridge, Massachusetts, where Kirkwood was acclaimed by some as a second Kepler. Needless to say, neither Kirkwood, nor Nichol, nor anyone else was able to give a physical explanation of the Law, or to serve evidence that it really confirmed the nebular hypothesis.

For all his enthusiastic endorsement of Kirkwood's Law, Nichol's discussion of the nebular hypothesis gave unwitting evidence of apparently widespread dissatisfaction with it. Yet objections to the hypothesis did not seem to have been translated from oral remarks into printed statements, let alone into detailed essays. Nichol referred to 'recent disputes' which arose, he wrote, from the clash between those who simply raised the nebular hypothesis to the rank of definite truth and those who insisted that it could be taken for truth only if it gave a rigorous explanation of all phenomena of the solar system. Nichol once more insisted that Laplace had offered only a hypothesis, but he could not help adding that its essential truth was not to be doubted. 'Mountains', he quoted an aphorism of de Saussure, 'should not be viewed through a microscope'.[205]

While Nichol's parry expressed well the fact that in science the advance toward truth is gradual, the fact remained that the same advance presupposed an even closer look at the intrinsic value of theories and conclusions. That even from a distance the nebular hypothesis showed at least one gigantic shortcoming was yet to be perceived. When it became evident during the 1850's that Laplace had bequeathed a flagrant error to posterity with his claim that the rings of Saturn were solid, the value of his nebular hypothesis did not seem to suffer a severe blow. Following the discovery of a darkish third ring in November 1850,[206] G. P. Bond, one of the discoverers at Harvard College Observatory, concluded on the basis of his continued observations that the ring was liquid.[207] That Laplace's claim about the stability of a solid ring of irregular shape was wholly inapplicable to Saturn was argued shortly afterwards by B. Peirce, also of Harvard.[208] He characterised Laplace's statement as a 'careless suggestion' in which 'he was blindly followed by his successors'.[209] Peirce, who submitted the idea of a fluid ring whose centre of gravity was kept stable by the satellites of Saturn, remarked at the same time that were the ring gaseous, its ultimate condensation into a single planet would be conceivable. 'Under this modification', he wrote, 'the nebular hypothesis may possibly be free from some of the objections with which it has been justly assailed'.[210] The written record contains little if anything about those objections. The nebular hypothesis appeared either too nebulous for concrete objections, or too authoritative to invite sustained criticism.

With the publication in 1853 of O. Struve's carefully documented memoir on the history of all observations of Saturn's rings,[211] interest was enhanced to such a degree that the stability of Saturn's rings was selected as the topic of the first Adams Prize to be given in 1856. The chief conclusion of James Clerk Maxwell's prize-winning essay[212] was that 'the only system of rings

which can exist is one composed of an indefinite number of unconnected particles, revolving round the planet with different velocities according to their
respective distances'.[213] As an observational evidence of his analysis Maxwell
referred to the transparency of the recently discovered darkish or innermost
ring. The essay promptly established him as a leader in mathematical physics.
No immediate echo followed his words contained in the introductory part of
his essay: 'I have shewn that a destructive tendency actually exists, but that by
the revolution of the ring it is converted into the condition of dynamical
stability'. This phrase followed his remark that the collapse of a ring like
Saturn's into one or more satellites was 'one of the leading doctrines of the
"nebular theory" of the formation of planetary systems', and that 'we are
familiar with the actual breaking up of fluid rings under the action of
"capillary" force, in the beautiful experiments of M. Plateau'.[214]

Maxwell's roundabout way of urging caution about the nebular hypothesis would have probably been lost on Nichol who referred to the Adams Prize
competition still to be adjudged when he completed his article on Saturn's
rings in his *A Cyclopaedia of the Physical Sciences*. In that article Nichol made
much of Bond's and Peirce's work in support of a non-solid ring, while playing
down at the same time Laplace's erroneous claim that the rings were solid. For
Nichol those rings still represented the perfect evidence of that rare possibility
specified by Laplace, when the outer zone of the shrinking solar nebula retained a perfect homogeneity preventing thereby its coalescence into a
planet.[215] No reference to the latest evidence against Laplace or his theory was
needed by Daniel Brewster of the University of Edinburgh to justify his
violent denunciation of the nebular hypothesis. Like his countryman, Nichol,
Brewster too excelled in optics, but history largely remembers him as a bio-
grapher of Newton for whom Newton was the unqualified embodiment of all
virtues, scientific and otherwise. In discussing Newton's letters to Bentley,
Brewster declared: 'They show that the *Nebular hypothesis*, the dull and dangerous heresy of the age, is incompatible with the established laws of the material
universe, and that an omnipotent arm was required to give the planets their
position and motions in space, and a presiding intelligence to assign to them the
different functions they had to perform'.[216] Had Brewster not been blinded by
his idea of Newton's genius, he might have taken a closer look at the reasons
why Sir Isaac demanded an 'omnipotent arm' for the positioning of planets
into their orbits, reasons which related to their angular momentum.

Hollow glory

No trace of this quasi-religious respect for Sir Isaac tainted the forty pages of an
essay, 'Recent Astronomy, and the Nebular Hypothesis', in the July 1855 issue
of the *Westminster Review*.[217] The anonymity of its author was an open secret.
Herbert Spencer, soon to embark on his ambitious project of bringing into one
mighty synthesis the whole kingdom of knowledge, had no formal qualifications to judge matters astronomical. He judged them and everything else anyhow, with the supreme confidence of one whose sight had already been firmly
set on his own system of the world. It was an evolutionary world going

through endless cycles of emergence and decay.[218] For the time being Spencer
spoke only of the emerging phase, a part of which was a theory of planetary
evolution, which differed but slightly from Laplace's theory. Unlike Laplace,
for whom comets were intruders into the planetary system, Spencer argued that
an evolutionary theory could easily make the comets an integral part of the
development of the system of planets. According to him comets formed as small
condensations in the outskirts of the primeval nebula somewhat earlier than its
development into a mass of which all parts rotated in the same plane and
direction. Apart from that Spencer followed closely the main ideas of Laplace's
theory about which he stated, probably with an eye on Nichol's publications,
that 'books of popular astronomy have familiarised even unscientific readers
with this conception'.[219] Spencer's unhesitating reference to Plateau's experiments as a confirmation of the theory certainly evidenced his firm belief in it.
The incisive analysis which Maxwell was to publish a year later on the dynamics of the rings of Saturn[220] was not to compete in popular effectiveness with
Spencer's facile account of each and every feature of the system of planets.
Spencer probably learned from Nichol's books about Comte's calculations
which he quoted with merely a touch of misgiving. This he allayed with
the remark that 'as a professor of mathematics, his [Comte's] authority is of
weight'.[221] Spencer, a Comtean in many respects, did not, as time went on,
believe with the same readiness other 'professors of mathematics' whose work
cast doubt after doubt on the nebular hypothesis during the remainder of his
long life. Needless to say, the cause of the rotation of the nebula as a whole was
no problem for Spencer. He traced it, as did Nichol, Laplace, and to some
extent Kant, to a slight asymmetry in the nebula prior to its condensation.

Spencer, who saw in the anomalies of the satellites of Uranus no real
difficulty for the nebular hypothesis, did not think that the solar system was
altogether without mystery. 'The problem of existence', he wrote in his concluding remarks, 'is not solved: it is simply removed further back'.[222] His next
phrase, 'the genesis of an atom is not easier to conceive than the genesis of a
planet', showed, however, that rhetoric could easily prevail on his critical
sense. About the genesis of atoms he knew nothing, while about the genesis of
planets he gave a discussion which seemed to know everything, except the
absolutely primordial beginning. For that beginning he still professed considerable awe in one specific sense. 'Creation by manufacture', he declared, 'is a
much lower thing than creation by evolution. A man can put together a
machine; but he cannot make a machine develope [sic] itself . On such a basis
one could even say with Spencer that 'the nebular hypothesis implies a First
Cause as much transcending "the mechanical God of Paley", as this does the
fetish of the savage'.[223]

Phrases like these perfectly fitted the British intellectual atmosphere ripe
for the publication of Darwin's Origin of Species. Indeed, discussion of the
nebular hypothesis and of planetary evolution in general, became a distinctly
Anglo-Saxon preserve thereafter. Owing to a general neglect of Laplace's
theory in France, her scientists did not yet notice the crucial barrier which
some of their numbers were to raise before long against the further advance of

the nebular hypothesis. J. Babinet, since 1840 a member of the Académie des Sciences, exaggerated not at all when stating in a lecture given in Paris on December 15, 1860, that the concluding chapter of Laplace's *Exposition*, 'one of the greatest honours to human intellect, did not in any way attract the attention of the public'.[224] What Babinet added was all first-hand information: 'The astronomers in their observatories, devoted exclusively to measuring angles and times, saw in the theory nothing for themselves . . . The mathematicians saw in it only hypotheses that did not come under the competence of their equations. As to the literary critics, the book did no more exist for them . . . than did the inscriptions of Babylon. I believe, I was one of those [few] who between 1816 and 1827, while Laplace was still alive, called attention in my lectures and conversations with scientists to his cosmogony as the only one that gives account of all that is contained in the solar system'.[225] It was irony itself that three months later a paper was read before the Académie des Sciences on a point of the cosmogony of Laplace which unwittingly opened a new chapter in the history of theories on planetary origins. The author of the paper was Babinet himself.

NOTES TO CHAPTER FIVE

1. The subtitle reads: *oder Versuch von der Verfassung und dem mechanischen Ursprunge des ganzen Weltgebäudes nach Newtonischen Grundsätzen abgehandelt* (Königsberg und Leipzig: bey Johann Friederich Petersen, 1755). The first 143 pages, in addition to the introductory LVI + VI' pages, are available in English translation by W. Hastie, *Kant's Cosmogony, as in his Essay on the Retardation of the Rotation of the Earth and in his Natural History and Theory of the Heavens* (Glasgow: James Maclehose and Sons, 1900). Pages 144–200 of the original contain Chapter VIII of Part II on the validity of mechanical philosophy in cosmology and the relatively short Part III on the denizens of other planets.

2. As reported in the biography of Kant by L. E. Borowski, *Darstellung des Lebens und Characters Immanuel Kants* (1804), a work approved by Kant himself. For its modern edition, see *Immanuel Kant: Sein Leben in Darstellungen von Zeitgenossen, Die Biographien von L. E. Borowski, R. B. Jachmann, und A. Ch. Wasianski*, edited by Felix Gross (Berlin: Deutsche Bibliothek [1912]). On the impounding of Kant's work, see p. 89.

3. In German libraries only six copies were known to exist in 1922, as reported in the standard study of Kant's scientific investigations by E. Adickes, *Kant als Naturforscher* (Berlin: W. De Gruyter, 1924–25), vol. 2, p. 207. Although Adickes convincingly established the fact that Kant's familiarity with physics and mathematics did not reach university level, he failed to perceive that precisely because of this Kant could not offer an explanation of the origin of the solar system matching the scientific attainments of the mid-1700's. Kant presented what already had been inadequate in Descartes' time, namely, a basically qualitative unfolding of some general presuppositions about matter and motion. That M. Knutzen, professor at Königsberg, who is usually referred to as Kant's private tutor in physical science, was not particularly suited for the task, will easily transpire from B. Erdmann's monograph, *Martin Knutzen und seine Zeit: Ein Beitrag zur Geschichte der Wolfischen Schule und insbesondere zur Entwicklungsgeschichte Kants* (Leipzig: Verlag von Leopold Voss, 1876; reprinted, Hildesheim: H. A. Gerstenberg, 1973). The seventh chapter deals with Knutzen's 'scientific' publications, but the only one which deserves that label is a small treatise on the famous comet of 1744. Knutzen gave no courses in mathematics and physics.

4. In the July 15, 1755 issue, p. 432.

5. *Kant's Cosmogony*, pp. 24–25 [XXIV]. Numbers in square brackets refer to pages in the original.

6. *Ibid.*, p. 26 [XXVIII].

7. *Ibid.*, p. 28 [XXXI–XXXII].

8. *Ibid.*, p. 28 [XXXII].

9. *Ibid.*, p. 29 [XXXV].

10. Kant quoted from the first volume of the German translation (*Uebersetzung der A lgemeinen Welthistorie . . . Genau durchgesehen und mit häufigen Anmerkungen vermehret von Siegmund Jacob Baumgarten* [Halle: bey J. J. Gebauer, 1743]) of the seven-volume *An Universal History from the Earliest Account of Time to the Present: Compiled from Original Authors and Illustrated with Maps, Cuts, Notes, Chronological and Other Tables* (London: printed for J. Batley, 1736–47). In the first volume the first section was on 'The Cosmogony, or Creation of the World' (pp. 1–52) and contained a short summary of Descartes' cosmogony (pp. 35–36). About Burnet it was mentioned that he planned to write a theory of the origin of stars and 'even of the solar system'. All that was offered on Whiston's theory was that it consisted in 'changing the course and orbit of the chaos [comet] into that of a planet'. Bentley was not referred to at all. Only conjectures can be made about the question whether Kant knew of Bentley's sermons and of Bentley's spirited rebuttal of the idea of the coalescence of a chaotic matter into planetary systems under the impact of gravitation, the very idea which Kant advocated. Kant had a marked interest in British authors, and copies of the many editions of Bentley's sermons were widely circulating in Germany during the first half of the eighteenth century. In addition, there appeared in Berlin, already in 1696, a Latin translation of Bentley's sermons, and a German translation in Hamburg in 1715. (See A. T. Bartholomew, *Richard Bentley, D.D. A Bibliography of his Works and of all the Literature called forth by his Acts or his Writings*, with an Introduction and Chronological Table by J. W. Clark [Cambridge, Bowes and Bowes, 1908], pp. 8–9). It should, however, be noted that Kant soon was to be referred to as 'the honest Kant', because of his care in identifying his sources. The reference to Bentley in Adickes' monograph (vol. 2, p. 210) suggests that Adickes was unfamiliar with the contents of Bentley's sermons, or else he would have readily spotted Kant's unwitting advocacy of what Bentley emphatically had rebutted.

11. See *Kant's Cosmogony*, p. 72 [25].

12. *ibid.*, p. 30 [XXXVII].

13. *Ibid.*, p. 35 [XLVI].

14. In three instalments in the January 1, 5 and 8 issues. For further details, see my *The Milky Way: An Elusive Road for Science* (New York: Science History Publications, 1972), pp. 196–97 and 214.

15. As first noted *ibid.*, p. 197.

16. *Kant's Cosmogony*, p. 36 [XLVIII].

17. *Ibid.*, p. 52 [VI']. Kant's correct explanation of the visual appearance of the Milky Way is largely responsible for the unsuspecting approvals which are customarily accorded to his explanation of the origin of the system of planets. A classic example of this can be found in the broadcast address given in 1954 by Karl R. Popper on the 150th anniversary of Kant's death. See Popper's *Conjectures and Refutations: The Growth of Scientific Knowledge* (2d ed.; New York: Harper & Row, 1968), p. 177.

18. *Ibid.*, p. 59 [9].

19. *Ibid.*, p. 67 [20].

20. *Ibid.*, p. 73 [26].

21. *Ibid.*, p. 75 [28].

22. *Ibid.*

23. *Ibid.*, p. 75 [29].

24. *Ibid.*, p. 76 [29].

25. *Ibid.*, p. 76 [30].

26. *Kant's Cosmogony*, p. 25 [XXV].

27. *Ibid.*, pp. 93–94 [51]. Since the detail, 640/650, or the ratio of the average density of planetary mass and of the sun's mass, came from Buffon's theory of planetary evolution, Kant must have been familiar with some standard objections to his own theory.

28. *Ibid.*, p. 77 [31].

29. *Ibid.*, p. 78 [32].

30. *Ibid.*

31. *Ibid.*, p. 79 [35].

32. *Ibid.*, p. 81 [36].

33. *Ibid.*, p. 80 [35].

34. *Ibid.*

35. *Ibid.*, p. 82 [37].

36. *Ibid.*, p. 80 [36].

37. *Ibid.*, pp. 83 and 87 [38 and 43].

38. *Ibid.*, p. 89 [45].

39. *Ibid.*, p. 91 [48]. The phrase is a classic example of the naïveté for which during the age of classical physics mechanical explanation was intelligibility itself. In that state of mind mere guesses about mechanical models were readily taken for complete solutions of exceedingly complicated problems. See on this Chapter Two, 'The World as a Mechanism', in my *The Relevance of Physics* (Chicago: University of Chicago Press, 1966).

40. *Ibid.*, p. 95 [51].

41. *Ibid.*, p. 95 [52].

42. *Ibid.*, p. 101 [59].

43. *Ibid.*, p. 107 [67].

44. *Ibid.*, p. 131 [96].

45. *Ibid.*, p. 114 [75].

46. *Ibid.*, p. 142 [110].

47. *Ibid.*, pp. 142–43 [110].

48. *Ibid.*, p. 154 [125].

49. *Allgemeine Naturgeschichte* (1755), pp. 173–200. This Third Part has very recently become available for the first time in English in my translation as an Appendix to *Cosmology, History and Theology*, edited by W. Yourgrau and A. D. Breck (New York: Plenum, 1977), pp. 387–403.

50. *Ibid.*, p. 187.

51. *Ibid.*, p. 189.

52. *Ibid.*, p. 190.

53. *Ibid.*, p. 185.

54. *Ibid.*, pp. 176–77.

55. See Abtheilung II, Betrachtung 7, 'Kosmogonie' in Kant's *Gesammelte Schriften*, vol. 2, (Berlin: G. Reimer, 1905), pp. 137–41.

56. The second and third editions appeared in 1770 and 1783.

57. *Ideen zur Philosophie der Geschichte der Menschheit*, Erster Theil (Riga und Leipzig: bey J. F. Hartknoch, 1785), pp. 4–5. Between 1785 and 1807 the work saw four editions.

58. The letter was dated November 13, 1765. See *Immanuel Kant: Briefwechsel*, with introduction, notes and indices by Otto Schöndörffer (Leipzig: Felix Meiner, 1924), vol. 1, pp. 36–40. For further details, see my *The Milky Way: An Elusive Road for Science* (New York: Science History Publications, 1972), pp. 198–200.

59. See my English translation of it, *Cosmological Letters on the Arrangement of the World-Edifice*, with Introduction and Notes (New York: Science History Publications, 1976). The anti-evolutionary view is particularly stressed in the Eighth Letter, pp. 98–99.

60. *Anleitung zur Kenntniss des gestirnten Himmels* (Berlin: bey Christian Friedrich Himburg, 1777), pp. 658–59.

61. *Kurzgefasste Erläuterung der Sternkunde und den [sic] dazu gehörigen Wissenschaften* (Berlin: bey Christian Friedrich Himburg, 1778), p. 512.

62. Concerning the *Erläuterung*, see the 2d edition (Berlin: bei Christian Friedrich Himburg, 1793), p. 734, and the 3d edition (Berlin: in der Himburgschen Buchhandlung, 1808), vol. 2, p. 321. As to the *Anleitung*, see for instance the 7th edition (Berlin: bei Christian Friedrich Himburg, 1801), p. 608, or the 9th edition, the last sent to press by Bode himself (Berlin: in der Nicolaischen Buchhandlung, 1823), p. 629.

63. The work appeared under the title, *William Herschel über den Bau des Himmels: Drey Abhandlungen aus dem englischen übersetzt. Nebst einem authentischen Auszug aus Kants allgemeiner Naturgeschichte und Theorie des Himmels* (Königsberg: bey Friedrich Nicolovius, 1791). The translator of Herschel's papers was G. M. Sommer, assistant librarian of the Kgl. Schloszbibliothek in Königsberg, while Gensichen was assistant supervisor of the student-body at the university there. The copious extracts from Kant's work amounting to almost forty pages are followed by Gensichen's assertions (pp. 201–204) of the priority and superiority of Kant's views on the Milky Way over those of Lambert.

64. 'Prüfung der Kantischen Hypothese von dem mechanischen Ursprung des Planetensystems', in *Philosophisches Archiv* (Berlin), Band I, Stück 2, (1972), pp. 1–36. For quotation, see p. 31.

65. *Ibid.*, pp. 35–36.

66. *Ibid.*, pp. 7–8.

67. *Ibid.*, p. 12. Schwab quoted from the French translation of Maclaurin's work which originally appeared in 1748.

68. *Ibid.*, p. 15.

69. *Ibid.*, p. 20.

70. *Ibid.*, pp. 3 and 24–25. The next year Schwab published in the same periodical (Band I, Stück 4, pp. 1–21) a detailed criticism of Kant's theory of the origin of Saturn's rings and of his derivation of the rotation-period of planets, noting with an eye on Gensichen that it is easier to admire than to evaluate a theory carefully. At the end Schwab took Herder to task for praising Kant's theory to the extent of equating it with contributions made by Copernicus and Newton. It took more than a hundred years before Schwab's stern though sound verdict was echoed in a prominent scientific context. In his Hitchcock Lectures given at the University of California in 1924, the Swedish mathematician and cosmologist, C. V. L. Charlier, chastised Kant for his ignorance of basic principles of Newtonian physics, decried the vogue of mentioning Kant's theory in the same breath with that of Laplace, and warned: 'As a *popular* treatise on cosmogony I consider the "Naturgeschichte" of Kant unsuitable and even dangerous as inviting feeble minds and minds uninstructed in natural philosophy to vain and fruitless speculations'. The text of Charlier's lectures was published in the *Publications of the Astronomical Society of the Pacific*, vol. XXXVII (1925); for quotation see p. 63.

71. Of these four editions, two were unauthorised. For further details, see Hastie, *Kant's Cosmogony*, pp. lviii–lix.

72. The flamboyant professor of physics at Göttingen, G. F. Lichtenberg, and the physician-astronomer of Bremen, W. Olbers, are a case in point. See on this my paper, 'Drei kosmologische Vorträge von Wilhelm Olbers', in *Nachrichten der Olbers-Gesellschaft*, Nr. 79, Oktober 1970, pp. 14–28.

73. The translation by J. K. Fr. Hauff appeared under the title, *Darstellung des Weltsystems* (Frankfurt a. M.: bey Varrentrapp und Wenner, 1797), and was reviewed in *Magazin für den neuesten Zustand der Naturkunde mit Rücksicht auf die dazu gehörigen Hülfswissenschaften* (Jena), edited by Johann Heinrich Voigt, 1 (1798): 173–76. It must be noted that in reporting the contents of Book V of the work the reviewer failed to mention that it contained a theory of the origin of the planetary system.

74. *Exposition du système du monde* (Paris: Imprimerie du Circle Social, An IV [1796]), a small octavo book in two volumes. For the chapter in question, see vol. 2, pp. 293–312. The following discussion of the evolution of Laplace's theory is in part identical with the text of my article, 'The Five Forms of Laplace's Cosmogony', *American Journal of Physics* 31 (1976): 4–11.

75. *Exposition* (1796), vol. 2, p. 294.

76. *Exposition* (1796), vol. 2. p. 294.
77. *Ibid.*, pp. 294–95.
78. *Ibid.*, p. 312.
79. *Ibid.*, p. 295.
80. *Ibid.*, p. 297.
81. *Ibid.*, p. 298.
82. As discussed in Chapter Four. See pp. 104–05.
83. Examples of this will be given in Chapter Six. See pp. 162 and 171.
84. *Exposition* (1796), vol. 2, p. 309.
85. 'An Account of the Discovery of Two Satellites revolving round the Georgian Planet', a paper read on February 15, 1787, before the Royal Society (*Philosophical Transactions* 77 [1787]: 125–29, see p. 128). A year later Herschel specified those deviations as 91 and 89 degrees respectively in his paper 'On the Georgian Planet and its Satellites' read before the Royal Society on May 22, 1788 (*Philosophical Transactions* 78 [1788]: 364–78, see p. 375).
86. Reference to that retrograde motion was part of the long title of his paper 'On the Discovery of four additional Satellites of the Georgian Sidus' read on December 14, 1797 (*Philosophical Transactions* 98 [1798]: 47–79; see p. 48). One wonders how could Laplace disregard Herschel's emphatic statement followed as it was by the comment: 'This seems to be a remarkable instance of the great variety that takes place among the movements of the heavenly bodies. Hitherto, all the planets and satellites of the solar system have been found to direct their course according to the order of the signs [of the zodiac]: even the diurnal or rotatory motions, not only of the primary planets, but also of the sun, and six of their secondaries or satellites, now are known to follow the same direction; but here we have two considerable celestial bodies completing their revolutions in a retrograde order'.
87. *Exposition* (1796), vol. 2, p. 296. In the decimal division of the circle which Laplace advocated the 100 degrees correspond to the customary 90 degrees.
88. *Ibid.*, p. 301.
89. *Ibid.*, p. 302.
90. *Ibid.*, pp. 302–03.
91. *Ibid.*, pp. 122–26.
92. *Ibid.*, p. 123.
93. What Laplace specifically showed in his celebrated memoir published in three parts in 1784–86 was that the actual decrease of the size of Jupiter's orbit and a similar increase of Saturn's orbit have periods of 929 years and that the eccentricities of planetary orbits always remain within tolerable limits.
94. *Exposition* (1796), vol. 2, p. 304.
95. See his *Esquisse d'un tableau historique des progrès de l'esprit humain*, in *Oeuvres de Condorcet* (Paris: Firmin Didot, 1847), vol. 6, p. 13.
96. *Exposition* (1796), vol. 2, p. 305.
97. *Ibid.*
98. *Ibid.* If, however, such was the case, there were enough 'opaque' bodies in the universe to trap all starlight and to make, in particular, impossible the visual appearance of the Milky Way, an implication which Laplace did not perceive.
99. For a brief discussion of that similarity, see C. W. Misner, K. S. Thorne and J. A. Wheeler, *Gravitation* (San Francisco: W. H. Freeman, 1973), pp. 623–24.
100. Through B. Merian's *Le Système du monde*, a condensation of Lambert's *Cosmologische Briefe*. For details, see Lambert, *Cosmological Letters on the Construction of the World Edifice*, p. 27.
101. *Exposition du système du monde* (Seconde édition, revue et augmentée par l'auteur; Paris: de l'Imprimerie de Crapelet, An VII [1799]) in one volume. For the last chapter, see pp. 341–51. The chapter on the atmosphere of celestial bodies was still Chapter ix of Book IV (pp. 245–47) and contained only two small stylistic changes with respect to the text in the first edition.

102. *Exposition* (1796), vol. 2, p. 303.

103. *Exposition* (1799), p. 343.

104. See note 85 above.

105. *Exposition* (1799), p. 347.

106. *Ibid.*, p. 346.

107. 'On the Utility of Speculative Inquiries' (1780), see *The Scientific Papers of Sir William Herschel*, edited by J. L. E. Dreyer (London: The Royal Society and the Royal Astronomical Society, 1912), vol. 1, p. lxxxi.

108. 'Catalogue of 500 new Nebulae and Clusters; with Remarks on the Construction of the Heavens' (1802), *ibid.*, vol. 2, p. 213.

109. Quoted in *The Herschel Chronicle*, edited by C. A. Lubbock (New York: The Macmillan Company, 1933), p. 310. That Laplace's theory of the evolution of the solar system was the crowning part of his lifelong effort to eliminate as much as possible 'final causes' from the study of celestial phenomena and of cosmological theories is the principal contention of J. Merleau-Ponty's article, 'Situation et rôle de l'hypothèse cosmogonique dans la pensée cosmologique de Laplace' (*Revue d'histoire des sciences* 29 [1976] : 21–49). Undoubtedly, it is a chief aim of science to divest quantitative connections among physical phenomena of their apparently mysterious character and to render in that sense any recurrence to a 'final cause' unnecessary. It is an entirely different question whether an infinite regress along the line of quantitative correlations is a satisfactory answer to the singularly specific character of material existence, a point already emphasised by W. Whewell (as noted later in this Chapter) in connection with Laplace's theory. For a more detailed discussion of this point see Lecture XVII ('Cosmic Singularity') in my Gifford Lectures given at the University of Edinburgh in 1975 and 1976 (*The Road of Science and the Ways to God* [Edinburgh: Scottish Academic Press, 1977]).

110. *Exposition du système du monde* (Troisième édition, revue et augmentée par l'auteur; Paris: Courcier, 1808). The work was now printed in two formats. One was small octavo in two volumes. For Chapter vi of Book V see vol. 2, pp. 379–99. The total length of the cosmogonical part approached 3500 words. The chapter on the atmosphere of celestial bodies (vol. 2, pp. 137–41) differed from that in the second edition only by a slight modification of the ratio of the mass of the moon to that of the earth. The other format was large octavo in one volume. See pp. 256–58 on the atmosphere of celestial bodies and pp. 388–94 on cosmogony.

111. *Exposition* (1808), vol. 2, p. 382.

112. *Briefwechsel zwischen W. Olbers und F. W. Bessel*, edited by Adolph Erman (Leipzig: Avenarius & Mendelssohn, 1852), vol. 1, p. 337.

113. For details, see my articles, 'The Early History of the Titius-Bode Law', *American Journal of Physics* 40 (1972) : 1014–23, and 'Das Titius-Bodesche Gesetz im Licht der Originaltexte', *Nachrichten der Olbers Gesellschaft* (Bremen) Nr. 86, Oktober 1972, pp. 1–8.

114. *Exposition* (1808), vol. 2, p. 396.

115. *Ibid.*, p. 388.

116. As will be discussed in Chapter Seven. See pp. 187–90.

117. *Exposition* (1808), vol. 2, p. 388. In his first essay on Saturn's rings (1787) Laplace merely implied their solidity. See his 'Mémoire sur la théorie de l'anneau de Saturne', in *Histoire de l'Académie Royale des Sciences, Année MDCCLXXXVII* (Paris: Imprimerie Royale, 1789), pp. 249–67. The solidity of rings was clearly asserted in the *Traité de mécanique céleste* (Paris: de l'Imprimerie de Crapelet, chez J. B. M. Duprat, an VII [1798]) vol. 2, pp. 165 and 373.

118. As will be discussed later in this Chapter. See pp. 142–43.

119. Both reviews appeared in widely circulating and prominent organs. That by Gauss in the *Göttingische Anzeigen von gelehrten Sachen*, July 23, 1808 (see Gauss, *Werke* [Göttingen: Königliche Gesellschaft der Wissenschaften, 1874], vol. 6, pp. 500–03). The review by Biot in *Le Moniteur universel*, 13 avril, 1808, Nr. 104, p. 410, col. 3 – p. 412, col. 2.

120. *Le Moniteur universel*, 7 juillet 1812, Nr. 189, p. 740, col. 3.

121. *Exposition du système du monde* (Quatrième édition, revue et augmentée par l'auteur; Paris:

Mme V^c Courcier, 1813). The format was large octavo in one volume. For the chapter in question see pp. 427–48. The total length of the section on cosmogony was about twice as long as in the first edition. The additions mostly concerned the nebulae and the future progress of astronomy. Laplace now explicitly referred to Herschel, a move which amounted to adding Herschel's authority and renown to the nebular hypothesis – a curious move in view of the meeting between Laplace and Herschel in 1802 reported above and of Laplace's continued disregard of Herschel's findings about the retrograde motion of two satellites of Uranus.

122. *Exposition* (1813), pp. 431–32.
123. *Ibid.*, p. 432.
124. *Ibid.*, p. 433.
125. *Ibid.*
126. *Ibid.*
127. *Ibid.*, p. 434.
128. *Ibid.*
129. *Ibid.*
130. *Ibid.*, p. 435. Laplace recalled as another possibility the explosion of a planet and referred to Olbers as the original proponent of this theory to explain the origin of asteroids.
131. *Ibid.*, p. 436.
132. *Ibid.*
133. *Ibid.*, p. 437.
134. *Ibid.*
135. *Ibid.*, p. 439.
136. *Ibid.*, p. 438.
137. *Ibid.*, p. 439.
138. *Ibid.*, p. 440.
139. *Ibid.*, p. 442.
140. *Ibid.*, p. 443.
141. *Ibid.*
142. *Ibid.*
143. *Ibid.*
144. *Ibid.*
145. *Ibid.*, p. 438.
146. *Ibid.*, p. 442.
147. *Ibid.*, p. 444.
148. *Exposition du système du monde* (Cinquième édition, revue et augmentée par l'auteur; Paris: Bachelier, Janvier 1824). This edition was published both in large octavo in one volume and in small octavo in two volumes by the same publisher. Only the large octavo indicated the month of publication. The fifth edition was followed by a sixth, published in Paris and Bruxelles a few months after Laplace's death on March 5, 1827, but it should be considered as a reprint of the fifth. For further details on this sixth edition and on another sixth edition published in 1835, see my 'The Five Forms of Laplace's Cosmogony', quoted in note 74 above.
149. In the issue of May 29, 1813, p. 581 col. 2 – p. 582 col. 1. The reviewer compared the *Exposition* to Newton's *Principia* and declared that because of its greater abundance of scientific facts the *Exposition* will last longer than the *Principia!*
150. *Bruno oder über das göttliche und natürliche Princip der Dinge. Ein Gespräch*, in *Sämmtliche Werke* (Stuttgart: Cotta, 1859), vol. 4, p. 271.
151. In his *Dissertatio philosophica de orbitis planetarum* (1801), in *Sämmtliche Werke* (Stuttgart: F. Frommans Verlag, 1927), vol. 1, pp. 28–29.
152. In 1816, in his *Vorlesungen über die Geschichte der Philosophie*, in *Sämmtliche Werke*, vol. 17, p. 282.
153. Throughout the three editions (1817, 1827, 1830) of his *Encyclopädie der philosophischen Wissenschaften im Grundrisse*, in *Werke*, vol. 9, p. 180. In that work Hegel recognized the existence of the asteroid belt and the basic need for observations.

154. *Johann Samuel Traugott Gehler's Physikalisches Wörterbuch* neu bearbeitet von Brandes. Gmelin. Horner. Muncke. Pfaff; Siebenter Band, Erste Abtheilung N-Pn (Leipzig: bei E. B. Schwickert, 1833), pp. 582–89.

155. *Ibid.*, p. 546.

156. *Ibid.*, p. 589.

157. Vienna: Im Verlage von J. G. Heubner, 1825, vol. 2, pp. 273–95.

158. *Die Wunder des Himmels oder gemeinfassliche Darstellung des Weltsystems.* Dritter Theil (Stuttgart: Hoffman'sche Verlagsbuchhandlung, 1836), pp. 190–206.

159. Stuttgart: Hoffman'sche Verlagsbuchhandlung, 1842, Dritter Theil, p. 640.

160. *Parerga und Paralipomena: Kleine philosophische Schriften*, edited by Julius Frauenstädt (3d ed.; Leipzig: F. A. Brockhaus, 1874), vol. 2, pp. 143–48.

161. *Ibid.*, pp. 143 and 145.

162. References are to the English translation in Hermann von Helmholtz, *Popular Scientific Lectures*, selected and introduced by Morris Kline (New York: Dover Publications Inc., 1962), pp. 59–92.

163. *Ibid.*, p. 77.

164. *Ibid.*, p. 76.

165. *Ibid.*, p. 75.

166. *Ibid.*

167. *Ibid.*, p. 76.

168. *Ibid.*

169. *Kosmos: Entwurf einer physischen Weltbeschreibung* (Stuttgart: J. C. Cotta'scher Verlag, 1845), vol. 1, p. 95.

170. *Ibid.*, pp. 98–99.

171. F. Arago, *Biographies of Distinguished Scientific Men*, translated by W. H. Smyth and others, First Series (Boston: Tickner and Fields, 1859), p. 358.

172. *Ibid.*, p. 363.

173. *Ibid.*

174. *Ibid.*

175. *Ibid.*

176. Extracts from that memoir were published in *L'Institut, Journal Général des Sociétés et Travaux Scientifiques de la France et de l'Étranger* (Paris), 3 (No. 90, 28 janvier, 1835): 31–33. For quotation see p. 31.

177. In essence, Comte solved the equation $Gr^2/d^2 = 4\pi^2 d/x^2$ for x, that is, for the period of rotation of a particle on the equatorial edge of the solar nebula when it extended to a distance d, equal to any actual planetary distance (r is the actual radius of the sun).

178. *Cours de philosophie positive*, Vol. II, *La philosophie astronomique et la philosophie de la physique* (Paris: Bachelier, 1835), pp. 285–300. For a discussion of Comte's positivist utilisation of the stability of the planetary system, see my *The Relevance of Physics* (Chicago: University of Chicago Press, 1966), pp. 470–72.

179. In his memoir quoted above, p. 31.

180. 'Mémoire sur les phénomènes que présente une masse liquide libre et soustraite à l'action de la pésanteur', in *Nouveaux Mémoires de l'Académie Royale des Sciences et Belles-lettres de Bruxelles*, vol. XVI (Bruxelles: M. Hayez, 1843), 34 pp.

181. *Ibid.*, p. 27.

182. *Ibid.*, p. 26. The typical size of the globe of oil was 6 cm in diameter and that of the central disk 2 cm.

183. *Ibid.*

184. *Ibid.*

185. Sir John F. W. Herschel, *Outlines of Astronomy* (London: Longman, Brown, Green, and Longmans, 1849), pp. 599–600.

186. *History of Physical Astronomy from the Earliest Ages to the Middle of the Nineteenth Century* (London: Henry G. Bohn, 1852).

187. *Ibid.*, pp. 282–86.

188. *Philosophy of the Inductive Sciences* (2d ed.; London: John W. Parker, 1847), vol. 2, pp.

418–19) and *History of the Inductive Sciences* (3d ed.; London: John W. Parker, 1857), vol. 2, p. 229.

189. See the American edition, Philadelphia: Carey, Lea & Blanchard, 1833, p. 149.

190. References are to the second edition published the same year, with an Introduction by E. Hitchcock (Boston: D. Lothrop & Co., 1853).

191. *Ibid.*, p. 232

192. *Ibid.*, p. 233.

193. *Ibid.*

194. *Ibid.*, p. 223.

195. Edinburgh: William Tait, 1837, p. 179. The work, which went through seven editions in seven years, did more than any other publication to turn the nebular hypothesis into a widely known facet of astronomical lore.

196. *Ibid.*, p. 155.

197. *Ibid.*, p. 164.

198. This phrase can be found in the first American edition (New York: H. A. Chapin & Co., 1840), p. 99.

199. First Edinburgh edition, p. 179.

200. *Ibid.*, p. 186.

201. *Views of Astronomy: Seven Lectures delivered before the Mercantile Library Association of New York in the Months of January and February, 1848* (New York: Greeley & McElrath, 1848). The text of this forty-page-long, very densely printed booklet was, in the words of the title page, 'reported for the New York Tribune by Oliver Dyer, phonographic writer'. For quotation, see p. 29.

202. Edinburgh: John Johnstone, 1848. Nichol's declaration, that the theory is 'hypothetical in its foundation' and 'also hypothetical throughout' (p. 120) and that 'it is not likely . . . that anyone who now well considers its logical position, will endeavour to erect it into a truth' (p. 132), conflicted with his enthusiastic presentation of evidences and considerations supporting it, and with his handling of the question of the satellites of Uranus. He saw in that question the 'only one' (p. 125 note) possible objection to the hypothesis, an objection 'which seems a grave logical question at the outset . . . neither to be overlooked, nor reasoned away'. The gravity of these remarks was quickly offset by the 'probability of the case', which 'certainly favour the idea that it [the retrograde motion of those satellites] is [due to] a disturbance' (p. 125 note)

203. London: H. Bailliere, 1850. 'Supplementary Note to Chapter VI', pp. *245–*249.

204. The mathematical form of Kirkwood's law is $n^2/n'^2 = D^3/D'^3$ where n and n' refer to the number of rotation of any planet during one revolution in its orbit, and D and D' refer to any planet's sphere of action in terms of its width which is defined as the distance from the zone of zero gravitation on the inner side of the orbit of a planet to a similar zone on the outer side of its orbit. The documents pertaining to the background, formulation, disclosure, and the first reception of Kirkwood's Law are given in full in the *Proceedings of the American Association for the Advancement of Science. Second Meeting, held at Cambridge [Mass.] August, 1849* (Boston: Henry Flanders & Co., 1850), pp. 207–21, 'Fifth Day, August 18, 1849, Section of Mathematics, Physics, and Astronomy'. Excerpts from these documents were also published about the same time under the title, 'On a New Analogy in the Periods of Rotation of the Primary Planets, discovered by Daniel Kirkwood, of Pottsville, Pennsylvania', in *The American Journal of Science and Arts* 9 (1850): 395–99. The impact of Kirkwood's Law on American men of science can be gauged from the fact that the four other papers read on August 18 received only a bare mention in the *Proceedings*. According to Ronald L. Numbers ('The American Kepler: Daniel Kirkwood and his Analogy', *Journal for the History of Astronomy* 4 [1973]: 13–21) American scientists felt at that time that Laplace's nebular hypothesis was highly questionable (they did not put on record their reasons for this) and that it received a powerful support through Kirkwood's Law. Yet ultimately it was their belief in cosmic evolution, and not in a particular form of planetary genesis, that was bolstered through the finding of Kirkwood. It certainly did not prompt them to publish anything of significance on the

nebular hypothesis. No impact seems to have been made in continental Europe by Kirkwood's Law. In England David Brewster brushed aside the Law on July 31, 1850, in his Presidential Address at the Edinburgh meeting of the British Association with the following remark, the motivation of which will be discussed shortly: 'The American astronomers regard this law as amounting to a demonstration of the nebular hypothesis of Laplace; but we venture to say that this opinion will not be adopted by the astronomers of England'. *Report of the Twentieth Meeting of the British Association for the Advancement of Science* (London: John Murray, 1851), p. xxxv.

205. *The Planetary System*, p. *247 note. There was no discussion of the nebular hypothesis in Nichol's *The Phenomena and Order of the Solar System* (1838; New York: Dayton & Newman, 1842), and in his *Contemplations on the Solar System* (2nd ed.; Edinburgh: William Tait, 1844).

206. For details, see A. F. O'D. Alexander, *The Planet Saturn: A History of Observations, Theory and Discovery* (New York: The Macmillan Company, 1962), pp. 176–78.

207. G. P. Bond, 'On the Rings of Saturn', *The Astronomical Journal* 2 (May 28, 1851): 5–8 and 8–10.

208. B. Peirce, 'On the Constitution of Saturn's Ring', *The Astronomical Journal* 2 (June 16, 1851): 17–19.

209. *Ibid.*, p. 17.

210. *Ibid.*, p. 19. As a concluding remark Peirce warned against 'rashness and extravagance in the immediate presence of the Creator', and 'against venturing too boldly into so remote and obscure field of speculation as that of the mode of creation which was adopted by the Divine Geometer' (*ibid.*).

211. O. Struve, 'Sur les dimensions des anneaux de Saturn', read on Nov. 14, 1851, in *Recueil de Mémoires présentés à l'Académie des Sciences par les astronomes de Poulkova, ou offerts à l'Observatoire central par d'autres astronomes du pays*, Premier Volume (St.-Pétersbourg: Imprimerie de l'Académie Impériale des Sciences, 1853), pp. 349–86.

212. 'On the Stability of the Motion of Saturn's Rings', in *The Scientific Papers of James Clerk Maxwell*, edited by W. D. Niven (Cambridge: Cambridge University Press, 1890), vol. I, pp. 288–377. Maxwell had already submitted his essay when he learned of a paper of Peirce, 'On the Adams Prize-Problem for 1856' (*The Astronomical Journal* 4 [Sept. 5, 1855]: 110–12), in which Peirce first argued against the solidity of rings and then tried to establish their fluidity. In that paper Peirce made no reference to the nebular hypothesis.

213. 'On the Stability of the Motion of Saturn's Rings', p. 373.

214. *Ibid.*, p. 295.

215. J. P. Nichol and others, *A Cyclopaedia of the Physical Sciences* (London: R. Griffin, 1857), p. 663. See also p. 529.

216. *Memoirs of the Life, Writings, and Discoveries of Sir Isaac Newton*, by Sir David Brewster (Edinburgh: Thomas Constable and Co., 1855), vol. II, p. 131.

217. [H. Spencer], 'Recent Astronomy, and the Nebular Hypothesis', *Westminster Review* 70 (July 1858): 185–225. The article was purported to be a review of the *Oeuvres de Laplace*, of Arago's *Popular Astronomy*, of two major works of the younger Herschel and of *Recent Progress of Astronomy: especially in the United States* by E. Loomis of Yale, but it contained largely Spencer's own discussion of the question of 'island universes', and of the nebular hypothesis. The latter took up three-fourths of the essay.

218. As set forth programmatically in his *First Principles*. For further details, see my *Science and Creation: From Eternal Cycles to an Oscillating Universe* (Edinburgh: Scottish Academic Press, 1974), pp. 295–96.

219. 'Recent Astronomy, and the Nebular Hypothesis', p. 201.

220. As will be discussed in the next chapter.

221. *Ibid.*, p. 202.

222. *Ibid.*, p. 225.

223. *Ibid.*

224. 'Cosmogonie de Laplace', in J. Babinet, *Études et lectures sur les sciences d'observation et leurs applications pratiques* (Paris: Mallet-Bachelier, 1865), vol. VII, pp. 105–06.

225. *Ibid.*, p. 106.

CHAPTER SIX

THE ANGULAR BARRIER

A barrier raised and lowered

'It is impossible to admit that these planets [the Earth and Neptune] were formed of the mass of the sun spread out to the [respective] planetary orbits . . . One has to conclude that if the entire sun was spread out to the planetary orbits, it had a rotational movement too slow to let the centrifugal force overcome the gravitational force and cause thereby the leaving behind of an equatorial ring isolated from the total mass'. So stated J. Babinet on March 18, 1861, before the Académie des Sciences in a paper on a point concerning Laplace's cosmogony.[1] Babinet reached his conclusion on the basis of a relatively simple computation which he carried out to test not Laplace's theory but an interpretation of it. According to that interpretation, which in Babinet's words was held by 'several persons', the primitive solar atmosphere was formed by the spreading of the sun's mass to the orbit of the most distant planet and perhaps somewhat beyond it. Babinet claimed that Laplace was explicit on the point that when the planets were formed they drew their masses from the solar atmosphere distinct from the solar nucleus.

Had Babinet checked the various editions of the *Exposition*, he might have easily noticed that, from the fourth edition on, Laplace envisaged a primitive condition of the sun with all its mass diluted into a spherical cloud of such low density as to be hardly perceptible.[2] Laplace presented this as a condition which preceded the one in which the solar atmosphere had already a bright nucleus in resemblance of some nebulae. It was that more recent phase of the sun's development which was given in the first three editions, where he started out with a widely diffused atmosphere surrounding the sun.[3] But in the fourth edition Laplace failed to specify the correlation between the formation of the planets and the existence of a nucleus in the primitive solar atmosphere. The interpretation, which Babinet meant to discredit, was not, therefore, wholly out of line with Laplace's statements, a circumstance which escaped Babinet. At any rate, he was not to be bogged down in the tedious work of comparing the texts of the various editions of the *Exposition* when mathematics could readily dispose of a model which, in his belief, was wrongly attributed to Laplace. He must have felt that Laplace could not propose a mathematically wrong model for the evolution of the system of planets.

The mathematics was certainly straightforward for the case of a dilute spherical mass which remained homogeneous as it kept shrinking. For such a case the conservation of angular momentum demanded that the sun's mass if spread out to the earth's orbit should have a period of rotation equal to 1,162,000 days or 3,181 years.[4] Spreading out the sun's mass to the orbit of Neptune would have implied the slowing down of the rotation of the sun's mass from its present value of about 25 days to 'more than 27 thousand centuries'.[5] From all this Babinet drew three conclusions. In the first he simply stated the impossibility of the formation of planets from the mass of the sun spread out to the various planetary orbits. In the second he called attention, no doubt with an eye on the sun's present slow rotation, to the 'extremely weak' rotational motion which had to exist around the many centres of gravity in that 'universal cosmic material' giving rise to as many stars. The third conclusion stated that 'if the entire sun had been diffused to the planetary orbits, it would have had a rotational movement much too weak to let the centrifugal force equal the gravity and to permit thereby the leaving behind of an equatorial ring isolated from the total mass'.[6]

Although Babinet's paper came to a close with a reference to the formation of rings in Laplace's theory, he did not mention Maxwell's memoir on Saturn's rings. Herbert Spencer, the first to discuss following the publication of Babinet's paper the merits of the nebular hypothesis, referred to Maxwell, but the reference merely served evidence that no new discovery could weaken the wide variety of verities, which he now set forth as the 'first principles' of each and every facet of cosmic and human evolution. Old statements too fared badly in Spencer's hands. According to him 'Laplace assumed that the ring would rupture at one place only; and would then collapse on itself'.[7] This was to attribute to Laplace a notion he did not voice as such. Nor was Spencer's rendering of Maxwell's findings the epitome of objectivity. The assumption of the breaking of the ring in one place is a questionable one, Spencer wrote, adding, that 'such at least I know to be the opinion of an authority second to none among those now living'.[8]

Spencer meant Maxwell, for he added in the same breath that the ring must consist of a material composed of many parts.[9] Could then this reference of Spencer to Maxwell be interpreted as something contrary to the truth of Laplace's theory? Not at all, if one was aware of the validity of one of Spencer's basic tools of explanation, the principle of the instability of the homogeneous: 'It is still inferable from the instability of the homogeneous, that the ultimate result which Laplace predicted would take place. For even supposing the masses of nebulous matter, into which each ring separated, were so equal in their sizes and distances as to attract each other with exactly equal forces (which is infinitely improbable); yet the unequal action of external disturbing forces would inevitably destroy their equilibrium – there would be one or more points at which adjacent masses would begin to part company. Separation once commenced, would with ever-accelerating speed lead to a grouping of the masses. And obviously a like result would eventually take place with the groups thus formed; until they at length aggregated into a single mass'.[10]

Almost a hundred pages later, another basic category of Spencer, the principle of equilibration, was also made to support the nebular hypothesis and in particular the formation of planets from rings: 'This ring must progress towards a moving equilibrium of a more complete kind, during the dissipation of that motion which maintained its particles in a diffused form: leaving at length a planetary body, attended perhaps by a group of minor bodies, severally having residuary relative motions that are no longer resisted by sensible media; and there is thus constituted an *equilibrium mobile* that is all but absolutely perfect'.[11] To the last word there was added a note in which Spencer provided unwitting evidence that his brave reasoning was an 'absolutely perfect' farce.

In the note Spencer spoke of a recent approval by Sir David Brewster[12] of a 'calculation by M. Babinet, to the effect that on the hypothesis of nebular genesis, the matter of the Sun, when it filled the Earth's orbit, must have taken 3,181 years to rotate; and that therefore the hypothesis cannot be true'.[13] To parry Brewster's remark Spencer now equated Babinet's calculations with those of Comte, which he quoted with approval only four years earlier but which he now recognised as based on *petitio principii*. Babinet's calculations were based, according to Spencer, 'on two assumptions, both of which are gratuitous, and one of them totally inconsistent with the doctrine to be tested'. True, exception could be taken to Babinet's calculations concerning the 'current supposition respecting the Sun's density', and concerning the rotation of all parts of the nebula with the same angular velocity. But Spencer merely revealed his amateurism in physical science (at the age of fifteen he quit formal training to supplement it for a few years with railroad engineering), when he wrote in a way of conclusion that 'if (as is implied in the nebular hypothesis, rationally understood) this spheroid resulted from the concentration of far more widely diffused matter, the angular velocity of its equatorial portion would obviously be immensely greater than that of its central portion'.[14] With such an oversight of the elementary laws of rotational dynamics he was never to catch a glimpse of the insurmountable barrier posed to Laplace's theory by Babinet's attention to the question of angular momentum.

New-World enthusiasm

Babinet was not mentioned by Daniel Kirkwood who in the 1860's waged his spirited crusade on behalf of the 'truth' of Laplace's theory from the chair of professor of astronomy at the University of Indiana, a position which came to the once shy, amateur astronomer through the enthusiasm by which some of his countrymen had greeted his discovery of the 'Law of the Sphere of Action'. Kirkwood's paper 'On the Nebular Hypothesis'[15] certainly evidenced the partisan thinking that can be generated by addiction to a theory. In taking up the question of the heavy tilt of the plane of the satellites of Uranus and their retrograde motions, Kirkwood looked for a solution in the great varieties of motions of the many parts into which the rings around Neptune broke up before those parts coalesced into its satellites. The nebulat theory was nebulous enough to provide convenient *ad hoc* solutions, but expediences of that kind were in the long run to undermine its credibility.

More importantly, the same bias could effectively prevent a recognition of what was soon to emerge as the most crucial feature of the solar system, the distribution in it of the angular momentum. Kirkwood's paper 'On Certain

Harmonies of the Solar System' from 1864[16] was a classic illustration of this. There he gave a detailed, quantitative development of his law of correlation which he called 'analogy'. His principal conclusion was that 'in the solar nebula, as well as in each of the gaseous planets, the ratio of the revolving or equatorial radius to the radius of gyration varied throughout the entire process of condensation: in other words, that the rate of variation of density from surface to centre was constantly changing'.[17] He illustrated this with two considerations. One was that the solar mass, if spread out homogeneously to the earth's orbit, would be rotating in 5,457 instead of 365 days. The other brought Kirkwood even closer to the moment of truth: 'If we compute the values of the principal radius of gyration when the spheroid extended to the present planetary orbits respectively, we find that the solar mass had reached a high degree of central condensation before the epoch of Neptune's separation'.[18] Fom this Kirkwood might have logically proceeded to the actual computation of the densities in the nebular spheroid at the actual planetary distances. Comparing these results with the actual masses of the planets might have in turn alerted him to the question of the distribution of angular momentum. But with his firm faith in the nebular theory and in his 'analogy' he bogged down in a tedious effort to reconcile his 'analogy' with the fact that the period of Saturn's rotation was not much greater than that of Jupiter. He also wasted his energy on trying to establish some correspondence between the size of the 'sphere of action' of each planet and their relative distances and did the same with respect to the satellites of Saturn and Jupiter.

In a note to his paper Kirkwood expressed his eager anticipation[19] of seeing in print a paper 'On the Nebular Hypothesis' by David Trowbridge[20] of Perry City, New York, the contents of which came to his attention shortly after he had completed his own article. The principal thrust of Trowbridge appeared to be aimed at deriving specific figures for the relative densities in the nebular spheroid, and represented in that respect a distinct advance over Kirkwood's efforts. Thus Trowbridge came even closer than Kirkwood did to the recognition of the paramount importance of the distribution of angular momentum in the solar system. Trowbridge's failure to sight the crucial issue evidenced once more and somewhat more keenly the blinding effect of one's engrossment with one's favourite theory. The near-miss by Trowbridge was all the more curious as from the very outset he tried to show that in all probability the primitive nebula was homogeneous. But from this he inferred that separate nuclei of condensation could in such a nebula readily evolve and ultimately issue in a rotational movement around the common centre of gravity. He emphasised this with an eye on nebulae giving rise to star clusters.[21] He failed to point out that there was at that time no evidence of that rotational movement which he postulated in clusters of stars. It equally escaped him that he did not provide even a tentative explanation for the rotation of that smallest portion of the nebula which presumably developed into a solar system. He merely stated that 'a fluid mass which does not rotate on an axis must ultimately become spherical in form . . . but if it had a rotatory motion, it can never become a sphere, although it may approximate one in form'.[22] It was in the analysis of the subsequent development of that rotating liquid spheroid where, according to him, 'we must look for that proof of the truth of the nebular hypothesis which will be in any great degree satisfactory'.[23] His anticipations were as illusory as were his conclusions specific.

Trowbridge's starting point was a very oblate rotating spheroid, a 'mostly aeriform fluid body',[24] with a mass equal to that of the sun. He felt that he could discount the mass of the planets which amounted only to 1/738th to the sun's mass. Clearly, he wanted to be very rigorous about the 'Principle of the Conservation of Aeras', or the principle of the conservation of angular momentum, in that spheroid. He perceived none of the irony that by the same token he ruled out of court the question of the total angular momentum in the actual planetary system, and of its true distribution among the various constituents of that system. It also escaped him that the features of that rotating spheroid kept pointing in that very same direction. First, he computed in that spheroid the radii of gyration 'at the time when it abandoned the several rings from which the planets were formed'.[25] At that moment the rings were supposed to have a period of rotation equal to that of the respective planets. The values, in millions of miles, were as follows: 0.46, 0.74, 0.95, 1.31, 2.21, 3.29, 5.18, 8.75, 12.26. On the basis of this he derived the following values for the relative densities in the successive layers of the rotating spheroid, again starting with Mercury and including the ring of asteroids: 27.1, 3.1, 1.0=earth, 0.23, 0.019, 0.003, 0.00037, 0.000032, 0.0000148. As was noted above, the key to all these calculations was the principle of the conservation of the angular momentum, but Trowbridge failed to recall this as he mulled over the startling result that the ratio of densities of the ring of Mercury and that of Neptune was 2,000,000 to 1. A little reflection on the respective masses and angular momenta of Mercury and Neptune could have given him a second thought on the matter. Two decades passed before this was pointed out.

What came in the meantime to be offered on the nebular theory of planetary origins hardly gave evidence of much acumen, although some of the scientists involved were well known for their incisive creativeness. Trowbridge himself saw the chief merit of his investigation in that it accounted 'in a very satisfactory way for Prof. Kirkwood's "spheres of attraction", in his beautiful "Analogy" '.[26] Kirkwood was also mentioned with high praise by Gustavus Hinrichs of Iowa State University, author of some of the most obscurantist articles that saw print in the 1860's in a leading scientific journal. He echoed Kant in advocating the existence of a repulsive force in addition to the attractive one. To this he added that there should also be a slight inequality between these two forces. Once this was admitted, a 'rotation of the nebulous matter' could be spoken of as 'direct consequence of this inequality'. The rest followed with bewildering ease: 'by attraction the [nebulous] matter acquires a globular form, the effected rotation produces a flattening of the globe, and from this moment the axis of rotation will remain stationary. By continued attraction the size diminishes, rotation increases in velocity, the flattening becomes greater, a ring is formed, producing a planet with its satellites, the whole system having motions and configurations, which conform to those observed in the actual world'.[27]

This is not the place to follow Hinrichs' attempts to solve such obvious difficulties of the nebular hypothesis as the small density of Saturn and the retrograde motion of the satellites of Neptune. A paper, in which the resistance due to the ether provided a key to deriving the figures for the density, rotation, and even the relative age of each and every planet and satellite system, had to be suspect from the very start. For the historian Hinrichs is significant only because he served evidence of the distrust of some leading scientists towards the nebular theory. He quoted Brewster's sharp censure of the theory and claimed that the foregoing difficulties brought it 'into disrepute'.[28] In Hinrichs' words even 'Plateau's researches on the equilibrium of fluids did but revive this theory for a moment'.[29] Hinrichs' subsequent publications[30] also evidenced the already typical pattern of taking all data about planets into consideration except the amount of angular momentum carried by each and all of them. Needless to say, the basic differences between Kant's and Laplace's theories remained largely ignored.

Old-World bias

The situation hardly improved when the nebular theory was discussed by first-rate scientists and scientific writers. Richard A. Proctor was an epitome of the latter though not on the merits of what he offered in this connection. He took up the topic in his monograph, *Saturn and its System*.[31] The results of Maxwell's study of the rings of Saturn were carefully given by Proctor who made no secret of their devastating bearing on Laplace's conclusions that the rings were solid. Proctor also recalled Maxwell's arguments against the possibility of considering the rings as a continuous liquid or vaporous body.[32] For all that Proctor claimed that 'no part of the solar system affords so striking an argument in favour of Laplace's Nebular Hypothesis as the Saturnian system of rings'.[33] Reasons for this were not given by Proctor who rather tried to answer objections to the nebular hypothesis. Proctor's answer to the objection having to do with the phraseology of Genesis need not detain us here. Another objection had to do with the opposition 'of modern science to the idea of a vast nebulous globe maintained in a state of extensive tenuity by intense heat'.[34] A solution to this could be had, according to Proctor, by assuming that the primary stuff in cosmic spaces consisted of very small meteoritic particles. Proctor failed to recognise that in a sense this amounted to substituting Kant's ideas for those of Laplace. The heat of the sun and of the planets could then be imagined as resulting from the impact of those particles. This advantage was, however, off-set by the basic weaknesses of Kant's assumptions of which Proctor seemed to be wholly unaware.

The third difficulty faced by Proctor concerned the retrograde motion of the satellites of Uranus. His answer typified the wishful thinking of all adepts of a 'favourite' theory. It mattered little that it was, in Proctor's words, an 'altered' form of the one proposed by Laplace.[35] Once the formation of planets was pictured as the result of endless agglomeration and collisions of larger and larger pieces of meteoritic matter, anything could happen: 'In the successive collisions through which each globe may be conceived to have been formed

. . . we appear to have a sufficiently plausible explanation of the peculiarity in question'.[36] This was a peculiar manner of giving away the game. The new explanation of one particular irregularity left without explanation the prevailing regularity in the solar system. For if in the case of the satellites of Neptune a most irregular feature was the result, why did the same mechanism produce only regular effects in all other cases? These latter cases were rather numerous in the solar system. Proctor himself set great store by the fact that the probability was less than 1 in 4,835,700,000,000,000,000,000,000 that all the 83 known asteroids should revolve in one direction on the basis of chance. But were not the collisions random, which produced the anomaly of the satellites of Neptune?

The facile oversight by Proctor of all this was typical in the sense that the same inconsistency had already plagued for generations all theorising about the origin of planetary systems. Proctor also became part of that pattern by his analysis of the habitability of the system of Saturn. Whatever indications science provided on that score strongly militated against the survival of any earthlike being on any or all parts of that system. These indications were duly reported by Proctor and in great detail.[37] He held out nevertheless the possibility that 'arrangements unknown to us prevail on Saturn which may render other parts [away from its equatorial zone] of his surface habitable as we should understand the term'.[38] To make the point more credible he quoted the younger Herschel, the great authority of British astronomy of the time, who claimed that details of planetary surfaces 'which convey to our mind only images of horror, may be in reality theatres of . . . beneficent contrivance'.[39] Science, however much, could not necessarily dispel the illusion that one could eat one's cake and still have it.

Admiration for Laplace's theory and a strong advocacy of a universe inhabited throughout went hand in hand also in the case of A. Secchi, famed for his work on the solar atmosphere and on the classification of stars on the basis of their spectra. His advocacy lacked solid grounds to the same extent as there was an almost complete absence of any critical remark in Secchi's account of Laplace's theory. Originally a part of his lectures on the sun, given in 1867, it appeared in final form in 1870 in his Le soleil.[40] Secchi took Plateau's experiments as a confirmation of Laplace's theory without any qualification.[41] He ignored the satellites of Uranus and mentioned only some asteroids as departures in the solar system from a common plane of orbiting.[42] Laplace's theory, according to Secchi, 'explains a mass of circumstances that are intimately tied together and could not be explained otherwise'.[43] Of the difficulties of the theory he gave at most an inkling. Actually, he equated the explanatory value of the nebular theory with Newton's formulation of the inverse square law of gravitation. As the latter explained the principal correlations among the members of the solar system, the former explained, so Secchi put forward his astonishing claim, 'the tangential impulsion [in the planetary system] and all its physical circumstances of the second order'.[44]

German campaign for Kant and countercampaign

While this was wholly unfounded, especially with respect to that 'tangential impulsion' or angular momentum, Secchi was not wide off target in stating that the 'scientists of the day were almost unanimous' in ascribing the system of planets to the condensation of a primitive nebula. Where unanimity was patently lacking concerned the suddenly arising campaign to turn Kant into the undisputed originator of the theory. The leader was J. F. C. Zöllner, the first occupant of a chair of astrophysics at the University of Leipzig. According to Zöllner, Laplace only repeated Kant as far as the nebular hypothesis was concerned. Zöllner claimed this over some twenty pages in a now classic monograph on stellar photometry.[45] Zöllner's strongly chauvinistic motivations could hardly lead to an objective evaluation of the respective texts which he quoted in far greater detail[46] than anyone who had discussed before him the possible dependence of Laplace on Kant. A good illustration of Zöllner's myopia was his jubilation over the fact that Kant 'derived' a period of rotation of 10 hours for the ring of Saturn.[47] If anyone, then Zöllner should have known that the alleged 'derivation' was as inconclusive as were patently wrong some other 'derivations' by Kant of various numerical details of the planetary systems. Perhaps the only valuable part in Zöllner's discourse was his remark that Laplace assumed but tacitly the rotation of the primitive nebula and that Kant was mistaken in his efforts to derive the same rotation from gravitational attraction alone.[48] Zöllner almost implied that Laplace was guilty of plagiarism, and he turned[49] against Laplace the words 'vérité, justice', which in the first three editions of the *Exposition* were printed boldface at the end of the section on nebular theory. About Laplace's opportunism, if not chameleon-type behaviour, much could be said, but did chauvinism ever promote either truth or justice?

Whether Helmholtz was familiar with Zöllner's dicta on the 'Kant-Laplace' theory is of no importance. Helmholtz's belief in the close identity between their views was already almost two decades old when in 1871 he gave, shortly before he moved from Heidelberg to the University of Berlin, a much publicised address 'On the Origin of the Planetary System',[50] which he also read a few weeks later in Cologne. Although at the very outset Helmholtz promised a lecture on 'the hypothesis of Kant and Laplace', only one-tenth of his lengthy discourse had a direct bearing on it. Helmholtz's carelessness about the historical record could not have been greater. According to him, in the view of Kant and Laplace 'our system was originally a chaotic ball of nebulous matter. . . . This ball, when it had become detached from the nebulous balls of adjacent fixed stars, possessed a slow movement of rotation'.[51] Helmholtz entertained as little doubt about the historical truth of the 'Kant-Laplace' theory as he did about its physical truth. He mentioned the exceptional behaviour of the satellites of Uranus, but not as something that required a major effort for its reconciliation with the theory.

Several years later Helmholtz added to his lecture a note[52] in which the famed Scottish physicist, W. Thomson, the future Lord Kelvin, was upheld against some remarks of Zöllner. These remarks, made in 1872 in the latter's cosmogonical book on comets,[53] had nothing to do with the nebular theory. The book itself might have been very effective to unmask the heavy misconceptions about the 'Kant-Laplace' theory, as Zöllner now gave in parallel columns long excerpts from the statements of both Kant and Laplace on planetary origins.[54] These texts, though incomplete and somewhat garbled, could have revealed at one stroke to any unbiased reader the fallacy of the alleged identity of the two theories. But in the First Reich, which was on its way to scientific pre-eminence, there was developing a strong craving for past glories as well. Neo-Kantianism too was on its rise and the sage of Königsberg soon became celebrated as a scientific genius. Objective scholarship could, therefore, find only limited place in the sundry studies that were published in Germany on the 'Kant-Laplace' theory during the following four to five decades.[55] Their authors did not come from the ranks of prominent scientists.

A better fate in the German academic world should have been the share of Emil A. Budde, a privatdozent at the University of Bonn, who worked from 1878 onwards in Paris, Rome, and Constantinople as correspondent of the *Kölnische Zeitung*. In the booklet, which he published in 1872 on Zöllner's book on comets,[56] he displayed a keen critical sense with respect to Kant's theory, the like of which one could not find in the comments on it of such prominent colleagues of his as Helmholtz and Zöllner. Budde's starting point was Babinet's paper which he was the first among German scientists to note and in which he correctly saw a proof that the actual planetary system could not be formed from the successive detachment of rings of a condensing solar mass.[57] He was also to the point with his insistence that Kant's assumption of a specific density distribution in the original chaos pressupposed in a sense what was to be explained.[58] In Budde's view it was more in keeping with astronomical observations to take meteorites as the original substratum of the cosmogonical process, and he referred in this connection to the estimates of V. Schiaparelli, according to whom the amount of meteorites in the solar system was of the same order as the mass of the large planets.[59] Another interesting remark of Budde in this connection was his singling out Robert Mayer as the first to emphasise the constant fall of meteorites on the sun and the planets.[60] On the really crucial point, the origin of angular momentum in the solar system, Budde did not go beyond either Kant or Laplace. According to Budde any impediment of the free motion of meteorites was to issue in their agglomeration into larger and larger masses, the motions of which, when projected at a central plane, had to result in a uniform circular motion.[61]

French loyalty to Laplace

Budde's emphatic reference to Babinet should make appear rather frustrating the failure of Édouard Albert Roche to mention Babinet in his long memoir which he read on June 29, 1873, before the Montpellier Academy of Sciences, on the constitution and origin of the solar system.[62] Roche's unawareness of Babinet might at once suggest that his memoir, the first full-fledged mathematical treatment of various phases of Laplace's theory, was not to correct the

mistaken trust in it. Roche offered his memoir to show how his study of the
question would 'confirm, specify, or modify Laplace's theory'.[63] The ninety
pages of Roche's memoir indicated no substantial modification of Laplace's
theory. Its most sensitive postulates and phases were consistently sidestepped by
Roche. He declared without any comment that 'when a ring is broken and
transformed afterwards into a spheroidal mass, this, in turn, can give rise to a
planet accompanied by satellites'.[64] Again, by taking recourse to 'convenient
conditions', he felt that he had said enough on the point that the spheroidal
mass had to have a good measure of stability to secure its transformation into a
planet.[65] Attention to minor details used up much useful space in the long
memoir of Roche, who bogged down in such questions as the law underlying
the relative distances of satellites, the origin of zodiacal light, the structure of
the earth, and the source of the sun's energy. More noteworthy was his claim
that the marked difference between the density and rate of rotation of the inner
and outer planets indicated a major change in the manner of the condensation
of the sun.[66] The best part of Roche's memoir concerned Saturn's rings about
which he had already shown in 1849 that they exemplified the law which later
became known as Roche's limit.[67] The law stated that a liquid satellite, having
the same density as the planet itself, would inevitably break into small pieces if
it were closer to the planet than 2.44 times its radius.

Roche, whose principal departure from Laplace's theory consisted in his
advocacy of the formation of rings *within* the primitive solar or planetary
atmospheres, had occasional second thoughts. He himself wondered how the
moon could have originated from the turbulent, primitive terrestrial atmosphere.[68] In recalling at the end of his memoir Laplace's diffidence in all that is
not the result of calculus, Roche took the view that such a distrust should ani-
mate above all those who dared to take up the topic after Laplace.[69] Roche
himself displayed no diffidence as he discussed the condensation of the sun's
atmosphere, the very first topic of his memoir. His starting point was almost
platitudinous. A solar atmosphere revolving once in every $25\frac{1}{2}$ days (the actual
rate of rotation of the sun) could not extend beyond 37 solar radii, or somewhat less than one fifth of the earth-sun distance. Thus, Roche went on, the
condensation of the solar mass extending to the farthest planet had to be accompanied by a decrease in its moment of inertia and by an increase of its rate of
rotation. Equating now the centrifugal force with the gravitational attraction
in the rotating ring just detaching itself from the sun, Roche noted that the
equation was equivalent to Kepler's Third Law. To this he added that 'this
agreement between a fundamental law of the system of the world and the
hypothesis of Laplace does not constitute a proof of his hypothesis, but is a condition which every cosmogonical theory must satisfy'.[70] Had Roche shown
more of the diffidence he claimed,[71] he might have caught a glimpse of the fact
that, in connection with the angular momentum, theories on the origin of solar
system had to satisfy far more than the foregoing simple requirement.

Vagaries from America

A fair measure of that diffidence would have certainly stood in good stead Pliny E. Chase, professor of philosophy at Haverford College, near Philadelphia. It was irony itself that, while Roche's reputable work saw but little circulation, Chase's essay was carried to the four corners of the world in the pages of the prestigious *Philosophical Magazine* in eleven instalments between September 1875 and December 1878. Chase wrote with an assurance for which neither any part nor the whole of the planetary system seemed to be unexplainable. Even when his dicta were not utterly obscurantist, they could still be fairly arcane. True, his starting point was not without some interest. His series of the products of the $\frac{4}{3}$rd power of 15, 24, 30, 42, 105, 168, 280 and 392 with the 'unit distance', which he defined as $\frac{9}{4}$th of the sun's radius, gave with remarkable accuracy the planetary distances.[72] About the derivation of the foregoing series running from 15 to 392 he gave no explicit information. From the extension of the same series he derived his 'harmonic anticipation of planetary or asteroidal matter at a distance from the sun equivalent to 0.269 of Earth's mean radius vector'.[73] A confirmation of that 'anticipation' came, according to him, from 'the sun-spot observations of De La Rue, Stewart, and Loewy . . . and the recent papers of Leverrier'.[74] The reference to Leverrier meant the mythical intramercurial planet, Vulcan, which figured prominently in Chase's communications when they came to a close three years later, although Vulcan was still being searched for in vain.

Here only a few glimpses can be given of what Chase offered in those three years. His was an almost morbid preoccupation with finding numerical correlations among data of planetary sizes, distances, masses, motions, resulting in his 'π-geometrical series',[75] 'triangular-numbers',[76] and sundry harmonics of pendular oscillations. On the basis of the latter he claimed 'a noteworthy numerical correspondence between the seven rupturing-nodes within the planetary belt and the seven condensation-falls from α Centauri to $\frac{\pi}{2}$ Sun'.[77] Chase was certainly not guilty of the charge, which he laid at the door of astronomers. Their views, he wrote, 'respecting the mode of action in world-building, have been various and vague'.[78] In a sense Chase was anything but vague, although his flood of data, often specified to the fifth decimal, could impress only the amateur. As to ideas underlying that flood of data Chase offered a thinly veiled number-mysticism. One could hardly expect from one infatuated with number 9 as a 'cosmical factor'[79] to touch on the question of angular momentum in the solar system even in a communication entitled 'Radiation and Rotation'.[80] His efforts would have at best deserved the classic remark of Wolfgang Pauli about a scientific paper he was asked to evaluate: 'It is not right; worse still, it is not even wrong'.

To the eagerness of Chase to prove everything about the solar system even the most unexpected turn of events presented no obstacle. In his ninth communication, from early 1878, he naturally claimed that his jungle of numerical sequences gave a perfect account of the bafflingly short orbital period of Phobos, the inner satellite of Mars. His 'proof' given in an eighteen-line-long sentence came to a close with the following words: 'all [these data] point to increments of wave-velocity as sources of interior nebular rupture, giving a new meaning to Herschel's doctrine of "subsidence," and making the inner moon of Mars a confirmation, rather than a formidable objection, to the nebular hypothesis'.[81] For Chase the nebular hypothesis of Laplace was an ultimate truth which held the key of explanation to everything connected with it. Chase's last communication ended with confident assertions about the inverse relationship between density and orbital period for the satellites of Mars of which he expected more to be discovered.[82]

Blow from on high

Phobos and Demos, the two satellites of Mars, were discovered in August 1877 by Asaph Hall of the Naval Observatory in Washington. The discovery seemed to be an uncanny connivance on nature's part to cast further doubts on Laplace's theory. In August of that year Mars was approaching a favourable opposition, but Hall decided to take up the search for Martian satellites only at the prodding of his wife.[83] He knew that after Herschel failed to sight satellites around Mars in 1783, only one astronomer, H. L. d'Arrest of Copenhagen, saw it worthwhile to carry out in 1864 systematic observations to that end. On the nights of August 17 and 18, Hall not only found two satellites, but to his great surprise the inner one seemed to orbit around Mars in about 7 hours. At first Hall believed that the unusually short period might have been due to his combining the sightings of several inner satellites into one. Careful observations on the nights of August 20 and 21 convinced him that 'there was in fact but one inner moon, which made its revolution around the primary in less than one-third the time of the primary's rotation, a case unique in our solar system'.[84] The case was equally 'unique' for Laplace's theory in which a satellite, if it indeed arose from a ring detached from the planet, must have had a period of orbiting greater than the period of rotation of the planet itself. Of this point no mention was made either by Hall or by his famed colleague, Simon Newcomb, who in November 1877 submitted more exact data on the motion of the newly sighted Martian moons.[85]

The problem posed for the nebular theory by Phobos was ignored by Newcomb also in his famous textbook, *Popular Astronomy*.[86] What Newcomb offered there on the nebular theory mirrored the almost schizophrenic mentality by which the nebular hypothesis was adhered to in spite of the growing recognition of its difficulties. One of the major modifications which Newcomb himself advocated for Laplace's theory consisted in picturing the formation of planets as having started with the one closest to the sun. Also, according to Newcomb, a broad and flat disk had first to form in the equatorial plane of the rotating nebula before it could divide into separate rings. Again, Newcomb emphasised that vaporous rings could not have enough cohesion to be thrown off as single units from the equatorial belt of a rotating nebular globe. Furthermore, he found no explanation on the basis of Laplace's theory for the shrinkage of the nebular atmosphere to half of its radius after the separation of a ring, if the 'double progression' of planetary distances was to be accounted for. One can today feel only frustration on finding Newcomb praise Laplace for presenting the sun as much older than the planets. Otherwise, Newcomb remarked, if the sun had contracted from a nebulous mass extending even to the orbit of Mercury, the solar shape would have become lens-like and not spherical. Most importantly, in such a case 'it would have been impossible to account for the slow rotation of the sun upon his axis'.[87] Was it not precisely on that slow rotation, or small angular momentum, that the triumphant march of Laplace's theory was to come to a grinding halt, and was not Newcomb himself pointing out that Kant was wholly wrong in trying to derive the rotation of the system from gravitational attraction alone?

In the eyes of Newcomb the greatest difficulty for the nebular hypothesis was the alleged coalescence of the material of a ring into a single planet. He admitted that Laplace's theory was 'not mathematical enough to be conclusive'.[88] Double stars revolving in highly eccentric orbits and ring-shaped swarms of small particles, rather than the planets and their highly specific system, represented for him the logical outcome of the nebular hypothesis. Still he clung to it as the only theory 'indicated by the general tendencies of the laws of nature' and which 'has not been proved inconsistent with any fact'.[89] It was with disdain that he brushed aside those who wanted to see it proved rigorously, as if one could ever observe the shrinking of the sun or the condensation of a nebula into stars and planetary systems. Actually, a rigorous test of the theory was incomparably closer than a nebula and could have been carried out with only pencil and pad in hand. The test was rooted in both a law and a fact of nature, namely, the conservation of angular momentum and its actual distribution in the solar system.

Blow from down under

Contrary to the general expectations, the days of fortune for Laplace's theory were now running out fast, although the blows against it resembled shadow boxing. This subconscious reluctance to formulate the fatal blow in specifics should appear tantalising in retrospect at least. In this connection one need only consider the eleven lectures which A. W. Bickerton, professor of chemistry at Canterbury College of New Zealand University in Christchurch, gave between July 1878 and November 1880 before the Canterbury Philosophical Society.[90] Chemistry was only slightly less removed from celestial mechanics than was the Canterbury Philosophical Society from the leading scientific Academies. Also, the nebular hypothesis was anything but a topic closely allied with the question of what Bickerton described as 'the sudden appearance of stars in various regions of the sky'.[91] That he did not call them novae should give some foretaste of that mixture of amateurism and insight which characterised his lectures throughout. Equally odd should appear the fact that the partial collision between two dark bodies, which he advocated as a solution, became in his hands an all-purpose mechanism to explain the major phenomena of the heavens from nebulae through variable stars to planets. Concerning variable stars he believed that in the partial collision between two dark bodies (stars) 'a cylindrical or curved slice has been cut out of each' and as a result 'we may have the molten interior of the body exposed to view'.[92] Due to the rotation ensuing from the shearing impact, this flaming interior appeared periodically to an onlooker on earth. If Bickerton's second lecture,[93] in which he promised a new explanation of the solar system, had any merit, it consisted in his relentless references to a long-forgotten theme, namely, that the origin of planetary systems should be sought in collision processes. Little did Bickerton guess that the theme of collisions would see before long a powerful revival in planetary theories.

Bickerton's contribution to that revival seems to be minimal. Undoubtedly, there was something original in the diagrams [*Illustrations XXIII* and *XXIV*] attached to his fourth lecture.[94] There he gave a schematic illustration of the ten steps through which the partial impact could lead to the formation of a ring-type nebula, or to a planetary ring. The clue to the process consisted in the formation of a third body from those parts of the two colliding dark stars that were torn off in their shearing collision. It could be made plausible that the third body not only received rotation through the couple of the shearing forces, but that it became also elongated due to the gravitational pull of the two receding dark stars. It was that elongated product which, because of its rotation, developed into a spiral. Cooling in turn made this spiral contract so that its arms ultimately formed a ring around the central nucleus.

Bickerton's faith in, if not infatuation with, the theory of '*Partial* Impact' (the italics were his) could hardly have been greater. In his fifth lecture, entitled 'On the Genesis of Worlds and Systems', he spoke of it as the 'perfect Aladdin's lamp' and promised to unfold its magic possibilities.[95] He did not entirely prove disappointing. Soon he was talking of Laplace's theory, advising his listeners that already many objections to it were hinted at in his previous lectures. Now he was disclosing one more: 'According to Laplace the surface of the nebulous sun should go on getting faster and faster as ring after ring was thrown off; but, as a matter of fact, the sun's energy of rotation is only one fifty-thousandth part of that necessary to throw off a ring, and those among us who believe in the conservation of energy ask, where is that energy gone?'[96] What Bickerton said implied a major thrust in the right direction, but he failed to advance to incisive details. At the least, he should have said something more about that factor of 'one fifty-thousandth'. For the rest of his lecture, he made only a few generic references to the need of satisfying the principle of the conservation of energy and by implication that of angular momentum. Bickerton's interests were in a wholly different direction. It concerned the cyclic rebuilding of worlds through endless collisions between dark stars which he pictured as the necessary formations from the cosmic ashes.

The intimation of a barrier

Captivated by the idea of eternal returns, Bickerton made only a passing reference in his eighth lecture[97] given on August 8, 1880, to H. Faye's recent critique of Laplace's theory. The reference proved the alertness of somewhat eccentric minds to anything that might support their pet theories. Bickerton possibly received in the last mail before his lecture the copies of *Comptes rendus* containing the text of Faye's papers read on March 15 and 22 before the Académie des Sciences.[98] Engrossed with his own theory he had no eyes for spotting the new avenue which Faye's papers opened for a better understanding of the question of the origin of the system of planets. This is not to suggest that Faye himself had exploited in full the possibilities contained in his formula for the evaluation of angular momentum of the solar nebula which he assumed to be equal in mass to that of the sun. His interest centered on the great disproportion between the centrifugal force and the gravitational force in the rotating solar nebula. The disproportion, which in the case of the sun is 1 over 28,000, had, according to Faye's estimate, never been appreciably smaller while the solar nebula kept contracting. This meant that 'a sun thus constituted and corresponding more to our modern ideas [about it] than to those of Laplace, could have never left behind the smallest portion of its mass by contracting in such a manner as to arrive to its present condition'.[99]

At the start of his second paper Faye spoke of the 'mechanical [kinetic] energy' of the rotating primitive nebula which resided 'entirely in its rotation'.[100] After the contraction of the nebula had run its course, the 'former energy remains in its integrity, but this time one finds it almost entirely in the circulations'.[101] However, Faye left undeveloped this cryptic phrase which might have led him to recognise that by far the largest portion of the total angular momentum of the solar system was embodied in the orbiting and rotations of the planets and their satellites. For all his insistence on the grave difficulties of Laplace's theory, Faye was far from ready to propose a radically different substitute for it. The main point made in his second paper was in fact a contribution to a corrected form of Laplace's theory. The retrograde rotations of Neptune and Uranus[102] now seemed to be a consequence of Faye's foregoing formula. When used to express the tangential velocity in the rotating nebula, the formula indicated a maximum velocity at a point lying somewhat beyond midway toward the periphery. Faye did not hesitate to conclude: 'Thus the nebular mass is divided during the entire process of its contraction into two very different regions: 1. the outer part, where the rings in giving rise to planets impress on them a retrograde rotation, as shown by Uranus and Neptune; 2. the interior part, where all the planets have a forward rotation as in the case of Saturn and Jupiter'.[103] As subsequent discoveries were to show, Saturn had one satellite and Jupiter two with retrograde motion.

The witness of the moon and planets

The solar system contained uncanny evidences against the nebular hypothesis even in the earth-moon region. The details of this were unfolded with impressive exactness by George H. Darwin, but equally astonishing should appear his failure to see the real import of his investigations. His appointment in 1883 as Plumian professor of astronomy at Cambridge came largely as a recognition of his monumental study of the changes in the motion of the earth and the moon due to tidal friction.[104] In 1879 Darwin reached the conclusion that the moon might have been steadily receding from the earth for the past 54 million years and that the moon's original formation had taken place not at its present distance, as demanded by Laplace's theory, but at about 5,000 miles, or 1/50th of that distance. This certainly meant a sharp conflict with Laplace's theory, but Darwin developed no doubts about its basic correctness. On the contrary, he believed that his analysis held the key to the explanation of the case of Phobos and that it also permitted considering the ring of Saturn as a 'satellite in the course of formation'.[105] At the same time, Darwin could not be more aware of the fact that his calculations rested 'on the principle of the conservation of the moment of momentum [angular momentum] . . . in tracing the parallel changes in the moon and the earth'.[106] The case could not be different for a similarly rigorous treatment of the development of the orbits of the planets and of their satellites. It appeared immediately to Darwin that his theory implied that 'before the birth of the satellites the planets occupied very much larger volumes, and possessed much more moment of momentum than they do now'.[107]

Darwin was, therefore, led by the inner logic of his investigations to the calculation of the orbital and rotational angular momenta of each member of the planetary system. He made public his results, possibly the most overdue feat in the history of science, in 1881, in his memoir 'On the Tidal Friction of a Planet attended by Several Satellites, and on the Evolution of the Solar System'.[108] The following table[109] contains the values of orbital momenta in units for which Darwin chose the earth-moon mass, the earth-sun distance, and the mean solar day:

Mercury	.00079
Venus	.01309
Earth	.01720
Mars	.00253
Jupiter	13.469
Saturn	5.456
Uranus	1.323
Neptune	1.806
Total	22.088

Compared with this total, the sun's angular momentum due to its rotation was a very small amount, lying between 0.444 and 0.679, corresponding to about 1/50th of the total orbital angular momentum of the planets. Darwin also showed that the rotational angular momenta of the planets and their satellites were negligible quantities with the exception of the earth, where the orbital momentum exceeded the rotational only by a factor of five. The extreme smallness of the rotational angular momenta of the planets and of their satellites meant the death-knell for the application of the tidal theory as a basic mechanism in the formation of the planets. Or as Darwin put it: 'If therefore the orbits of the planets round the sun have been considerably enlarged during the evolution of the system, by the friction of the tides raised in the planets by the sun, the primitive rotational momentum of the planetary bodies must have been thousands of times greater than at present . . . But it does not seem probable that the planetary masses ever possessed such an enormous amount of rotational momentum, and therefore it is not possible that tidal friction has affected the dimensions of the planetary orbits considerably'.[110]

At this point the consideration developed by Babinet might have become a decisive help for Darwin to take a fresh look at the lopsided distribution of the angular momentum between the planets and the sun. Unfortunately Babinet's name was not among the references in Darwin's memoir. It came to a conclusion by a detailed list of the modifications which the influence of tidal friction entailed for the nebular hypothesis. Although some of the modifications were in Darwin's words 'of considerable importance', they afforded 'no grounds for the rejection of the nebular hypothesis'.[111] This might have been true with respect to the tidal forces, but hardly of the principle of the conservation of angular momentum which governed not only their action but also the principal mechanism in Laplace's theory.

A sober appraisal and continued oversight

The evidences disclosed by Faye's and Darwin's investigations were not in fact needed to prompt one to take a dim view of Laplace's theory. Its old difficulties, and especially the one originally intimated by Babinet, were enough for that purpose, provided they were matched with a healthy measure of scepticism toward scientific idols. The case was well illustrated by J. B. Stallo's dismissal in 1882 of the nebular hypothesis as practically untenable. He did this in his work, *The Concepts and Theories of Modern Physics*,[112] which represented the first broad scrutiny of the various assumptions and theories that dominated late-19th-century physics. Among those theories was the nebular hypothesis. Stallo's discussion of the question was not only distinguished by his awareness of some basic differences between the ideas of Kant and Laplace. He also realised the real bearing of Babinet's remarks to which his attention was called by Budde's essay. According to Stallo the difficulty posed by the angular momentum was a 'formidable' one.[113] After quoting several paragraphs from Babinet's paper Stallo concluded: 'The discrepancies here brought to light between the actual orbital periods of the planets and the corresponding periods found by calculation in accordance with the postulates of the nebular hypothesis, are so enormous that there appears to be no possibility of accounting for them by the assumption of a progressive contraction of the orbits of the several planets since their projection, and the consequent quickening of their orbital motions.'[114]

It was in the same vein that Stallo spoke of the 'radical inconsistency'[115] of the short orbiting period of the inner satellite of Mars with the nebular hypothesis. It could not be saved from the teeth of angular momentum, so Stallo reasoned, either by a recourse to tidal friction or to a retardation in the 'aethereal medium'.[116] As to the latter, he not only found its effectiveness for the purpose insufficient, but also expressed grave doubts as to its existence. With respect to the tidal friction, it could at best make the orbiting of the inner satellite of Mars equal to the period of rotation of the planet itself. Stallo himself felt that the heavy accumulation of the difficulties of the nebular hypothesis was now beginning to be 'extensively realised'.[117]

He patently overstated the case. The theory of 'meteoric agglomeration,' in which Stallo saw a reliable substitute to Laplace's theory,[118] neither had following nor could it cope with the question of angular momentum. This latter point had completely escaped Stallo who, in close dependence on Budde's essay, reported at length and without a single word of criticism the reasoning of Laplace about the emergence of a rotational movement in a principal plane from the chaotic motion of many small particles.[119] Stallo set, as Budde did, great store by the point that a chaotic agglomeration of momenta could not be perfectly symmetrical, for in that case no rotation could develop in the system as a whole. With Budde, Stallo failed to realize that the laws of probability were equally against assuming that the agglomeration was so asymmetrical as to contain in germ (in the projection of all the momenta to a principal plane) the cause and measure of the uniform rotation and orbiting of much of the solar system. For all of Stallo's words, the barrier of angular momentum, to say nothing of its peculiar distribution, remained as intact as ever.

As to the meteoric hypothesis, it hardly gained momentum during the next six years. In 1888, in the third edition of his work, Stallo still spoke of the nebular hypothesis as one that 'has attained very general acceptance'.[120] Such was indeed the case. Those who in the 1880's kept giving their vote to Laplace's theory by trying to remedy its defects were at one with Stallo in the latter's strange oversight of a most important paper published in 1884, which will be discussed shortly. They were mostly French scientists who ignored Stallo's strictures of Laplace's theory, although the French translation of Stallo's work was reprinted four times between 1884 and 1905. They also failed to perceive the true implications of Darwin's work. The first to be mentioned is Faye who never realised that it was the question of angular momentum that vitiated both Laplace's theory and his own efforts to construct a viable modification of it. In a lecture given at the Sorbonne on March 15, 1884, on the formation of the universe and of the solar system,[121] Faye singled out the retrograde motion of the satellites of Uranus and Neptune as the factor which 'reduced to nothing the celebrated hypothesis of Laplace'.[122] Such was a strange evaluation of the general belief in the truth of Laplace's theory but in line with Faye's pronounced inability to write scientific history, remote or recent. More forgivable was Faye's contention, generally shared during the closing decades of the 19th century, that nebulae and especially ring-like nebulae were closely analogous to the nebula which coalesced into the solar system.

To cope with the retrograde motion of the satellites of Uranus and Neptune, Faye called attention to the fact that in a rotating, homogeneous, spherical nebula the tangential velocities increased linearly with distance from the centre. If concentric rings were slowly developing within such nebula, the outer part of each ring would have a greater tangential velocity than the inner part, and consequently the globe developed from the ring would rotate in the forward direction. If, however, a considerable central condensation had already taken place in the nebula, the tangential velocity in the rings would decrease with the square of distance from the centre. As a result, planets formed from such rings would have retrograde rotation. That the retrograde rotation of the satellites of Uranus and Neptune (which in turn were indices of the sense of rotation of the two planets) could be more complicated did not occur to Faye, who gave a detailed diagram of his 'explanation' [*Illustration XXV*], which forcefully suggested that the case was clarity itself. He had no second thoughts in admitting that a consequence of his theory was that the earth (and by implication all planets with forward rotation) was 'much older than the sun'.[123] Otherwise, Faye added, 'the stars would rise in the west and would set in the east; the moon itself, like the satellites of Uranus and Neptune, would have a retrograde motion'.[124] A splashy argument but only to convince the unwary.

Faye's lecture came to a close with an homage to Descartes whom Faye apotheosised as 'le sieur *Cogito*', or 'Mister Thought'.[125] Such a note was sadly out of place. It escaped Faye that Descartes had hardly shown himself the epitome of rigorous thinking in his theory of planetary origins. In Faye's account of Descartes' theory there was not even a hint that planets were former stars, that sunspots played a crucial role, and that comets too were former stars.[126] It was merely rhetoric, not rigorous weighing of the pros and cons, to claim in 1884, as Faye did, that Laplace's theory was one 'of the highest manifestations of the human mind'.[127] Faye displayed no respect for the facts of scientific history as he insisted, to exculpate Laplace, that observational work on the motion of satellites of Uranus had not yet been completed in Laplace's time.[128] Finally, Faye's own previous work should have suggested to him that it was the abdication of rigour to suggest that all other phases of Laplace's theory were unobjectionable.

How oblivious one could grow to obvious difficulties because of undue attachment to one's favourite theory can be seen in the chapter which followed the text of Faye's lecture. There he found support for his views in the conflict of current estimates of the sun's age and of paleontological time that elapsed since the solidification of the earth's crust. The latter was listed by Faye as about 100 million years (75 millions for the Primary, 19 millions for the Secondary, 6 millions for the Tertiary), while the former corresponded to about 20 million years according to calculations in vogue since Helmholtz first submitted them in 1854 on the basis that the sun obtained its heat from contracting from the orbit of Neptune.[129] The earth, therefore, could be rightfully considered much older than the sun if Faye's theory was true, and Faye hastened to strengthen its truth by dressing it in a mathematical formula.[130] It consisted of two terms of which the first decreased and the second increased in value as the contraction of the solar nebula went on. In the first part of the contraction the first term was dominant, permitting the evolution in the still largely homogeneous solar nebula of planets with forward rotation. In the second part of the contraction the two terms played equal roles with the resulting formation of the outermost planets (satellites) with retrograde motion. The sun itself was the product of the third phase characterised with the prominence of the second term. In a final note Faye attributed the formation of comets to local disturbances, thinking that the last objection to the theory of Laplace, who considered comets intruders, had thereby been eliminated.[131]

A barrier to stay

As if by fate and irony, the most crucial objection to Laplace's theory came to be unfolded about the time when Faye's book began circulating. To compound the irony, the objection was equally fatal to Laplace's theory and to its modification by Faye, whose idea implied that not 1/700th but a considerable part of the material of the entire solar system lay outside the orbit of an earth already fully formed. Faye paid no attention the fact that the subsequent shifting of that material within the earth's orbit demanded a thorough redistribution of the angular momentum, the very topic of a short paper by Maurice Fouché, a mathematician of twenty-nine. Partly because Fouché had not published anything until then, his paper[132] was presented on November 24, 1884, to the Académie des Sciences by Charles J. E. Wolf, astronomer of the Paris Observatory and since 1883 member of the Académie. Fouché's originality was in part due to his unawareness of Darwin's and Babinet's work. He gave no details about the considerations that let him obtain in terms of the earth's mean distance from the sun, of the sun's mass, and of the mean sidereal day the following figures for the total angular momentum of the sun and of the planets respectively: $2\pi \times 0.000,000,353,8$ and $2\pi \times 0.000,009,611,6$.[133] The ratio, 1 to 28.5, of the two figures differed considerably from the ratio 1 to 50 as given by Darwin but still sufficiently high to prove an all-important point. Fouché, however, left it to his readers to realise the obvious, namely, that whereas the sun's mass constituted more than 99 percent of the total mass of the solar system, almost the very opposite was true of the total angular momentum. More than 97 percent of it resided in the planets whose mass was puny indeed when compared with that of the sun.

The brevity and directness of Fouché's paper presented a great advantage over the technically forbidding pages of Darwin's memoir. Most importantly, Fouché stated explicitly at the very outset the real thrust of his paper. It represented in his words 'a very serious difficulty against Laplace's theory'.[134] The difficulty had also a direct bearing on the various modifications of that theory. In fact, the very first sentence of Fouché's paper referred to Laplace, Faye, and Wolf. The latter will be discussed shortly. According to Fouché's disclosure, the importance of a quantitative evaluation of the total angular momentum in the solar system and its distribution between the sun and the planets struck him during a conversation with Camille Flammarion, the famed populariser of astronomy during the latter half of the 19th century. Fouché then looked up the literature and concluded wrongly enough that the point had not been raised beforehand. It had at best been vaguely intimated, so Fouché remarked, in Trowbridge's insistence in 1864 on the necessity of an early concentration of much of the matter of the solar nebula in the sun. That the rather obscure work of Trowbridge came within Fouché's ken, but not Darwin's memoir and Babinet's paper, should appear astonishing. Even more so should appear the fact that what Fouché did, could and should have been done a number of times during the previous hundred years, first by Laplace himself and afterwards by any of the prominent scientists who had blindly espoused Laplace's theory, a glittering superstructure resting on deceptively fragile foundations.

Fouché did his best to drive home his historic lesson by unfolding various aspects of his basic calculation. First, he took the case of a homogeneous, ellipsoidal nebula equal in mass to that of the sun, but extending to the orbit of Neptune and rotating with the planet's angular velocity. The angular momentum of such a nebula would have been $2\pi \times 0.00604$, or more than 600 times greater than the combined value of the angular momentum of the planets and of the sun. 'One may therefore see', he added, 'what an enormous condensation one should accept [in the nebula] to reduce its moment of inertia [angular momentum] to one six-hundredth of what would have been its value in the case of homogeneity'.[135] Then Fouché took up the case of an ellipsoidal nebula with its mass divided into two parts: the central part carrying a mass endowed with the present value of the sun's angular momentum and an outer part, which he called the primitive solar atmosphere, having an angular momentum equal to that of all planets and satellites. From this it was possible to infer that the total mass of that atmosphere exceeded but slightly the total mass of planets and satellites. Therefore, he wrote, 'the whole atmosphere of the nebular mass should have successively turned into planets, which is very difficult to admit'.[136] It was no help to assume still undetected quantities of matter in the solar system. Neither this device nor the assigning of very high density to the outer regions of that atmosphere could be invoked, Fouché noted, 'without increasing at the same time the moment of inertia of the total mass'.[137] This was a sagacious remark. The angular momentum was, indeed, for Laplace's theory and for all its modifications a barrier around which no detours could be charted.

Insensitivity to the barrier

The immediate reaction to Fouché's paper was insensitivity itself. Wolf, who in 1884 began to publish a series of articles on the cosmogonies and planetary theories of Kant, Laplace, and Faye in the *Bulletin astronomique*, summed up in a very brief footnote the contents of Fouché's paper. In the summary there was no explicit reference to the angular momentum, only to the 'conservation of the quantities of movement'.[138] Equally perfunctory and uninformative was the report on Fouché's paper in the same, December 1884, issue of the *Bulletin*. In a five-line summary of it the question of angular momentum was completely ignored.[139] Wolf himself had ample time to do better justice to Fouché's paper when in late 1885 he prepared his articles for publication as a book which also contained Wolf's critique of Darwin's theory of tides and a translation by Wolf into French of Kant's *Allgemeine Naturgeschichte*. Wolf's previous footnote on Fouché was carried into the book without any change or amplification.[140] Ironically, the footnote came shortly after Wolf listed the objections that had been made against Laplace's theory. According to these objections no rings could form in the process as specified by Laplace; the many parts of a ring could not coalesce into one planet; the planets, if formed from the rings, should have had a retrograde rotation; the first satellite of Mars and the inner ring of Saturn were closer to those planets than permitted by Laplace's theory, which did not permit either the retrograde rotation of the satellites of Uranus and Neptune.[141] It wholly escaped Wolf that all these objections were dwarfed by comparison with the importance of the question of the distribution of angular momentum in the solar system.

As to Faye, he published in 1885 his work in a new edition which contained fifty more pages than the first, with the chapter on the evolution of the solar system completely rewritten.[142] The changes concerned only style not substance. The problem of angular momentum, together with the papers of Babinet, Darwin, and Fouché, remained non-existent for Faye. He had so little doubt about the correctness of his theory anchored in his two-term formula that he now listed in great detail and in a schematic form the progress of the formation of the solar system. No wonder that he presented further details about the geological record to the point of asserting: 'Lest we should close our eyes and reject the inconveniencing data in order to reduce the duration of the great phenomena of natural history on our globe, one has to conclude that our globe is older than the sun; in other words, the first rays of the sun, as it was being born, had to illuminate an earth already consolidated, already shaped by waters under the sole influence of the central heat'.[143] Rarely in the history of science were the eyes of a scientist more disastrously closed and rarely were 'inconveniencing data' more effectively ignored.[144]

After Faye and Wolf a few words may be in order about Flammarion, whose conversation with Fouché was recorded by the latter as the spark that led him to an evaluation of the distribution of angular momentum in the solar system and of its bearing on the nebular hypothesis. Flammarion, perhaps the most popular populariser of astronomical lore of all times, had hardly a receptive frame of mind to ponder the implications of Fouché's results. They did not enter into the later editions of his *Astronomie populaire*, first published in 1877.[145] Clearly, Flammarion did not wish to part with such a deceptively facile 'explanation' of the formation of planetary systems as was Laplace's theory. He needed it as a scientific justification of his belief that the universe was populated throughout. Although at first Flammarion earned his living as a practising astronomer, he soon began to reap huge royalties from books of which probably the most representative was *La pluralité des mondes habités*.[146] Its numerous and rapid reprints made the youthful author quickly famous in an age which eagerly let itself be led by the nose as long as the nose-ring had the glitter of scientific words. The extremely cold temperatures on the distant planets were as readily talked away by Flammarion as were the scorching clouds enveloping Mercury and Venus. His quick success emboldened him to announce the impending publication of his *Les mondes imaginaires et les mondes réels*, which promised in its subtitle a trip across the skies with a critique of all 'human, scientific and literary' theories about the inhabitants of other planets.[147] The haughty brushing aside by Flammarion of the flights of fancy of Cyrano de Bergérac, Fontenelle, Kant, and of others was a diversionary tactic on his part. He wanted to create the impression that his ideas on the denizens of other planets and planetary systems had been tested in the cauldron of criticism. It was also Flammarion who parleyed into a huge book the sundry speculations on Martians who caught the imagination of the public following V. Schiaparelli's much publicised reports of canals on Mars.[148] No wonder that in the end Flammarion became a prolific populariser of parapsychic phenomena and theories in which he saw as few problems as he did in the world of planets.

Psychology and planetary theories were not wholly disparate aspects even in the case of such a reputable scientist as Lord Kelvin. He took up the question of planetary evolution in a lecture 'On the Origin of the Sun's Heat'[149] given at the Royal Institution on January 21, 1887. Once the explanation of the sun's heat by gravitational contraction and in terms of the kinetic theory of gases was done, Kelvin could turn to what he called 'the most interesting part of the subject – the early history of the Sun'.[150] He was to answer the question, 'What was the condition of the sun's matter before it came together and became hot?'[151] The question could be answered, according to Kelvin, by various assumptions. One was the collision of two solid masses, each with the earth's density but with half of the sun's diameter, falling towards one another from a distance twice the separation of the earth from the sun. The other was the collision in the centre of 29 million moons, originally placed on the surface of a spherical space with a radius 100 times greater than the earth–sun distance. In both cases the collision would have produced a very hot gaseous nebula expanding into a sphere with a radius forty times the radius of the earth's orbit. Thus one would have the primitive nebula with which Laplace started.[152]

Kelvin also gave credit to Laplace[153] for assuming the nebula to have a definite quantity of angular rotation. He took special pains to emphasise the importance of considering the question of angular momentum, and in an aside he explained the old expression for it, or 'moment of momentum'. He even calculated that in order to produce the present rotation of the sun, the two colliding masses should have had a velocity of two metres per second in a direction perpendicular to the line joining their centres.[154] As to the model of 29 million moons, he merely stated that they should have been endowed with a small rotation, obviously in the same direction. This way the angular momentum of the solar system could be readily accounted for. In all this, Kelvin ignored the Frenchmen, Fouché and Babinet, as well as Darwin, his countryman. Kelvin might have, of course, realised the importance of the point independently. At any rate, he was oddly original as he remarked: 'The moment of momentum of the whole solar system is about eighteen times that of the sun's rotation; seventeen-eighteenths being Jupiter's and one-eighteenth the sun's, the other bodies being not worth taking into account in the reckoning of moment of momentum'.[155]

Such was a distinctly careless procedure, as the contributions of Saturn, Uranus, and Neptune could not be ignored, but it matched the carefree veneration which Kelvin accorded to the nebular theory and to its Laplacian form. In Kelvin's Victorian rhetoric the nebular theory of the solar system, 'founded on the natural history of the stellar universe, as observed by the elder Herschel, and completed in detail by the profound dynamical judgment and imaginative genius of Laplace, seems converted by thermodynamics into a necessary truth'.[156] Rarely was the actual course of development so badly misread. One wonders what was the momentary connection with science of a scientist who wrote without the slightest evidence of second thoughts: 'There may in reality be nothing more of mystery or of difficulty in the automatic progress of the solar system from cold matter diffused through space, to its present manifest order and beauty, lighted and warmed by its brilliant sun, than there is in the winding up of a clock, and letting it go till it stops'. To this he added in an equally stupefying note that 'a watch spring is much farther beyond our understanding than is a gaseous nebula'.[157]

Nebulae for foil

Kelvin's address came to a close with a short discussion of meteors and meteorites. The last phase of respectability for Laplace's theory owed much to a sudden rise of interest in them. The chief worker in the field, J. Norman Lockyer, created quite a sensation when in late 1887 he claimed that the famous 'nebulium' line in the spectrum of Orion nebula 'completely' coincided with a line which he observed in the spectrum of magnesium coming from vaporised aerolitic meteorites.[158] He saw in this a confirmation of the view that swarms of meteorites constituted the prime matter out of which nebulae, primitive solar atmospheres, and planetary systems developed.[159] The one who gave closer and exact scrutiny to the merits of the meteoritic hypothesis was Darwin. He did this in a memoir presented on July 12, 1888, to the Royal Society.[160] Unfortunately, the memoir was restricted to the investigation of the question whether a swarm of meteorites, when treated as a gas whose molecules are in constant elastic collisions, could be considered essentially equivalent to a fluid sphere in which hydrodynamical pressure is the chief physical factor.

While the elimination by Darwin of the question of the origin of rotation in the meteoritic nebulae was in itself justified,[161] it was also symptomatic of the insensitivity which prevailed in the closing decades of the 19th century to distribution of angular momentum in the solar system. The measure of that insensitivity betrayed itself both in short statements and in lengthy appraisals. As to the former, one may recall the words of Oliver Lodge carried to countless readers of his *Pioneers of Science*, first published in 1893. According to him, the nebular hypothesis which 'since the time of Laplace . . . has had ups and downs in credence, sometimes being largely believed in, sometimes being almost ignored, . . . holds the field with perhaps greater probability of ultimate triumph than has ever before seemed to belong to it'. This hardly meritorious evaluation of the true merits of the nebular hypothesis was then followed by the assertion that Kant was its original formulator and that it should be known 'rather by his name than by that of Laplace'.[162]

With respect to lengthy appraisals the most noteworthy case was provided in 1896 in the new, considerably revised edition by Spencer of his 1858 essay.[163] He still showed awareness only of Babinet's paper but not of those by Fouché and Darwin. Spencer's solution to the objection contained in Babinet's calculation consisted in the claim[164] that the original solar nebula had to have a much greater angular velocity on its periphery than in its central region, because the molecules forming the periphery were attracted from much farther regions and thus acquired greater velocity. The turning of that linear velocity into a rotational one was once more assigned by Spencer to that old and dubious device, the necessarily slight asymmetry of the original mass of gas contracting into the primitive solar atmosphere. To make more palatable the idea of a spherical mass of gas, in which the angular velocity *increased* with distance from the centre, Spencer referred to spiral nebulae. He merely proved that somersaults in scientific reasoning easily go hand in hand with turning inside out what is obvious in scientific evidence.

References to various nebulae as evidences of the truth of nebular hypothesis formed the concluding chapter of the book which R. du Ligondès published in 1897 on the origin of the solar system.[165] Its many diagrams and mathematical formulae could give at first sight the impression that one had at hand a rigorously argued analysis of the question, but the actual situation could not have been less satisfactory. The formulae and diagrams largely concerned the principal specification which du Ligondès introduced into the nebular hypothesis. According to him, the gravitational condensation of the primitive nebula resulted in a central mass surrounded by a very wide and flat gaseous disk in which circular lines of minimum density had to develop causing its rupture into concentric rings [*Illustration XXVI*]. That these rings 'rapidly condensed' into planets[166] was only one of the numerous cases in which a crucial phase of the nebular hypothesis was taken by du Ligondès as self-explanatory. Needless to say, such was also the case with the problem of the origin of uniform rotation in the primitive nebula. Du Ligondès simply paraphrased what Laplace kept proclaiming authoritatively in the pages of his *Exposition*, namely, that the slight dissymmetry in the motion of the particles of the original nebula must have yielded a net maximum amount of rotational motion along one specific plane on the basis of the principle of the conservation of areas, that is, of angular momentum.[167]

The year, 1897, was made memorable in the history of theories on the origin of planetary systems through two lectures. One was part of a lecture series on the role of tides in the solar system given by Darwin, the foremost authority in the field, at the Lowell Institute in Boston. There he claimed that there was 'good reason for believing that the Nebular Hypothesis presents a true statement in outline of the origin of the solar system, and of the planetary subsystems'.[168] For principal proof he referred to photographs taken recently of nebulae, 'in which we can almost see the process in action'.[169] Far clearer than those photographs should have appeared to Darwin his own, now almost two-decade-old calculations on the distribution of angular momentum in the solar system. It should, indeed, give a sobering second thought about the true nature of the development of science that Laplace's simplistic handling of the question had gone unchallenged for sixty years, and that for another four decades it could largely be ignored, even by such a master of celestial dynamics as Darwin, that any meaningful advance in the explanation of the origin of the solar system was blocked by a formidable barrier. For, even if one simply assumed an originally rotating spherical nebula, the ultimate distribution within it of the quantity of rotation, or angular momentum, was still to be explained. By strange, almost symbolic coincidence, in the same year of 1897, which marked practically the centenary of the original formulation of the nebular hypothesis, a cautious objection was voiced against it in a lecture given by someone professionally strange to the field. Unlike previous objections against the nebular hypothesis, this grew ultimately into a major disclaimer marking, as will be seen in the next chapter, a turning point in the history of speculations on planetary origins.

NOTES TO CHAPTER SIX

1. 'Note sur un point de la Cosmogonie de Laplace', *Comptes rendus* 52 (1861): 481–84.

2. 4th ed. (1813), p. 431.

3. 1st ed. (1796), vol. 2, p. 301; 2d ed. (1799), p. 345; 3d ed. (1808), vol. 2, p. 386. Complete references to these editions were given in the preceding chapter.

4. 'Note sur un point de la Cosmogonie de Laplace', p. 483. Few things can give a layman better insight into the measure of neglect shown by prominent scientists to test quantitatively the nebular hypothesis than the simplicity of the equation for the conservation of angular momentum in a rotating nebula of uniform density, an equation by which the hypothesis can be tested and was tested by Babinet. In the equation $\omega'd'r'^5 = \omega dr^5$ the primed letters refer to the angular velocity ω', radius r' and density d' of the shrinking nebula and the unprimed letters express those quantities for the actual sun. Since the density d' for any radius is equal to dr^3/r'^3, the foregoing equation can be reduced to $\omega'r'^2 = \omega r^2$. Furthermore, since ω, r and r' (which can be any actual planetary distance) are known, the calculation of ω' for any planetary distance becomes an exercise in elementary arithmetic.

5. *Ibid.*

6. *Ibid.*, p. 484.

7. *First Principles* (London: Williams & Norgate, 1862), p. 365. It was the first volume and general foundation of his *System of Synthetic Philosophy*.

8. *Ibid.*

9. *Ibid.*, p. 366.

10. *Ibid.*

11. *Ibid.*, p. 449.

12. I have not been able to verify this statement of Brewster in his papers published within the year separating the publication of Babinet's paper and the *First Principles* of Spencer.

13. *Ibid.*

14. *Ibid.* This footnote was reprinted without change in the 3d (1870, p. 491) and 4th editions (1882, p. 491).

15. *American Journal of Science and Arts* 30 (1860): 161–81

16. *American Journal of Science and Arts* 38 (1864): 1–18.

17. *Ibid.*, p. 5.

18. *Ibid.*

19. *Ibid.*

20. *American Journal of Science and Arts* 38 (1864): 344–60 and 39 (1865): 25–43.

21. *Ibid.*, p. 346.

22. *Ibid.*, p. 349.

23. *Ibid.*

24. *Ibid.*, p. 350.

25. *Ibid.*, p. 351.

26. *Ibid.*, p. 358.

27. 'The Density, Rotation, and Relative Age of the Planets', *American Journal of Science and Arts* 37 (1864): 53.

28. *Ibid.*, p. 48.

29. *Ibid.*

30. 'Introduction to Mathematical Principles of the Nebular Theory or Planetology', *American Journal of Science and Arts* 39 (1865): 46–58, 134–50, 276–86.

31. *Saturn and its System* (London: Longman, Green, Longman, Roberts, & Green, 1865).

32. *Ibid.*, pp. 114–16.

33. *Ibid.*, p. 201.

34. *Ibid.*, p. 202.

35. *Ibid.*, p. 203.

36. *Ibid.*, p. 203 note.

37. *Ibid.*, Chapter VII 'The Habitability of Saturn', pp. 156–85.

38. *Ibid.*, p. 185.

39. *Ibid.* Contrary to Proctor's claim the quotation was not from Herschel's *Outlines of Astronomy*, p. 286, but from his *A Treatise on Astronomy* (London: Longman, Bees, Orme, Brown, Green & Longman, 1833), p. 278.

40. *Le Soleil* (Paris: Gauthier-Villars, 1870).

41. *Ibid.*, p. 333.

42. *Ibid.*, p. 340.

43. *Ibid.*, p. 341.

44. *Ibid.*

45. *Photometrische Untersuchungen mit besonderer Rücksicht auf die physische Beschaffenheit der Himmelskörper* (Leipzig: W. Engelmann, 1865), pp. 214–31.

46. Zöllner quoted from the second edition of Laplace's *Exposition* without referring to the considerable changes which the text had undergone in the 4th edition.

47. *Ibid.*, pp. 226–27

48. *Ibid.*, pp. 230 and 225.

49. *Ibid.*, p. 231.

50. References are to the English translation in H. Helmholtz, *Popular Lectures on Scientific Subjects: Second Series*, translated by E. Atkinson (London: Longmans, Green & Co., 1881), pp. 139–95

51. *Ibid.*, pp. 173–74.

52. *Ibid.*, pp. 196–97.

53. *Ueber die Natur der Cometen: Beiträge zur Geschichte und Theorie der Erkenntniss* (Leipzig: Wilhelm Engelmann, 1872).

54. *Ibid.*, pp. 460–63.

55. Examples are *Kant und Newton* by K. Dieterich (Tubingen: H. Laupp, 1876), *Kant oder Laplace* by A. Meydenbauer (Marburg: 1880), and the article 'Kosmogonie' in *Lexikon der Astronomie*, edited by H. Gretschel (Leipzig: Verlag des Bibliographischen Instituts, 1882), pp. 283–91. In this article the nebular hypothesis was entirely credited to Kant with only a short paragraph (p. 289) reserved for Laplace.

56. *Zur Kosmologie der Gegenwart: Bemerkungen zu J. C. F. Zöllner's Buch Ueber die Natur der Kometen* (Bonn: Eduard Weber's Buchhandlung, 1872). In this 70-page-long brochure the nebular hypothesis was dealt with on pp. 21–37.

57. *Ibid.*, p. 23.

58. *Ibid.*, p. 32.

59. *Ibid.*, p. 29.

60. *Ibid.*, p. 28. Robert Mayer sought in this process the source of the origin of the sun's heat in his brochure, *Beiträge zur Dynamik des Himmels in populärer Darstellung* (Heilbronn: Verlag von Johann Ulrich Landherr, 1848; see reprint edition in *Die Mechanik der Wärme in gesammelten Schriften von Robert Mayer*, edited with notes by J. J. Weyrauch [Stuttgart: Verlag der J. G. Cotta'schen Buchhandlung Nachfolger, 1893], pp. 151–223). Although Mayer insisted at the outset that the source of the sun's heat is closely tied to 'the organization of the planetary system', he offered no speculation about its origin and

evolution. The absence of such speculations was also evidenced by the extensive notes of Weyrauch, who referred only to Laplace's *Traité de mécanique céleste*, but not to his *Exposition.* Sporadic references in the literature to Mayer as the founder of the meteoric hypothesis of the origin of planetary system are without foundation.

61. *Ibid.*, p. 30.

62. 'Essai sur la constitution et l'origine du système solaire', *Académie des Sciences et Lettres de Montpellier. Mémoires de la Section des Sciences*, vol. 8 (Montpellier: Boehm et Fils, 1872–75), pp. 235–324.

63. *Ibid.*, p. 235.

64. *Ibid.*, p. 246.

65. *Ibid.*, p. 259.

66. *Ibid.*, p. 267.

67. *Ibid.*, pp. 289–301. Roche's original memoir, 'La figure d'une masse fluide soumise à l'attraction d'un point éloigné', *Académie des Sciences et Lettres de Montpellier. Mémoires de la Section des Sciences*, vol. 1 (Montpellier: Boehm, 1847–50), pp. 243–62 and 333–48, was read on June 18, 1849. On Saturn, see p. 258. Roche discussed his topic without any reference to the question of the origin of the solar system.

68. *Ibid.*, p. 284.

69. *Ibid.*, p. 324.

70. *Ibid.*, p. 237.

71. *Ibid.*, p. 324.

72. 'Cosmical Activity of Light', *Philosophical Magazine* 50 (1875): 252.

73. *Ibid.*, p. 252 note.

74. *Ibid.*

75. ' 'In the Beginning'. – I. Mass and Position', *Phil. Mag.*, 51 (1875): 318.

76. 'On the Nebular Hypothesis. – II. Interaction', *Phil. Mag.*, 1 (1876): 507.

77. 'On the Nebular Hypothesis. – VI. Momentum and *Vis viva*', *Phil. Mag.*, 4 (1877): 295.

78. 'On the Nebular Hypothesis. – VIII. Criteria', *Phil. Mag.*, 5 (1878): 362.

79. 'On the Nebular Hypothesis. – X. Predictions', *Phil. Mag.*, 6 (1878): 452.

80. 'On the Nebular Hypothesis. – IX. Radiation and Rotation', *Phil. Mag.*, 6 (1878): 128–32.

81. 'On the Nebular Hypothesis. – VIII. Criteria', p. 363.

82. 'On the Nebular Hypothesis. – X. Predictions', p. 454.

83. *Observations and Orbits of the Satellites of Mars. With Data for Ephemerides for 1879* (Washington: Government Printing Office, 1878), p. 5.

84. *Ibid.*, p. 6.

85. An oversight all the more curious, because the moving of Phobos across the Martian sky at least twice during one Martian day is a truly spectacular phenomenon.

86. *Popular Astronomy* (New York: Harper & Brothers, 1878), pp. 493–99.

87. *Ibid.*, p. 497. Newcomb's treatment of the nebular hypothesis remained unchanged even in the 4th revised edition (1882), pp. 505–11.

88. *Ibid.*, p. 497 (4th ed., p. 509).

89. *Ibid.*

90. Published in *Transactions and Proceedings of the New Zealand Institute*, vols. XI–XIII, 1878–80.

91. 'On Temporary and Variable Stars', *Transactions* 9 (1878): 118.

92. *Ibid.*, p. 123.

93. 'Partial Impact: A Possible Explanation of the Origin of the Solar System, Comets, and Other Phenomena of the Universe', *Transactions* 11 (1878): 125–32.

94. 'Partial Impact: On the General Problem of Stellar Collision', *Transactions* 12 (1879): 181–86.

95. *Ibid.*, p. 191.

96. *Ibid.*, p. 194.

97. 'The Origin of the Solar System', *Transactions* 13 (1880): 154–59. For reference on Faye, see p. 154.

98. 'Sur l'hypothèse de Laplace', *Comptes rendus* 90 (1880): 566–71 and 'Sur l'origine du système solaire', *Comptes rendus* 90 (1880): 637–43.

99. *Ibid.*, p. 571. Faye derived for the ratio of the centrifugal and gravitational forces the formula $\beta'^2 R'/28,000\beta^2 R$, where β was a function of α and n, a small fraction and a small positive number, respectively. This meant that the ratio could not be at any time much greater than at present.

100. *Ibid.*, p. 637.

101. *Ibid.*

102. Because the orbiting of Triton, the inner satellite of Neptune, is retrograde, the rotation of the planet itself was at that time assumed to be such. The inclination (98 degrees) of the axis of Uranus to its orbital plane makes its rotation retrograde.

103. 'Sur l'origine du système solaire', p. 640.

104. 'On the Precession of a Viscous Spheroid, and on the Remote History of the Earth' (1879), in *Scientific Papers by Sir George Howard Darwin*, vol. 2, *Tidal Friction and Cosmogony* (Cambridge: University Press, 1908), pp. 36–139.

105. *Ibid.*, p. 131.

106. *Ibid.*, p. 130.

107. *Ibid.*, p. 133.

108. *Ibid.*, pp. 406–58.

109. *Ibid.*, p. 438.

110. *Ibid.*, p. 445.

111. *Ibid.*, p. 458.

112. *The Concepts and Theories of Modern Physics*, published as Volume XXXVIII of the International Scientific Series by D. Appleton Co., in New York, in November 1881. References are to the reprint edition in 1882. With the exception of a subclose no changes were made in Chapter XV, 'Cosmological and Cosmogenetic Speculations – The Nebular Hypothesis', in the second, revised edition, published in 1884.

113. *Ibid.*, p. 281.

114. *Ibid.*, p. 283.

115. *Ibid.*, p. 284.

116. *Ibid.*, p. 285.

117. *Ibid.*, p. 286.

118. *Ibid.*

119. *Ibid.*, pp. 288–89.

120. It was edited in 1960 with introduction and notes by P. W. Bridgman (Cambridge, Mass.: The Belknap Press of Harvard University Press); see p. 286.

121. The lecture was published later that year as Chapter XII of Faye's *Sur l'origine du monde: Théories cosmogoniques des anciens et des modernes* (Paris: Gauthier-Villars, 1884), a work which gave platitudinous summaries of the cosmogonies of the ancients (Moses, Plato, Aristotle, Lucretius, Ovid) and of the moderns (Descartes, Newton, Kant, Laplace).

122. *Ibid.*, p. 189.

123. *Ibid.*, p. 192.

124. *Ibid.*

125. *Ibid.*, p. 197.

126. *Ibid.*, p. 186.

127. *Ibid.*, p. 135.

128. *Ibid.*, p. 134.

129. *Ibid.*, p. 200. For details on the late-19th-century conflict between physicists (astronomers) and geologists (evolutionists) concerning the age of the sun and of the crust of the earth, see my *The Relevance of Physics* (Chicago: University of Chicago Press, 1966), pp. 301–305.

130. *Ibid.*, p. 202. The two-term formula was $ar + \dfrac{b}{r^2}$ with a decreasing and b increasing as the contraction of the nebula went on.

131. *Ibid.*, pp. 204–05.

132. 'Sur la condensation de la nébuleuse solaire dans l'hypothèse de Laplace', *Comptes rendus* 99 (1884): 903–06. A short and noncommittal notice of Fouché's paper appeared almost immediately in the December issue of the *Bulletin astronomique* 1 (1884): 615–16.

133. *Ibid.*, pp. 904–05. The units in which Fouché calculated angular momentum were the mean earth–sun distance, the sun's mass, and the mean solar day.

134. *Ibid.*, p. 906.

135. *Ibid.*, p. 905.

136. *Ibid.*, p. 906.

137. *Ibid.*

138. 'Les hypothèses cosmogoniques', *Bulletin astronomique* 1 (1884): 313–19, 431–39, 478–89, 583–96; 2 (1885): 69–78, 222–32, 318–30. For the footnote on Fouché, see pp. 591–92.

139. *Ibid.*, pp. 615–16. The wording of the summary clearly indicated that Wolf was its author.

140. *Les hypothèses cosmogoniques: Examen des théories scientifiques modernes sur l'origine des mondes, suivi de la traduction de la Théorie du Ciel de Kant* (Paris: Gauthier-Villars, 1886).

141. *Ibid.*, p. 35.

142. H. Faye, *Sur l'origine du monde* (2d ed.; Paris: Gauthier-Villars, 1885); on the evolution of the system of planets, see pp. 262–78.

143. *Ibid.*, p. 281.

144. No technical details were added in Faye's summary of his cosmology and planetary theory which he prepared for 'scientific readers' under the title, 'La formation de l'univers et du monde solaire', in *Annuaire pour l'an 1885 publié par le Bureau des Longitudes* (Paris: Gauthier-Villars, 1885), pp. 757–804. See especially pp. 789–800. Eleven years later Faye published a third edition of his *Sur l'origine du monde* (Paris: Gauthier-Villars, 1896) which was practically identical with the second edition.

145. *Astronomie populaire* (Paris: C. Marpon et E. Flammarion). By 1911 more than one hundred thousand copies of it had been sold, not counting translations.

146. *La pluralité des mondes habités* (Paris: Mallet-Bachelier, 1862).

147. First published in 1865 (Paris: Didier) and was already in its 12th printing by 1874.

148. *Le planète Mars et ses conditions d'habitabilité: Synthèse générale de toutes observations* (Paris: Gauthier-Villars et Fils, 1892), a volume of 608 large octavo pages.

149. 'On the Origin of the Sun's Heat', in Sir William Thomson, *Popular Lectures and Addresses* (2d ed.; London: Macmillan & Co., 1891), vol. 1, pp. 376–429.

150. *Ibid.*, p. 410.

151. *Ibid.*, p. 411.

152. *Ibid.*, p. 414.

153. *Ibid.*, p. 421.

154. *Ibid.*, p. 416.

155. *Ibid.*, p. 420.

156. *Ibid.*, pp. 421–22.

157. *Ibid.*, p. 422.

158. 'Researches on Meteorites', *Nature* 37 (1887): 55–61 and 80–87. See especially p. 86.

159. For a fully documented form of the various stages of Lockyer's investigations and speculations on this, see his *The Meteoric Hypothesis: A Statement of the Results of a Spectroscopic Inquiry into the Origin of Cosmical Systems* (London: Macmillan & Co., 1890).

160. 'On the Mechanical Conditions of a Swarm of Meteorites, and on the Theories of Cosmogony', *Proceedings of the Royal Society* (London) 45 (1889): 3–16.

161.. About rotation Darwin said merely the following: 'A meteor swarm is subject to gaseous viscosity, which is greater the more widely diffused is the swarm. In consequence of this a widely extended swarm, if in rotation, will revolve like a rigid body without relative motion (other than agitation) of its parts. Later in the history the viscosity will probably not suffice to secure uniformity of rotation, and the central portion will revolve more rapidly than the outside' (*ibid.*, p. 15).

162. *The Pioneers of Science and the Development of their Scientific Theories* (London: Macmillan, 1893), p. 267. The same passage remained unchanged even in the last, revised edition (1926) reprinted by Dover (New York, n.d.), p. 267.

163. 'The Nebular Hypothesis', in *Essays Scientific, Political, and Speculative* Volume I (New York: D. Appleton & Co., 1899), pp. 108–81.

164. *Ibid.*, p. 121.

165. *Formation mécanique du système du monde* (Paris: Gauthier-Villars et Fils, 1897). The almost 200-page-long book was prefaced by T. Moreux's 38-page-long résumé of du Ligondès' theory.

166. *Ibid.*, p. 57.

167. *Ibid.*, pp. 17–18.

168. *The Tides and Kindred Phenomena in the Solar System* (Boston: Houghton, Mifflin and Company, 1898), p. 338.

169. *Ibid.*, p. 339.

CHAPTER SEVEN

COLLISIONS REVISITED

A geologist's puzzle

Looking back on the strange detours that led him to the recognition of the elliptical orbit of planets, Kepler had more than one reason to marvel at his intellectual journey. The path which speculations on planetary origins were to follow afterwards witnessed the truth of his remark that roads to discoveries were as surprising as the discoveries themselves. Elements of surprise marked every aspect of the process by which it became gradually recognised that the Laplacian theory had to be replaced by another approach. To be sure, astronomers had by 1900 made speculations on planetary origins their own preserve. It was certainly not expected that the new chapter in the story would soon be opened by an outsider, a geologist, Thomas Chrowder Chamberlin. In addition, he was already in his fifties, hardly young enough to chart new avenues in his own field, glacial geology, let alone far outside it. Yet it was the study of glaciers that led him to a drastic reconsideration of the whole status of the question of planetary origins.

About that strange road over glaciers to planets Chamberlin disclosed as much as he left untold.[1] His puzzles might have originated as early as the 1880's when he served as a U.S. geologist in charge of the glacial division. He had already been for five years chairman of the Department of Geology at the University of Chicago when, in 1897, he first vented his misgivings about Laplace's theory before a major scientific forum. The occasion was the meeting of the British Association for the Advancement of Science in Toronto, the date August 20, 1897. His studies of glaciers impelled him, Chamberlin argued in a long address,[2] to picture the remote climatic history of the earth as an alternation of hot and cold periods. This seemed to him to be at variance with the commonly accepted form of the Laplacian hypothesis, according to which the earth passed from a hot, gaseous condition to a steadily cooler liquid and solid state. Chamberlin admitted that 'even following the general line of the nebular hypothesis a cold [primitive] earth is hypothetically possible'.[3] His mistrust in the hypothesis was, however, too strong to let him marshal arguments in support of it rather than against it. Of the whole texture of the hypothesis he singled out one specific phase, the separation of the moon's material from the earth in the form of a ring. At that point the earth's radius was supposedly equal to the present earth–moon distance, and the centrifugal force just slightly greater than the gravitational attraction, to permit the detachment of the ring. Chamberlin wondered if in the process the equatorial atmosphere of the earth should not have flown off with the moon-ring, and added the portentous remark: 'The subsequent contraction of the earth should apparently have accelerated its rotation to such an extent that the retention of the outer equatorial atmosphere would be put in jeopardy'.[4]

This was all that had at least an implicit bearing in Chamberlin's discourse on the question of angular momentum, the crucial issue of the nebular hypothesis. His main line of attack rested with the kinetic theory of gases and he presented long tables of the mean velocities of various molecules at various temperatures. The tables showed that much of the earth's atmosphere would have escaped if it passed through a temperature range from 4000° C to 3000° C.[5] He touched but lightly on the bearing of the same consideration on another and more telling aspect of the Laplacian theory as he remarked: 'It would seem, therefore, that unless the argument from molecular velocities is radically and grievously at fault the hypothesis of a gaseous earth-moon ring is untenable unless a degree of tenuity be assumed which separates the molecules beyond the limits of effective kinetic relations'.[6]

From the historical viewpoint more interest was carried in Chamberlin's repeated acknowledgments of the collaboration by Forest Ray Moulton, a young astronomer at the University of Chicago. The new planetary theory was to bear the names of these two, but nothing indicated that either of them had by 1897 sighted its salient features. In the summary of Chamberlin's address, which Moulton composed for his fellow astronomers a few months later, there was no reference to the question of angular momentum.[7] Moulton certainly saw the matter with the eyes of an astronomer as he gave emphasis to some general remarks of Chamberlin on the advantages of the 'meteoroidal hypothesis' as an alternative to the Laplacian theory. In Moulton's words the meteoroidal hypothesis meant that 'the aggregation of discrete meteorites revolving around the Sun unto one mass would be a slow process, and since they would all be revolving in the same direction the kinetic energy impact would be small, so it is plain that the collection into a body, such as the Earth, might take place only without developing at any time a high temperature'.[8] This served one more evidence that espousing the meteoroidal hypothesis did not imply a recognition of the problem of angular momentum, in line with what was already noted in the previous chapter. Clearly, then, nothing extraordinary could be expected from Moulton's hopeful conclusion that 'on the ground common to geology and astronomy, as has been the case with chemistry and physics, many discoveries may be made which shall shed a new light on both'.[9]

A joint salvo by Chamberlin and Moulton

Actually, judging by the results, an unusually creative exchange of ideas did take place between Chamberlin and Moulton during the next three years. The first fruits of their discussions were published almost simultaneously in early 1900 in two separate papers. They were remarkable not only by their incisiveness but also by the absence of unnecessary duplication of the material presented. The paper by Moulton[10] listed in three groups the objections that could be raised against the Laplacian theory. The first group contained the objections based on a comparison between the observed facts and the conditions implied for the nebula by the theory. It had, of course, already been known for some time that a rotating, homogeneous nebular mass, being in hydrodynamical equilibrium, could not give rise to retrograde motions, or to significant deviations from a common plane of rotation and from circular orbits. There was a distinct touch of originality in Moulton's pointing out that the planetary masses, if they indeed coalesced from respective rings, were also at variance with the presumed homogeneity of the nebular mass. It was again that homogeneity which did not seem to be borne out, according to Moulton, by the motion of the rings of Saturn if these indeed were detached from an originally homogeneous planet.

Moulton based his second group of objections on the consideration whether the generally assumed initial conditions of the nebula could have led to some of the major features of the planetary system. As a most natural case of these conditions one could assume the dilution of the solar mass to the orbit of Neptune. At the surface of such a nebular globe the escape velocity would have been such as to imply the permanent loss of much of the lighter elements. The case appeared even more telling when compared with the actual atmosphere of some of the planets: 'If such concentrated bodies', Moulton asked, 'as the Moon and Mercury are unable to hold atmospheres at present temperatures, is there any ground for supposing that the Earth-Moon ring, or the ring of Venus, or of Mars, could have held any of the atmospheric gases, or water vapour, at the [much higher] temperatures necessarily assigned them by the Laplacian theory?'[11] Moulton advanced similar reasoning against the possibility of the formation of quasi-permanent gaseous rings, a pivotal condition in Laplace's theory for the eventual formation of planets.

The pitfalls lying hidden in the ring model for Laplace's theory were now exploited by Moulton with relentless resolve. He argued that from a highly diluted spheroid rotating uniformly, the matter would be separated continuously and not in distinct rings at great intervals. A strong point in all this was that Moulton invariably assumed conditions most favourable for Laplace's theory. But the latter appeared unsatisfactory at each and every step considered. Moulton investigated in detail whether a planet already in its final stage of formation from the ring could attract to itself the small remains of that ring. The mathematical analysis showed that the planet could gather to itself only that part of the ring which extended 60 degrees on each side of the planet.[12] Equally unfavourable for the Laplacian theory was Moulton's investigation of the question whether the formation of a planet could get under way in that ring which, according to the hypothesis, had to be homogeneous and of very low density. Here Moulton set great store by the extreme smallness of gravitational attraction between minute particles of matter as the case had to be in an essentially gaseous ring. Also grave question marks appeared about the stability of a large globe, if it formed itself within that ring. Threatening the stability of such a globe was the questionable stability of the ring itself in respect to the attractive forces of the central body. Moulton reached the conclusion, along lines paralleling those of Roche,[13] that for the stability of the ring its density had to be three times greater when it separated from the nebula than was the density of the latter. Another conclusion of Moulton was that if the ring and the central mass were of the same density, the ring as such could never be closer to the centre of the solar nebula than 1.35 times the latter's radius. In Moulton's own summary, 'the discussion of these limits shows that a Laplacian ring could not have contracted into a planet, and that the condition of the solar nebula must have been one of great heterogeneity instead of homogeneity in concentric layers'.[14]

Moulton's third category of objections rested on a 'comparison of those properties of the supposed initial system with the one now existing, which are invariant under all changes resulting from the action of internal forces'.[15] This third section was much shorter than the first two, as Moulton discussed only one such invariant factor, the angular momentum. He did it with full knowledge of Darwin's memoir of 1881, which he prominently quoted in the introductory part of his paper. He did not mention Babinet and Fouché, although his argument had much in common with the principal point made by these two. What Moulton offered was not without ambiguities. He failed to specify the units in which he made his calculations, nor did he make it clear what he meant by the phrase that 'the planets present no difficulties', because 'the moment of momentum of the present system can be determined with almost perfect accuracy'.[16] Actually, it was clear that such an accuracy was wanting in his figures. He computed the sun's moment of momentum or M by considering it homogeneous, while his figures of M for the various extensions of the solar nebula were based on assuming the density to be a specific function of distance from the centre. However, the inaccuracies were too small to invalidate his claim that if the nebula 'contracted with the law of density always that adopted above, the centrifugal force would not equal the centripetal until it had shrunk far within *Mercury's* orbit'.[17] The basis for this claim was set forth in the following Table:

When the nebula extended to Neptune's orbit,	$M = 32.176$
When the nebula extended to Jupiter's orbit,	$M = 13.250$
When the nebula extended to the Earth's orbit,	$M = 5.690$
When the nebula extended to Mercury's orbit,	$M = 3.400$
In the system at present,	$M = 0.151$

The enormous difference between the first and last figures was a far cry from what one could expect if the Laplacian theory was dynamically sound. The difference could not be ascribed to uncertainties about the true variation of density in that rotating nebula, or to simplifications in the mathematical technique. The difference pointed, in Moulton's words, 'to a mode of development quite different from, and much more complicated than, that postulated in the nebular theory under discussion'.[18] The few words offered by Moulton about the new 'mode' struck the keynote of the new phase which was to open in the history of theories on the origin of planetary systems. According to Moulton the original solar nebula was not only 'heterogeneous to a degree not heretofore considered as being probable', but that it also 'may have been in a state more like that exhibited in the remarkable photographs of spiral nebulae recently made by Professor [James E.] Keeler'.[19] As further development showed, the spirals soon explicitly played the role of a major, if not foremost, evidence in support of the new 'mode' or theory.

The only striking similarity between the paper of Moulton and that of Chamberlin, which was to complement the former, consisted in Chamberlin's concluding allusion to Keeler and to its photographs of spiral nebulae. More of this later. The originality of Chamberlin's paper was secured by Moulton's leaving to his senior colleague the detailed discussion of the question of angular momentum. The reason for this was Moulton's acknowledgment of the general feature of the close collaboration between Chamberlin and himself during the previous three years. During that time, Moulton wrote, their exchange of ideas was so frequent and close that it did not seem possible 'to divide the responsibility for the various methods of attacks, and the manner of carrying them out'. Still he felt that by and large it was Chamberlin's role to employ 'his keen perception of physical relations, and his exceptional powers of invention', whereas it fell mainly to him [Moulton] to elaborate mathematically Chamberlin's insights.[20] The recognition of the decisive impact of the question of angular momentum seemed, therefore, to have come from Chamberlin, as his paper,[21] published concurrently with Moulton's, was largely devoted to it. Actually, as they disclosed it years later, the credit for the recognition belonged to Moulton.[22]

With Chamberlin a full exploitation of the question of angular momentum was of paramount importance. He thought that if the hot Laplacian nebula was replaced by a cold globular swarm of meteorites, the objections made against the theory on the basis of the kinetic theory of gases would lose much of their force. Chamberlin made no secret of his heavy reliance on Darwin's memoir of 1881 and on Moulton's computations, but he offered several independent considerations which helped much in bringing out in full force the bearing of angular momentum [M. of M. = moment of momentum = angular momentum] on the truth of Laplace's theory. By converting Moulton's data into Darwin's units Chamberlin presented the following Table[23] for a comparison between the successive nebular stages and the corresponding parts of the present system:

	Nebular M. of M.	Present M. of M.	Ratios
Neptunian stage,	4,848.055	22.76661	213 to 1
Jovian ,,	1,996.420	14.18161	141 to 1
Terrestrial ,,	857.330	0.71008	1,208 to 1
Mercurial ,,	512.290	0.67979	754 to 1

The striking irregularity in the sequence of ratios strongly indicated, so Chamberlin argued, that the source of trouble was not with the assumptions underlying the computation, but in the wholly unsatisfactory character of the Laplacian theory as far as the conservation of angular momentum was concerned. The generally accepted estimate of the densities of globular gaseous nebulae also seemed to suggest that the conflict between the Laplacian theory and the data of angular momentum could not be caused by a 'radical error in the law of density'.[24]

On the basis of the Laplacian theory it was almost inevitable to assume that, as Chamberlin put it, 'there would be some systematic and rational relationship between the masses separated from time to time and the moments of momenta of these masses, for the separation was due to a common progressive cause, the acceleration of rotation'.[25] The data arranged by Chamberlin in the following Table[26] told an entirely different story:

Ring	Percentage of mass of its nebula	Percentage of M. of M.	Percentage of M. of M. reduced to basis of .00001% of nebular mass
Neptunian	0.00507	7.93	.0156
Uranian	0.00454	6.31	.0139
Saturnian	0.02852	27.78	.0098
Jovian	0.09530	94.97	.00996
Martian	0.0000323	0.36	.1099
Terrestrial	0.0003160	2.42	.0766
Venus	0.0002495	1.89	.0755
Mercurial	0.0000205	0.12	.0566

The Table evidenced no systematic variation, and in the case of Jupiter it laid bare two particularly destructive details for the Laplacian theory. One of these was that high as was the ratio of Jupiter's angular momentum to the parent nebula, it was proportionately exceeded in almost all other cases. The other detail was that if the Laplacian theory was true then 'in the formation of the Jovian ring less than one thousandth of the mass carried away 95 percent of the moment of momentum'. The remark then could aptly be made by Chamberlin that 'the minor planets [Mars, Earth, Venus, Mercury] had a narrow escape from not being at all', since 'one nineteen-thousandth more of the mass thrown off with an equal proportion of rotational momentum would have exhausted the supply [of angular momentum]'.[27]

Darwin's work contained enough evidence to permit Chamberlin to answer in the negative the question whether tidal friction would transfer so much angular momentum from the sun to the planets. The data of angular momentum, he wrote, 'seem to require the assignment of some mode of origin by which the peripheral portion of the system acquired all but a trivial part of the mass'. The situation could be met, so he believed, by 'the possible formation of the system by the collision of a small nebula upon the outer portion of a large one, the smaller one having necessarily a high ratio of momentum to mass, while the larger one may have had little or no rotatory momentum, or even an adverse rotation'.[28]

Chamberlin made no secret of his reluctance to chart a course diametrically opposite to the one implied in the Laplacian theory which he characterised as 'perhaps the most beautiful and fascinating ever offered to the scientific public'.[29] He was also aware of the difficulties which the near-collision envisaged by him presented for an analysis in terms of celestial mechanics. The lack of theoretical justification could, however, be supplemented by appeal to factual evidence, or to use Chamberlin's words, by 'a purely naturalistic and inductive method'.[30] The evidence consisted of Keeler's photographs. According to Chamberlin they were no support for the Laplacian theory, because there was no trace on any of them of a subdivision of nebular masses into rings. But a great many of the photographs showed spirals, big and small, all of which, so Chamberlin thought, were produced by collision. He felt that the distribution of mass in the arms of spiral nebulae represented a stage closely akin to the actual formation of planets with orbits of increasingly small ellipticity. Before long he saw in at least some of the spirals the actual and overwhelming evidence of his own views of planetary formation.

The crucial role which was suddenly assigned in cosmological speculations to Keeler's photographs of spiral nebulae quickly met with his approval. In a paper published in the June 1900 issue of the *Astrophysical Journal* Keeler gave a detailed analysis of the characteristics of celestial photography with the Crossley reflector of Lick Observatory in California and emphasised the predominance of spirals among nebulae. More importantly, he presented his findings as having 'a direct bearing on many, if not all, questions concerning the cosmogony'. The example he mentioned in this connection was the spiral nebula. If it was indeed the normal form assumed by a contracting nebular mass, then the idea naturally offered itself, so Keeler wrote, that 'the solar system has been evolved from a spiral nebula' and not from a much less differentiated type implied in the Laplacian theory.[31] Since Keeler gave credit for these inferences to Chamberlin and Moulton, their confidence could only increase in the correctness of their interpretation of spiral nebulae. Reference to nebulae appeared in the very title of Chamberlin's next essay, his first probing into the details of the collision originating a planetary system like ours.[32]

By turning to the process of collision Chamberlin first wanted to secure the building material which, according to him, had to be a swarm of meteoritic particles. This material could be produced, he felt, by the relatively close approach of two stars, each with a planetary system, a process in which tidal forces would cause the breaking up of planetary bodies into many smaller ones. This procurement of material for future planetary systems from the partial or total destruction of already existing ones is worth noting. It revealed at one stroke Chamberlin's somewhat ambivalent approach to the crux of the new theory and also his addiction to a type of universe in which everything went through endless cycles of evolution and decay.[33] Near-approaches of planetary systems and the breaking up of comets in the vicinity of the sun turned out to be on closer analysis a meagre source for meteorites. One reason for this was the paucity of mass involved in such processes, the other the obvious rarity of them, a circumstance which Chamberlin recognized.

A far more ample source appeared to him in the near approach of a gaseous nebula in slow rotation to an extinct, solid star.[34] The case was much more complex, though, than the progressive elongation and rotation of two stars presented in a diagram [*Illustration XXVII*]. From this Chamberlin inferred the eventual development of long arms at opposite ends from the elongated nebula and their fragmentation into small knots. Most importantly, the dynamics of the case seemed to lead to a spiralling motion of the two arms, and this meant the imparting of a large amount of angular momentum to the material there. Chamberlin felt himself to be on firm ground as he declared: 'The effects of explosive projection combined with concurrent rotation must obviously give rise to a spiral form'.[35] Emboldened by his apparent success, he illustrated with similar diagrams the case of a close approach of 'two live suns', and the case of their partial collision.[36] Once more the formation of a spiral nebula was a foregone conclusion.

It remained to be shown that such near approaches were sufficiently frequent in the actual constitution of the realm of stars and nebulae. The answer to this problem depended on the proportion of 'live stars' to extinct ones. If, as Helmholtz suggested almost fifty years earlier, gravitational contraction was the only source of a star's energy, the active life of a star was at best a few hundred million years and, as a result, extinct stars outnumbered by far the active ones. Chamberlin, who was among the first to see in the newly discovered radioactivity a factor that could greatly lengthen a star's lifetime,[37] spoke again of it as a means of the cyclic rejuvenation of stars.[38] At any rate, there was still the 'inductive' approach, or the recourse to spiral nebulae. Speaking of them Chamberlin shared the then prevailing view for which all galaxies were systems much smaller than the Milky Way, distributed on its two sides within two imaginary hemispheres of which the Milky Way was believed to form the connecting plane.[39] In that picture the spirals could readily be taken as planetary systems in formation. Their apparent crowding around the galactic poles could also suggest that collisions producing them were not infrequent. In addition to quoting Keeler, Chamberlin enriched his essay with the photographs of six spirals, a procedure to which he was to give, as time went on, further emphasis.

The planetesimal theory

The existence of spirals, if taken as *de facto* illustrations of a specific theory, could only weaken interest in further scrutiny of the process of collision producing them. This was already in evidence in the lengthy research report, 'Fundamental Problems of Geology', which Chamberlin submitted on September 30, 1904, to the Carnegie Institution of Washington.[40] The title of the report bespoke his motivations. The third or concluding part of the essay consisted in unfolding the consequences of the new theory for questions of geology and it was the minutely descriptive style of a field-geologist that characterised the exposition throughout. Chamberlin's starting point was the origin of the meteoritic swarm which replaced the hot Laplacian nebula as the substratum of future planetary system. He found that such a swarm could not very likely assemble itself from a homogeneously diffused condition. The opposite route, the process of dispersion, offered three possibilities: dispersion by explosion, dispersion by collision, dispersion by tidal disruption. By the latter he meant the fragmentation of a planet under the gravitational pull of its sun.

As Chamberlin rejected all these cases as unsatisfactory approaches, he began to use, somewhat confusedly, the word 'planetesimal'. It meant for him not so much a state of matter different from the meteoritic one, but an organisation of meteorites into roughly circular, that is, 'planetesimal' orbits around a sun.[41] He spoke of the dispersion of a comet's head 'into the planetesimal state',[42] and described the yearly recurrence of the brilliant meteor showers in August and September as illustrations of the 'planetesimal mode of organisation'.[43] As he began to describe his now famous 'planetesimal hypothesis' of planetary origins,[44] he presented as its typical form the case in which 'the parent nebula of the solar system consisted of innumerable small bodies, planetesimals, revolving about a central gaseous mass, somewhat as planets today'.[45]

Chamberlin was not altogether felicitous with phrases as he tried to specify the extent of his theory. He left it to his reader to fathom the reasons why his theory (disregarding now the factor of near collision) was essentially different not only from the Laplacian nebular hypothesis but also 'from the meteoritic conception of Lockyer and Darwin'.[46] Clarity was wanting when Chamberlin stated that 'the hypothesis postulates no fundamental change in the system of dynamics after the nebula was once formed, but only an assemblage of the scattered material',[47] and that for the sake of concreteness that special case was chosen 'in which the nebula is supposed to have arisen from the dispersion of a sun as a result of close approach to another large body'.[48] The fogginess of the latter statement was only increased by what he added in the same breath: 'The case does not involve the origin of a star nor even the primary origin of the solar system, but rather is a rejuvenation and the origin of a new family of planets'.[49] Five pages later Chamberlin claimed that his theory 'confines itself to a supposed episode of the sun's history in which the present family of planets had its origin, and in the initiation of which a possible previous family may have been dispersed'.[50]

If such was indeed the case, then Chamberlin clearly stretched the meaning of the word geological when he claimed that the cosmological episode envisaged by his theory 'may be regarded as *geologic*, since it specially concerns the birth of the planet of which alone we have intimate knowledge'[51] (italics added). Actually, his theory stretched far beyond all 'geologic' epochs as he explicitly suggested (1) that the planetesimal nebula was the effect of the disruptively close approach of another sun to our ancestral sun which might have already had a planetary system of its own; (2) that it required another close approach by another sun to start the evolution of the present planets in the planetesimal nebula by producing in it two spiral arms with knots of condensation. It did not help clarity either that Chamberlin was wont to fall back on his 'inductive' method which, in fact, was heavily 'deductive' in character. In interpreting radioactivity as a source of stellar energy, he was on the right track, but his inferences from photographs of spiral nebulae and of solar protuberances had a touch of arbitrariness.

Concerning the spirals he now quoted for the first time a statement of Keeler in support of considering spirals of utmost importance in interpreting the evolution of planetary systems.[52] Chamberlin held out the hope that further refinement in observation would lead to the discovery of spiral nebulae 'that

represent the solar system more nearly in mass and proportion'.[53] He discussed
solar protuberances as models of spiral arms with obvious tendentiousness and
vagueness. According to Chamberlin these protuberances would shoot as far
as the actual limits of the planetary system, were it not 'for the retarding influence of the immense solar atmosphere'.[54] He also credited the sun's gravity
for breaking down 'the expansive potency of this prodigious elasticity'
embodied in the protuberances. In his always qualitative and often vague
reasoning about complex problems of celestial dynamics the identity of the
emergence of protuberances and of spiral arms remained above any doubt.[55]
He could, of course, claim with some justification that he had already analysed
the process in some detail, but the existence of spirals was unquestionably a
powerful prompting for him to consider the issue sufficiently explored.

Chamberlin now confidently turned to seemingly easy details such as the
amount of mass to be ejected from the sun. This could hardly appear a major
problem since the planets totalled less than one percent of the sun's mass. Also,
it could plausibly be assumed that the eruptions would take place in pulses
giving rise to knots (protoplanets) in the arms. The first outbursts would then
cause the separation of the lighter surface material, accounting thereby for the
lower density of the outer planets. The subsequent outbursts would in turn
provide the material from deeper and denser layers, as demanded by the higher
density of the inner planets. 'It is thus conceived', Chamberlin summarised the
first phase of planetary formation, 'that a spiral nebula, having two dominant
arms, opposite one another, each knotty from irregular pulsations, and
rotatory, the knots probably also rotatory, and attended by subordinate knots
and whirls, together with a general scattering of the larger part of the mass in
irregular nebulous form, would arise from the simple event of disruptive
approach'.[56] Bafflingly, Chamberlin disposed in one short paragraph of the
question of angular momentum without mentioning it explicitly, although it
was largely on that ground that he conducted his spirited campaign for the
replacement of the Laplacian theory by a new one. Recalling that three-body
problems do not admit exact solutions, he rested his case on 'special solutions'
which 'seem to justify the inference that effective rotation would arise'.[57]

It wholly escaped Chamberlin that it was a strange procedure to criticise
and reject the Laplacian theory by marshalling tables of angular momentum
and not to offer at the same time even approximate calculations about its
distribution between the central mass and the two main protuberances drawn
out of it by the passing star. This was all the more curious as Chamberlin
insisted on the point, undoubtedly with an eye on the question of angular
momentum, that in those protuberances, or spiral arms, there were knots
surrounded by subordinate knots, all rotating in the sense defined by the
bending of the spiral arms. He offered no specific proof for the crucial claim
that on the basis of his theory the bending of those arms had to be a sharp one,
securing both the circularity of planetary orbits and the proper distribution of
angular momentum. Instead, Chamberlin devoted lengthy paragraphs to such
specifications of the theory as the irregular or 'planetesimal' shape of the knots
in the arms and the supposedly small amount of gaseous matter there. He said

less about the more important point of the growth of those knots into planets. It was brought about, according to him, by the gathering up of the planetesimal particles through gravitational attraction as they orbited around the sun. He acknowledged that the dynamics of the 'close approach' most likely put the planetesimals into greatly varying and 'broadly elliptical' orbits. It now became his crucial task to explain how this situation could slowly transform itself into the basically circular paths of the planets. His surprisingly short explanation was that 'when two bodies in concentric elliptical orbits unite, their conjoined mass must move in an orbit that is intermediate between the two previous orbits, and this new orbit, in all cases investigated, is less eccentric than one of the previous orbits'.[58] Chamberlin failed to notice that since the almost circular planetary orbits represented one extreme in a gamut of eccentricities they could not become 'intermediate' in the sense intended by him.

Of the difficulties of his explanation Chamberlin recognised two. One concerned the evident slowness of the process, but he felt confident that in view of recent discoveries made about radioactive elements the cosmic past could safely be imagined as exceedingly vast. The other and more serious difficulty was connected with the direction of the rotation of most planets and satellites. In the Laplacian theory the gaseous ring was usually imagined to be rotating as a unit, with its outer parts moving faster, and thus the actual forward rotation seemed to follow readily. But in the foregoing aspect of the planetesimal hypothesis a retrograde rotation should have been the logical out-come, since all planetesimals were supposed to move independently, and consequently the inner ones must have had a greater orbital velocity than the outer ones. Chamberlin sought a way out of this dilemma by arguing that a 'planetesimal in a smaller elliptical orbit can come into contact with a planetary nucleus in a larger orbit *only* when a more or less aphelion portion of its orbit coincides with a more or less perihelion portion of the larger orbit of the nucleus, and a planetesimal in a larger orbit can come into contact with a planetary nucleus in a smaller orbit only when a more or less perihelion portion of its orbit coincides with a more or less aphelion portion of the nucleus' orbit'.[59] The 'graphical inspection' of the two kinds of coming into contact indicated to Chamberlin that 'the possibilities of overtake favourable to forward rotation exceed those favourable to retrograde rotation'.[60]

The fondness of the originator of a theory for his brainchild was visible in the manner in which Chamberlin sought confirmation for it in a comparison of the present eccentricities of the planets with the ones assigned to them on the basis of the theory. The latter values were obtained by taking into account the radius and mass of the respective planets, since according to the planete-simal hypothesis the greater was the mass collected by the planetary nucleus, the smaller its eccentricity had to become. The amount of mass was in turn proportional to the mean radius of the planet's orbit, which determined the volume of the ring from which the material of the planet collected itself. Curiously, there appeared to Chamberlin a favourable touch to his reasoning in the Table he submitted:[61]

Nucleus of	Assigned eccentricity	Present eccentricity
Mercury	0.25±	0.2
Venus	.21	.006
Earth	.2	.017
Mars	.28	.093
Asteroids (mean)	.33	.38 downward*
Jupiter	.336	.048
Saturn	.366	.056
Uranus	.37	.046
Neptune	.38±	.009

*mean about 0.15

In the same Table Chamberlin also saw the basis for a 'rational law' to supplant the purely algebraic relationship which was noticed by Bode concerning the relative distances of planets.[62] Chamberlin showed, however, full awareness of the obvious difficulties which his theory faced, when he took up the question of the number of permanent nuclei (protoplanets) developing in each planetary belt. He frankly admitted that the formation of a single planet in one belt was not a necessary outcome on the basis of his theory, though he argued laboriously that such was the plausible result.[63]

The last third of Chamberlin's essay, a discussion of the 'geologic' consequences of his theory, need not detain us here. For the history of our topic there was an interesting aside in the short progress report by Moulton whom Chamberlin engaged as a co-worker in carrying out the objective of his essay heavily supported by a grant from the Carnegie Institution of Washington. Moulton realised that in order to clarify the air about the genuine value of the 'nebular hypothesis' a careful study of the whole history of theories on planetary origins was to be made. He outlined it as follows: 'The first epoch reaches up to Laplace, the second consists of Laplace and the commentators on his work, including the modifications introduced by the theory of the conservation of energy; the third starts with Darwin's work on tidal evolution and reaches to our work in 1900'.[64] This worthwhile project failed to be carried out and gross inaccuracies about the history of theories of the origin of planetary systems kept their prominent place in the literature.

References to the history of the topic were wholly absent when Moulton's important paper, in which he presented the planetesimal theory along strictly technical lines, appeared in 1905.[65] His was a complete faith in the correctness of each line in a sweeping paragraph in which he claimed that once the spiral nebula was formed from the sun by the close approach of another star, practically all features of the planetary system followed as inevitable consequences and in a manner prescribed by the planetesimal hypothesis. Concerning the manner let it suffice here to recall two of the specifics listed by Moulton. One was the growth of the planet's mass as the cause of the increase of its rate of rotation and of the decrease of the eccentricity of its orbit. The other was, understandably enough, the claim that 'the greater part of the moment of momentum will belong to the planets'.[66]

Unfortunately, on this last and decisive point, the question of angular momentum, Moulton was sparing with proofs. First, his diagram [*Illustration XXVIII*] of the formation of the spiral arms indicated anything but a sharp curving[67] to make possible an ultimately circular orbit for the planets. Second, he offered no proof of the contention that 'it follows from the origin of the system that the remote planets should possess most of the moment of momentum of the system', and that it is an inevitable consequence of the spiral theory that Jupiter, which contains about one-tenth of 1 percent of the mass within the orbit of Saturn, should contain more than 95 percent of the angular momentum of the whole system. In connection with these figures Moulton quoted Chamberlin as if the latter had been the first to offer such data, and added that 'the whole question of the moment of momentum is a rock on which the ring theory breaks'.[68] As it turned out, the same rock beckoned calamity to the planetesimal theory as well. At any rate, it wholly escaped Moulton that a safe sailing for a planetary theory consisted in steering clear of two rocks, not one.

The other rock was constituted by the nearly circular planetary orbits, but this could hardly appear a threat as their derivation by Moulton seemed to have solid foundations from the viewpoint of celestial mechanics. No misgivings were voiced about the fact that the collisions between planetesimals and between planetesimals and protoplanets, on which the physical truth of Moulton's arguments ultimately rested, had not been investigated. After all, there were spiral arms which in some nebulae formed an almost perfect circle, and this could readily be taken as a fascinating evidence of what was to be proven. The extent of that fascination could not remain hidden to the reader of the second volume of a massive textbook on geology, co-authored by Chamberlin.[69] The volume opened with an eighty-page-long discussion of the remote origin of the earth. Half of that discussion was devoted to the planetesimal theory, illustrated with eleven striking photographs of spiral nebulae. Although Chamberlin called this discussion 'the first full statement' of the theory,[70] it added nothing really new to his earlier statements. The gain was largely in the illustrations and in Chamberlin's summary of the modification of Laplace's theory by Darwin and by Lockyer. One would have looked in vain for the names of Buffon, Kant, Babinet, and Fouché. The impact which spiral nebulae had on the thinking of Moulton was amply in evidence when his *Introduction to Astronomy* was published in 1906.[71] There, the twenty-five-page-long account of the theory was headed by the caption: 'The Spiral Nebula Hypothesis'. This new name was, according to Moulton, a most effective way to keep the planetesimal theory 'in sharp contrast with the Laplacian ring theory'.[72]

Anglo-Saxon appraisals

It should not, therefore, cause any surprise that admirers and opponents of the new theory gave a prominent place to the issue presented by spiral nebulae within the context of planetary evolution. As to admiring comments, a good and early case was a paper by R. G. Aitken, astronomer at Lick Observatory and well known for his work on binary stars. After reviewing the grave difficulties of the Laplacian theory, Aitken turned his attention to spiral nebulae as the first evidence which assures to the new theory 'a high degree of confidence'.[73] 'Celestial evidence' played an even more prominent part in J. E. Gore's favourable appraisal of the planetesimal theory. The evidence coming in from Lick and Yerkes Observatories seemed to put in doubt the very existence of ring nebulae and for Gore this meant that there 'is no warrant in the heavens' for the Laplacian hypothesis.[74] 'The heavens are strongly in favour of a new theory, a new cosmogony', he wrote with an eye on spiral nebulae, which formed, so he stated, 'a large proportion of the half million nebulae discovered with the Crossley reflector'.[75]

The spirals figured prominently in the partial endorsement of the planetesimal theory by the Swedish Nobel-laureate in chemistry, Svante Arrhenius, who tried to derive spirals and planetary systems alike from the partial collision of two stars [Illustration XXIX], a process which he did not analyse quantitatively.[76] This was only one of the glaring weaknesses in the numerous cosmological writings of Arrhenius. He was so strongly devoted to the idea of a perennial rejuvenation of matter and energy as to argue that spirals were cosmic places where entropy was decreasing! No such pretensions marred the support of W. Sutherland of the planetesimal theory. It came in 1911 in a note added in press to his article on 'Bode's Law and the Spiral Structure',[77] in which he showed that the actual planetary distances could be generated with good approximation from the properties of logarithmic spirals.

As could be expected, emphasis on spirals by the supporters of the planetesimal theory could not fail to provoke sharp reaction on the part of its opponents. The most notable of them was T. J. J. See, of the Naval Observatory of Mare Island, California, whose numerous memoirs displayed more prowess in details of mathematical analysis than in productive originality, as shown, for instance, by his recalculation in 1905 of the distribution of angular momentum in the planetary system.[78] He characterised as 'not only gratuitous, but unjustifiable and in the highest degree improbable' the recent suggestions that spiral nebulae were evidences of tidal disruption caused by close encounter between celestial bodies. According to him the near circularity of planetary orbits showed 'the absurdity' of such suggestions which he branded as 'simply misleading and mischievous'. He peremptorily claimed that 'there is not the slightest probability that our solar system was ever a part of a spiral nebula'.[79] Three years later, in 1909, he declared with the same impetuousity 'that most of the recent speculations on cosmogony are not worth the paper they are written on'.[80] Even if he had not identified those speculations as having been published in the Astrophysical Journal and supported by the Carnegie Institution, it was all too clear who were his chief targets.

See hardly made his diatribe any less offensive as he tried to set himself up as an authority in the matter by stating that he had been continuously working for the past twenty-five years on the topic.[81] When at long last the first evidence of that claim saw print in the same year,[82] it contained no reference to Chamberlin and Moulton. The irony of this had to appear all the more vivid, because in the theory of planetary evolution, which See now briefly outlined, the solar system, to quote his very words, 'was formed from a spiral nebula revolving and slowly coiling up under mechanical conditions'.[83] Equally ironical had to appear See's reference to the millions of spirals[84] playing the role of 'celestial evidence' for his theory. He called it the 'Theory of Resisting Medium' in line with the emphasis he put on the role of that medium in securing nearly circular orbits for the planets. The planets themselves had to be captured 'in the midst of the solar nebula',[85] since, according to See, 'Babinet's criterion' proved that they could not be detached from the protosun. See was not clear whether he meant by that 'midst' the centre of the nebular spiral or one of the many possible condensations in its arms. His confidence in his own reasoning knew no bounds. From the unexpected conclusion concerning the capture of planets, he wrote, 'there is absolutely no escape'.[86] The same unlimited confidence transpired from his interpretation of the presence of 8 planets, 25 satellites, and 625 asteroids (as the count then stood), or a total of 658 'small' bodies in the solar system. He saw in this the evidence that the chances were 2^{658} to 1, or 10^{66^6} to 1, that the original nebula would develop into a solar system with 'many small' bodies rather than into a double star.[87]

For all the bewilderment that could be occasioned by See's often intemperate and sweeping statements, his emphatic attention to the double problem of angular momentum and of nearly circular orbits should commend itself. Such a competent expert in the field as Sir George H. Darwin failed to call attention to the fact that the planetesimal theory contained no rigorously developed solution to that twofold problem. He took serious exception to the planetesimal theory[88] only on the question whether the accretion of large planetary bodies was feasible in the disk-shaped swarm of planetesimals into which the spiral arms supposedly developed. In the account by Chamberlin and Moulton mutual attraction and perturbation were not considered in order to prove with greater ease the all-important point that increasingly more circular orbits were to emerge from coalescence through collisions. A consummate master in the intricacies of celestial dynamics, Darwin felt that collision between mutually attracting bodies with separate orbits and different velocities was a most exceptional case. In his words, 'it would overtax the skill of the finest shot if the pheasant were so massive as to deflect the shot'.[89] The Martians of Wells' War of the Worlds were in Darwin's eyes imaginary enough to have the ability to hit the earth with their missiles. He strongly doubted that 'we poor terrestrial beings' would solve such a problem in ballistics in 'ten thousand years from now'.[90] As usual, prognostication concerning scientific progress once more proved to be the riskiest kind of prophecy.

French silence and German protest

Opposition to a theory can at times take the form of silent treatment and this is what seemed to happen as if by conspiracy in the land of Laplace. Between 1900 and 1920 Chamberlin's name not once appeared in the pages of the prestigious *Bulletin astronomique*, which carried in all its issues briefly annotated listings of articles published abroad in leading astronomical journals and even had on occasion a special section on the latest in cosmogony. During the same period Moulton's name occurred in the *Bulletin* more than a dozen times but only in connection with such papers of his that had nothing to do with the planetesimal theory![91] Émile Belot, chief engineer of the governmental department of manufacturing and the most prolific French author on planetary origins during the first quarter of the century, was, on the other hand, given in the *Bulletin* much attention for his theory which he characterised as a corrected form of the Cartesian vortices.[92] In the manner characteristic of somewhat eccentric thinkers he reduced all cosmic development to a particular process. His choice, collision, consisted in the penetration of a rotating protosun into a cosmic cloud. The collision, according to him, made the protosun pulsate while passing through the cloud and forced the detachment of some of the protosun's material. The result was the formation of funnel-like spiralling layers at the upper end of which the reflection of waves triggered by the pulsation produced condensations of matter, the protoplanets [*Illustration XXX*].

Such bizarre speculations were absent in the two chief studies of the origin of planets by French authors around that time. One of them was written by Charles André, director of the Observatory of Lyons, who claimed that of all theories (he made no mention of the planetesimal theory) that of Laplace was 'still the most solid'[93] and also 'close to the truth'.[94] André's claims were all the more baffling as he quoted Darwin's figures for the distribution of angular momentum in the planetary system.[95] According to André the figures merely indicated a point already made by Darwin, namely, that tidal forces were insufficient for the explanation of the formation of a system of planets like ours. The author of the other work in question was Henri Poincaré, possibly the most brilliant mathematical physicist of the early 20th century, who presented his *Leçons sur les hypothèses cosmogoniques* as a complete survey of theories on the origin of planetary systems that had been proposed since Laplace.[96] For Poincaré, who considered Laplace's theory the best in the field,[97] the question of the distribution of angular momentum constituted no major difficulty. He did not refer to Babinet, and the import of Fouché's paper consisted in his judgment in the need to postulate a massive condensation in the centre of the primitive solar nebula of Laplace.[98] Equally conspicuous should have appeared the absence of any reference in Poincaré's lectures to Chamberlin and Moulton, especially in view of the fact that Poincaré discussed the 'latest' on the topic, See's capture theory. About the latter Poincaré aptly noted that it was wholly unwarranted to suppose that practically all captures had taken place in the same direction and in the same plane.[99]

Neglect by Poincaré of the planetesimal theory sanctioned then an attitude in France as can be seen, for instance, in the monograph, which Alexandre Véronnet, astronomer in Strasbourg, published in 1914.[100] Needless to say, there was no mention of Chamberlin and Moulton in a short communication which K. Birkeland presented to the Académie des Sciences on November 4, 1912. His paper[101] represented the first specific application of the motion of charged particles in the sun's magnetic field as a factor as important as gravitation in the formation of the system of planets. What Birkeland offered was an odd mixture of fantasies, gross oversights, platitudes, and lucky guesses. As to the first, he attributed an electric potential of 600 million volts to the sun's surface. In computing the 'limiting circles' of charged particles emitted from the sun as a huge cathode, Birkeland took no account of the sun's magnetic field. He was clearly platitudinous in noting that the orbiting of negative particles should be retrograde if the sun's magnetisation was opposite to that of the earth. His lucky guess was that a major part of the total mass of the universe consisted in the ionised particles of interstellar dust clouds.

Reaction to the planetesimal theory in the land of Kant reflected the often sharp differences between Germany and France. Rejections of the theory were emphatic and immediate in Germany and were, not surprisingly, coupled with a defence of Kant's ideas.[102] No touch of chauvinism tainted the monograph which Friedrich Nölke published in 1908.[103] His presentation of Kant's theory was probably the most reliable that had appeared until then,[104] but Nölke gave a poor reading of history as he assigned to Chamberlin and Moulton the origin of attention to the question of the distribution of angular momentum.[105] In connection with their theory Nölke made an original and very destructive point. In the planetesimal theory planetary orbits closely approaching a circle developed from the increase in the mass of planets through collision with myriads of planetesimals. A doubling of the mass was, however, necessary to reduce by half (a measure still far from being satisfactory) the eccentricity of orbit, which originally had to be very great in the case of all planets because of the dynamics of their common origin. But in the planetesimal theory by far the largest portion of planetesimals consisted of hydrogen, the lightest element, which, according to the kinetic theory of gases, only planets with already massive bodies, such as Jupiter and Saturn, could accumulate and retain in sufficiently large quantities. In the case of the earth, where all the hydrogen in the ocean amounted to only 1/38,000th of the total mass, it was clearly impossible to derive its nearly circular orbit from the process in question, and the same had to be true of the even smaller planets, Mercury, Venus, and Mars.[106] It had, therefore, to be obvious that whatever the success of the planetesimal theory in steering clear of the rock of angular momentum, it crashed against the rock of circular orbits.

The tidal theory of Jeffreys and Jeans

This point was further underlined by H. Jeffreys who in 1916 called attention to the fact that collisions were far more frequent between planetesimals than between planetesimals and protoplanets.[107] The reason for this lay in the relatively much larger surface of bodies with very small mass as planetesimals were by definition. Collision between planetesimals meant, so Jeffreys argued, their volatilisation, a process which then inevitably led to the emergence of a Laplacian nebula, larger than the one preceding the near collision, but equally impotent to generate a planetary system. As Jeffreys' calculations showed, collisions between planetesimals were ten times more frequent than collisions between planetesimals and protoplanets even in the case when the diameter of the former was as much as one ten-thousandth of that of the protoplanets and when these were expected to increase their mass by only one thousandth. Actually, the planetesimal theory assumed planetesimals with much smaller diameter and implied a much larger increase of the mass of protoplanets. Thus, there could be little doubt 'that if the eccentricities were at all high at the commencement the planetesimal system must have turned to gas long before the planets could grow appreciably by accretion'.[108] Nothing was to be gained if one assumed that the colliding planetesimals first broke into solid fragments. In such a case the frequency of collisions between planetesimals was to increase until the conversion into a nebular state was complete. Or as Jeffreys put it, 'this alternative . . . only postpones the difficulty without avoiding it'.[109]

The solution to the problem of planetary origins was sought by Jeffreys in a recourse to Poincaré's model of a nebula with strong central condensation into which he introduced a rotation that increased from the centre. His was clearly an effort which tried to solve the problem of the simultaneous derivation of both the distribution of angular momentum and of circular orbits by positing the presence of both in the original nebula. The procedure meant, by Jeffreys' own admission, the abandonment of a reliance on spiral structure, or on any known physical formation as a model.[110] This must have cast immediate doubt on Jeffreys' speculation which received its death-knell when James H. Jeans showed[111] that no condensation can occur in a material as tenuous as the non-central regions of the protosun had to be in Poincaré's and Jeffreys' assumptions.

Almost simultaneously with Jeans' paper, Jeffreys gave in the July 1917 issue of *Science Progress* a qualitative support to the basic role of tidal action in the origin of planetary systems.[112] His paper must have appeared insignificant to anyone familiar with the massive memoir which Jeans read eight months earlier before the Royal Astronomical Society.[113] The originality of Jeans' memoir concerned the emergence of *one* huge filament of almost liquid mass from the sun under the influence of tidal forces generated by a passing star. The emergence of one huge filament instead of two spiral arms was as much a departure from the planetesimal theory as was the emphasis on *one* single rupture instead of repeated pulsations. Interesting elaborations on Jeans' conclusions were soon offered by Jeffreys. His analysis[114] of the effectiveness of the resisting medium in reducing the eccentricities of the planets showed that the development of the solar system implied a length of time comparable to the age of the earth as calculated from radioactivity.

A vastly detailed form of the new or tidal theory of planetary origins came in 1919 with the publication of Jeans' *Problems of Cosmogony and Stellar Dynamics.*[115] It represented a novel approach also in the sense that Jeans urged the limitation of discussions 'to the abstract problem of the behaviour of masses of astronomical matter under varying dynamical force'. This meant that 'the latest observational evidence' was not to be given undue importance in charting cosmogonical theories. For as Jeans noted: 'Up to the discovery of the spiral nebulae, most theories of cosmogony tried to prove that Saturn's rings (the most sensational astronomical objects then known) formed an intermediate stage in the evolution of planetary systems: since the discovery of spiral nebulae, the tendency has been to try to prove that the spiral nebulae form the link in question'.[116] A principal result of Jeans' investigations was that whereas a rotating mass with spiral arms formed a crucial stage in the evolution of almost all large-scale celestial phenomena, such as nebulae, star clusters, binary and multiple stars, the spirals were of no use in explaining the formation of planetary systems like ours.[117]

The success of Jeans' monograph (it earned for him the coveted Adams Prize for 1917) made his name almost synonymous with the theory which reigned supreme for the next two decades. Further investigations by Jeffreys did not create a similar stir when put together in a monograph in 1924.[118] Jeffreys himself stated that his theory was partly based on that of Jeans. More accurately, Jeffreys merely modified Jeans' theory on some not really essential points. Whereas according to Jeans the sun extended at the moment of its tidal disruption to the orbit of Neptune, in Jeffreys' estimates its original confines were well within the orbit of Mercury. Jeffreys also claimed that the encounter had to be an intermediate type between the 'slow' and the 'transitory' encounters envisaged by Jeans.[119] The closeness of their views indicated no appreciable change when in 1928 there appeared under the title *Astronomy and Cosmogony* a considerably enlarged version of Jeans' monograph,[120] and a year later a similarly enlarged new edition of Jeffreys' work.[121]

Of the several summaries of the tidal theory offered by Jeans over a period of almost three decades, one deserves special attention both because of its graphic character and of its lucidity supported by no mathematics. Its starting point is the steady increase in the brightness of a star heading toward our sun on an almost collision course:

> As we . . . watch the changing panorama of the sky somewhere between two and three thousand million years ago, we notice a star gradually increasing in brightness until it outshines all the others in brilliancy, and finally looks incomparably brighter than Sirius does now. It looks bright because it is very near rather than because it is intrinsically very bright; indeed, it has approached quite unusually near to the sun. And as we watch, it comes ever nearer; it is heading almost straight for the sun. It is no longer a mere point of light. We see it as a large disc. And now it has come so near that its mechanical effects are beginning to shew. Just as the moon, by its nearness to the earth, raises tides in our oceans, so this enormously more massive body is, by its nearness, raising tides in the fiery atmosphere of the sun. Because it is so much more massive than the moon, these tides are incomparably greater than those which the moon raises in the earth. They become so great that, at a point right under the star, the sun's atmosphere is drawn up to form a huge mountain, many thousands of miles high. This mountain travels over the surface of the sun, keeping always under the star which causes it, as this moves on its way through space. At the opposite point of the sun's

surface, another but smaller mountain keeps always opposite the main one. As the star approaches ever nearer, these tidal mountains continue to increase in height, until at last, when the other star is so near as to fill up a large part of the sky, a new feature enters. So far, the gravitational pull of the star has been drawing up the summit of the larger mountain in opposition to the gravitational pull of the sun, but the latter has always been the stronger. Now the second star comes so near that the balance suddenly swings over in the other direction; the second star outdoes the sun in gravitational pull, and the top of the mountain shoots off towards it. As this relieves the pressure on the lower parts of the mountain, these also shoot upwards, and then the parts below them, and so on, so that a whole stream of matter shoots out from the sun towards the second star. If this star came continually nearer to the sun, the end of the jet of matter would reach it in time, and the substance of the jet would join the two stars together like the bar of a dumb-bell.

Actually the other star is not heading directly towards the sun; after coming very near indeed, it finally passes on its way without actually colliding. As it recedes its tidal pull diminishes. No more matter is pulled off the sun, and the jet which has already come off forms a long filament of hot filmy gas suspended in space. In shape, it is rather like a cigar, pointed at its two ends. The point which is now furthest from the sun was originally the peak of the tidal mountain. The thick middle of the cigar consists of the matter which came off plentifully when the star was nearest, and its tidal pull was strongest. Finally, the pointed end nearest the sun is formed of the last thin dribble of matter which came off just before the tidal pull became too weak to draw any more matter away from the sun.

Even as we watch this cigar-shaped filament of fiery spray, it gradually cools and, as it does so, it condenses into detached separate drops, much as a cloud of steam condenses into drops of water. Yet these drops, like the filament itself, are colossal structures; their size is on the astronomical scale. Naturally they are biggest near the fat centre of the cigar, where the matter of the filament was most abundant, and are smallest at the two ends.

Finally, these detached drops of matter begin to move about in space as separate bodies. They do not fall back into the sun, because the pull of the other star, which we now see receding in the distance, has set them in motion; unless they happen to be moving directly towards the sun, they will not fall into it but describe orbits round it. This is a direct consequence of the law of gravitation, which was the same thousands of millions of years ago as it is today. Some of these orbits may be nearly circular while others are greatly elongated. As we watch the orbits for millions upon millions of years, we see them gradually and very slowly changing their shapes. The condensed drops of matter do not move in unobstructed paths, for the great cataclysm we have just witnessed has left space littered with its debris. The great drops must plough their way through this, and as they do so the shapes of their orbits gradually change, until at last, after thousands of millions of years, they move round the sun in almost circular orbits, just like the planets of today. And indeed these bodies are the planets; the dramatic spectacle we have just witnessed . . . is one which must inevitably happen in Nature whenever one star approaches close enough to another, and its final scene is so exactly like the solar system, that we have every reason to suppose that this is actually the way in which the planets came into being. So far as we can judge from their present arrangement and movements, it seems most likely that they were torn off the surface of the sun by the tidal pull of a passing star which happened to pass very unusually near to it some few thousands of millions of years ago.[122]

More noteworthy than the graphic and lucid character of this account of the tidal theory was the apparent lack of second thoughts on the part of its illustrious author concerning the emergence of nearly circular orbits. It was with similar confidence that Jeans assumed, fourteen years earlier, that the question had already been settled by See's studies of the change of highly eccentric orbits into closely circular ones due to the braking effect of resisting medium.[123] Jeans seemed to take lightly the fact that all parts, big and small alike, of the material torn off the sun by the passing star must initially have moved in a highly eccentric orbit if indeed the filament constituted one big lump of gaseous matter. For the reduction of such an orbit to a less eccentric one an interstellar matter of sufficiently high density was necessary for which, however, there was no observational evidence. Apart from this Jeans found that in addition to the problem of angular momentum (of which he was very much aware although his rendering of Babinet's objection was far from being accurate[124]) in the foregoing gaseous mass there was another and no less portentous problem. It derived from his analysis of large rotating gaseous bodies exemplifying the Laplacian nebula. His analysis showed[125] that regardless of the actual value of the angular momentum in our planetary system, no successive detachment of rings could take place in contracting nebulae comparable in mass to an average star. Such nebulae were to break into two parts, giving rise to binary stars, which are most typical features of the starry realm, but around which the formation of planetary systems even remotely resembling a stable system like ours is a well-nigh impossibility.

From this additional blow to the Laplacian hypothesis it also followed that the tidal theory worked only if one assumed a close encounter between two sufficiently compact stars whose outer layers were about to turn into the liquid state.[126] In other words, the phase of stellar evolution envisaged by Jeans for the effective production of planetary systems meant that much of the interstellar material had already been collected into the bodies of stars. Again, within the framework of Jeans' theory no sufficient braking material could be provided by that cigar-shaped filament torn off the less massive star through its close encounter with a more massive one. For as Jeans emphasised, this filament had to be such as to permit quick condensation within it of the planets.[127] Clearly then, not enough matter could be left for dissipation in interplanetary spaces to act as a braking medium for the production of nearly circular orbits. If the process of planetary formation outlined by Jeans had a really valuable detail, it consisted in its accounting for the variation of planetary masses with distance from the sun. The planets had to be smaller at the two ends of the filament and larger at its middle part in agreement with what is observed in our planetary system.

The detachment of one big filament, as advocated in Jeans' theory, also seemed to secure the possibility of a condensation into planetary bodies, since the mass of the filament gave rise to a sufficiently large gravitational attraction within itself, preventing thereby the early dissipation of much of its material. This was a distinct advantage over the planetesimal and the Laplacian theories in which there was no safeguard against the almost immediate dissipation of the detached material. Details of the process envisaged by Jeans remained, however, as conjectural as was his seemingly ingenious explanation of the formation of satellite systems around planets. He imagined that when the freshly formed and still partly liquid planets made their first approach to the sun on their originally very eccentric orbits, the sun exerted on them the very same tidal force, though on a much smaller scale, which was exerted on the sun by the passing star. On such a basis the formation of satellite systems had to be, in conformity with observation, more effective around the larger planets, since their solidification could not be as rapid as that of the smaller planets. The formation of the moon remained unaccounted for, but this was not considered a major fault since no other theory had shown any promise on this score.

The point where the first major inconsistency of Jeans' tidal theory was laid bare concerned the rotation of the planets and of the sun. In the tidal theory the rotation of the sun originated with the material which fell back on it after it acquired a transverse deflection through the pull of the passing star. Similarly, the rotation of planets came from the material falling back on them after they had undergone a tidal fission following their first perihelion passage. The amount of such material could readily be estimated from the known values of rotational angular momentum and Jeffreys found[128] that in the case of the sun the material in question had to be equal to that of Jupiter. This represented no major improbability, but it was impossible to admit that the material which supposedly fell back on Jupiter might have exceeded by a factor of 400 the total mass of its satellites. Much the same results were true for Mars, Saturn, Uranus, and Neptune. A possible solution of the problem consisted in assuming that the original radius of both the sun and the planets was larger by an order of magnitude, but this implied even greater inconsistencies. Jeffreys, therefore, proposed that the encounter between the sun and the star be pictured as a shearing collision rather than a close approach. His calculations showed that by making suitable assumptions about a good number of physical parameters the mechanism in question would lead to such values of the rotational angular momentum which were correct within the actually observed order of magnitude.[129]

The mirage of solution and its dissipation

In looking back on his achievement from the distance of a quarter of a century, Jeffreys recalled with a touch of nostalgia about the year 1929 that it 'was the first time that a theory of the origin of the solar system had made correct quantitative predictions, even in order of magnitude, and it looked as if the main outlines were established and the rest was a matter of working out details'.[130] The situation was far from being so satisfactory, but the prospect of having the final solution at hand effectively blocked for a while at least the sighting of major cracks. This was a typical pattern throughout the whole history of theories on planetary origins. It asserted itself with a striking force in the lifelong efforts which H. P. Berlage Jr., a Dutch meteorologist in Indonesia, launched about that time to formulate the long-sought explanation of the origin of the system of planets. It was also typical that his initial speculations did not lack some valuable details. They consisted in his insistence on taking into account the sun's electric field and its continuous emission of charged particles, and on a mathematical treatment of the theory. But one must certainly be perplexed by the fact that in an essay of 68 pages,[131] which was published in 1927·and came to a close with a bibliography of 94 items, one would have looked in vain for the names of Babinet, Fouché, and Moulton.[132] Nor in Berlage's study was there any discussion of the distribution of angular momentum in the solar system. From the integer multiples of electric charges and of masses of ionised molecules one could, of course, derive a mathematical formula which could somewhat arbitrarily be interpreted as evidence of distinct concentric rings of ions around the protosun. Yet in this derivation not only the magnetic field of the sun was not taken into account, but not even the centrifugal force.

Behind such a baffling waste of talent and effort there must have been undue preference for some particulars. This was disclosed three years later by Berlage in a frank phrase: 'I always insist on the point that Bode's law touches on the very root of the problem'.[133] Ten years later, in 1940, and after three more publications, in which he emphasised, again not without some originality, the presence of a broad flat disk and the role of viscosity within it as a starting point of planetary development, he was in a sense even more behind the times. He looked upon his work as an evidence that 'the solar system in successive stages of evolution, could be identified with a CARTESIAN whirl, with KANT's disc and the rings of LAPLACE: Even CHAMBERLIN and MOULTON's planetesimals were met at a certain stage, whereas for JEANS' foreign star there might have been work to do at the moment of conception. It might have started the evolution, leaving the rest to be done spontaneously. At any rate, this beautifully ordered structure, which is our planetary system, is essentially self-made'.[134]

What in particular should have revealed to any well-informed reader of this phrase the measure of its author's wishful thinking was his statement that all that the tidal theory could explain was the imparting to the matter torn from the sun 'sufficient moment of momentum around its primary'.[135] Actually, by 1940 it had become widely known that the tidal theory not only could not cope with the question of angular momentum, but that within its assumptions even the coalescence of the big filament into large planetary masses was highly questionable. This latter problem was sighted by Jeffreys himself, who, as was noted above, proposed the replacement of a near approach with a shearing collision. This, however, meant that the material detached from the sun came in great part from its interior layers and had very high temperature. The latter factor endangered the stability of the filament, whereas the former presented problems about the material composition of planets. Moreover, the shape of the detached material could no longer be assumed as resembling a cigar or an elongated cylinder. Most likely, it looked like a ribbon. Because of these new circumstances the question had to be re-examined whether sufficiently large liquid bodies, the cores of planets, could be formed in the detached material. While Jeffreys' investigations suggested the feasibility of the process,[136] a few years later L. Spitzer put forward considerations which indicated that the filament was most likely to dissipate itself into 'an enormously extended atmosphere of some sort around one of the two stars'.[137] Such an atmosphere seemed to be strongly reminiscent of the Laplacian nebula although with the advantage that it now could possess the requisite distribution of angular momentum, since its outer layers were rotating at a faster rate. Still, it must have appeared something of an impasse that the truth of the collision theory depended on the hardly promising question whether a Laplacian nebula could condense into planetary bodies.

That the question of angular momentum contained a fatal objection to all types of collisional theories of the origin of planetary systems was conclusively argued by Henry Norris Russell in a lecture series on the topic at the University of Virginia in 1934.[138] Russell's originality lay in his focusing attention of the angular momentum with respect to the minimum distance required between the sun and the passing star to permit the detachment of a sufficiently large amount of matter from the former through gravitational attraction. Assuming a star with average mass and velocity approaching the sun in a hyperbolic orbit, Russell found that the perihelion distance, which in this case is closely approximated by half of the semilatus rectum or semi-parameter p, of the hyperbolic orbit, could not be greater than one and a half million miles, or about 0.03 percent of the earth-sun distance. This, however, meant that the angular momentum of the star with respect to the sun was less than one tenth of the weighted average of the angular momentum per ton of all planets. Clearly, the passing star could not be the source of the planets' angular momentum. As Russell said tersely: 'To put so much angular momentum into the ejected material during the encounter would seem to be impossible'.[139] The data of the Table (see next page) which Russell offered[140] showed all too clearly how the difficulty increased as one moved from the innermost planet toward the more distant ones.

The dynamics of the encounter permitted no logical escape from the impasse. One could not reasonably assume that the passing star dragged the material of Jupiter and of Neptune 90 and 500 times respectively above the

Body	Semi-parameter	Angular momentum per ton	Ratio
Star	0.03	0.25	1.0
Mercury	0.37	0.61	2.4
Venus	0.72	0.85	3.4
Earth	1.00	1.00	4.0
Mars	1.51	1.24	5.0
Jupiter	5.19	2.28	9.1
Saturn	9.50	3.08	12.3
Uranus	19.11	4.39	17.6
Neptune	30.07	5.49	22.0
Weighted average		2.63	10.5

original height of the ejected material out of which they were formed. Again, the assumption that almost immediately after the tidal eruption they shifted their very eccentric orbits into nearly circular ones amounted to begging the question of the corresponding acquisition of angular momentum. The situation had to appear even worse for the grazing collision postulated by Jeffreys, since in that case the value for p for the star would have been only 0.01 astronomical unit and this implied a proportional increase of the relative amount of angular momentum per ton for the planets.

Russell's book was quickly reviewed in *Nature* by none other than Jeffreys, who wrote in connection with the question of angular momentum per ton: 'If this objection is valid, and I can see no reply at present, we have no satisfactory theory of the origin of the solar system'.[141] This meant the admission, to quote Jeffreys again, that 'the problem of the origin of planets is still unsolved'.[142] The other principal architect of the tidal theory, Jeans, left matters of technicality where they stood in 1928, and partly because of this Jeffreys, who kept for the rest of his life a vivid interest in the question of planetary origins, was already speaking as the principal authority in the field. His evaluation of the state of the art can, therefore, dispense with a detailed analysis of several minor theories proposed during the decade prior to 1935.[143] A really new approach was briefly outlined, though without enthusiasm, by Russell himself at the end of his lectures.[144] In order to steer clear of the rock of angular momentum, he suggested that the sun was originally a binary star and that the collision took place between its companion and a passing star. Assuming that the companion star revolved around the sun at a distance of about that of Saturn, it could appear perhaps not completely improbable that the breaking into parts of the companion star resulted in the formation of planetary system around the sun.

Actually, as was shown in detail shortly afterwards by R. A. Lyttleton, who in 1935–36 worked as a Procter Fellow with Russell in Princeton, the process was sound dynamically in three respects.[145] First, the sun's companion could easily be removed through its encounter with the passing star; second, through the same mechanism larger bodies (planets) could be left behind with 'widely varying angular momentum per unit mass and possibly causing them all to proceed *round* the Sun in the same general direction nearly in a plane' (italics added);[146] third, there seemed even to be some promise for the production of satellites through encounters between pairs of planets.[147] A trouble with all this lay in the vagueness of the word *round*. On this point Lyttleton did not show the measure of insight displayed by Russell who bluntly noted that 'the supposable evolution of the initial orbits into more nearly circular forms has also all the difficulties previously mentioned'.[148] This was as explicit an acknowledgment as there could be that authors of theories of the origin of the solar system should steer clear not of one, but of two major rocks. No wonder that Russell's lectures came to a close with a wishful look at the earlier stages of cosmic evolution when collisions between nebular condensations were still more numerous. He carried his vistas back to the explosion of Lemaître's hypothetical Primeval Atom at the start of the present expansion of the universe. After all, did not radioactive measurements and the analysis of the red-shift of expansion assign roughly the same age to the solar system and to the universe of expanding galaxies? Was it not tempting to think that in the primordial whirl of nebulae 'almost anything may have happened',[149] including the formation of planetary systems in very large numbers?

The lure of 'normalcy'

Tempting it had to be, if one were to judge this from the hardly disguised uneasiness that could be felt in Russell's acknowledgment of the extreme rarity of planetary systems on the basis of collisional theories. Astronomers were again reminded of this rarity, when in 1936 Z. Kopal, of the Stefanik Observatory in Prague, showed that the actual distances of the planets were too great to be produced by a single close encounter.[150] The need to postulate another such event with all the delicate parameters implied in it clearly increased the improbability of the formation of planetary systems beyond all imagination. By the mid-1930's the notion of the extreme rarity of planetary systems had been brought to the attention of circles far outreaching the astronomical profession. The chief role in this respect was played by Jeans. In his famous monograph of 1917 he had already noted that formation of planetary systems could no longer be viewed as a 'normal cosmogonic process',[151] since near-collisions producing it had to be extremely rare. Was it not in part this idea of the extreme rarity of planetary systems that made him state that 'the time for arriving at conclusions in cosmogony has not yet come'?[152] It mattered not, as subsequent developments showed, that there was no lasting ground for his claim that his tidal theory was 'the only theory which is not open to obvious and insuperable objections'.[153] About ten years later, in the second edition of that classic work, he re-emphasised, and in a categorical manner, the rarity of planetary systems. He spoke of them as 'freak-formations'[154] and estimated the chance of a star to form a system of planets as 1 in 6×10^{17} years.[155] This meant that an earth like ours would evolve once in every 500 million years in a typical galaxy like ours.[156] Three years later, in 1931, his graphic description of that unimaginably rare process exuded his full confidence in its truth: 'we have every reason to suppose that this is actually the way in which the planets came into being'.[157]

The rarity of planetary systems meant also the rarity of life in the universe. The most memorable remarks on this score came from Jeans himself. In his Halley Lecture of May 23, 1922, he gently lampooned the universally shared inference of the Copernican and Galilean revolution, on the basis of which one posited planets around each and every star with each and every planet being 'peopled by living beings similar to ourselves'.[158] Six years later, he phrased the same point with a singular sharpness: 'It does not at present look as though Nature had designed the universe primarily for life; the normal star and the normal nebula have nothing to do with life except making it impossible'.[159] The next year, in 1929, his famed *The Universe around Us* began to carry the same theme to millions of the readers of its uncounted editions, reprints, and translations. Life, he noted there, 'is limited to a tiny fraction of the universe',[160] and he took pains to correct a stereotyped illusion of modern intellectual convictions: 'The three centuries and more which have elapsed since Giordano Bruno expressed his belief in an infinite number of worlds have changed our conception of the universe almost beyond description, but they have not brought us appreciably nearer to understanding the relation of life to the universe'.[161]

On the heels of *The Universe around Us* came its sequel, *The Mysterious Universe*, the most widely read science popularisation of high quality written in many a year. The book opened, tellingly enough, with a vivid portrayal of the uniqueness of the physical process capable of producing a planetary system like ours. Jeans described the stars as ships voyaging in 'splendid isolation' in the immense ocean of space: 'They travel through a universe so spacious that it is an event of almost unimaginable rarity for a star to come anywhere near to another star'.[162] It was, of course, true that even in a universe consisting of a finite number of galaxies, as postulated by the latest cosmological models, the total number of stars was also 'unimaginably' large. Figures like 10^{21}, or a thousand million million million, were indeed beyond imagination, and they still could provide millions of millions of planets, even if the rate of collisions was only one in 500 million for any star over a period of two thousand million years. One could, in fact, say with Jeans that a new planetary system was born in the universe every few hours. Yet these figures given by Jeans in 1941, in his last major pronouncement on the topic,[163] could not conceal the fact that of these thousands of millions of planets only a fraction could be assumed to be similar to the earth in physical and chemical characteristics. Scattered through a finite though immense universe, those million or so earth-like planets, separated from one another by many thousands of light years, still represented an extremely rare phenomenon.

By 1941 Jeans was toward the end of a brilliant career. The tide which he helped rise in cosmology and planetary theory appeared to be ebbing. His tidal theory was no longer the talk of the day. Like all other theories before, it failed to give a satisfactory explanation of the system of planets. But whatever its shortcomings, it stemmed the tide, for a while at least, of pseudoscientific affirmations by some scientists about the presence of planetary systems and of planetary denizens in every nook and cranny of the universe. If such systems were very small in number, even smaller had to be the number of cosmic colonies of beings similar or superior to man. But the principle of plenitude still stood for 'normalcy' with its irresistible lure. A sign of this was the manner in which Harold Spencer Jones, the Astronomer Royal, exploited in 1940 Russell's nostalgic vision about the chaotic whorls dominating the cosmic scene shortly after the expansion of the universe got under way. What he wrote was a classic example of how the persistent failures could be cast in a favourable light by even more persistent preferences. On the basis of evidences provided by failures to construct a viable theory of planetary origins it seemed, Sir Harold wrote with an eye on the tidal theory and on Russell's modification of it, that 'such a combination of initial conditions was required to enable the system to come into existence that the system must be almost unique'.[164] Through fond contemplation of the gigantic turmoil of collisions and close approaches between stars in the erstwhile chaos the contour of evidences could be made less sharp and constraining. The magic device consisted in the otherwise plausible point that conditions in the primeval chaos were very different from those revealed by present-day observations. While on that basis one could write that 'we cannot draw any certain conclusion' about the manner in which planetary systems were formed, logic was hardly served in what was added in the same breath in the way of a grand conclusion: 'Though we are still unable to say in detail how the solar system originated, we are probably correct in concluding that it is by no means so exceptional as we had thought and that there are likely to be many other stars, in addition to the Sun, which are accompanied by systems of planets'.[165]

It was a rare and more creditable performance when Fred L. Whipple flatly stated a year later that there 'appear to be damning arguments against every theory so far proposed for the origin of the planets',[166] and avoided being involved in a dubious maze of hints about the presence of planetary systems everywhere in the universe. The 1940's had not run their course, when Jeffreys already noted in one of his usually authoritative surveys of latest theories that they reflected the desire for finding justification of looking upon planetary systems as 'normal' phenomena.[167] In the new theories the sought-after dynamics of planetary evolution followed once more the pattern of circles. In more than one sense.

NOTES TO CHAPTER SEVEN

1. He did this in a somewhat poetical style in the Introduction to his *The Origin of the Earth* (Chicago: University of Chicago Press, 1916, pp. 1–9), a popular summary of his investigations about the origin of planetary systems.
2. 'A Group of Hypothesis Bearing on Climatic Changes', *The Journal of Geology* 5 (1897): 653–83.
3. *Ibid.*, p. 668.
4. *Ibid.*, p. 666.
5. *Ibid.*, p. 665.
6. *Ibid.*, p. 668.

7. 'Professor Chamberlin on the Nebular Hypothesis', *Popular Astronomy* 5 (1897–98): 508–11.

8. *Ibid.*, p. 509.

9. *Ibid.*, p. 511.

10. 'An Attempt to Test the Nebular Hypothesis by an Appeal to the Laws of Dynamics', *The Astrophysical Journal* 11 (1900): 103–30.

11. *Ibid.*, p. 113.

12. *Ibid.*, p. 118.

13. *Ibid.*, p. 121. It should be indicative of the restricted circulation of Roche's memoir of 1873 that Moulton could not obtain a copy of it. His ideas of its contents were based, as had been the general case until today, on second-hand summaries and popularizing accounts.

14. *Ibid.*, p. 126.

15. *Ibid.*, p. 105.

16. *Ibid.*, p. 127.

17. *Ibid.*, p. 128.

18. *Ibid.*

19. *Ibid.*, p. 130. Keeler was director of Lick Observatory when he died in 1900, at the age of forty-three. His spotting of very large number of nebulae with the 36-inch Crossley reflector and the photographs he made of them had been a chief topic of conversation in astronomical circles as the nineteenth century came to a close. In November 1899 Keeler published a short note 'On the Predominance of Spiral Forms among the Nebulae' in the *Astronomische Nachrichten* (Nr. 3601; 151 [1899]: cols. 1–4), but it is unlikely that either Chamberlin or Moulton could use it for their papers printed in early 1900. In that paper Keeler estimated the number of nebulae, most of them spirals, within the reach of his telescope, at 120,000.

20. *Ibid.*, p. 106.

21. 'An Attempt to Test the Nebular Hypothesis by the Relations of Masses and Momenta', *The Journal of Geology* 8 (1900): 58–73.

22. The disclosure came in their joint paper, 'The Development of the Planetesimal Hypothesis', (*Science* 30 [1909]: 642–45), from which one could learn that once it occurred to Chamberlin that the slow rotation of the sun could not provide sufficient equatorial velocity in the solar nebula extended to Mercury's orbit for the separation of Mercury's ring, he called in Moulton for consultation. It was through that conference between the two that Moulton's attention was brought to the general problem of angular momentum in the nebular hypothesis. In 1916 Chamberlin further specified in his *The Origin of the Earth* (Chicago: University of Chicago Press, p. 50) that 'it was found later that Babinet had detected discrepancies of the same type many years previously'. To this Chamberlin very correctly added that Babinet did not see the objection fatal to Laplace's theory.

23. 'An Attempt to Test the Nebular Hypothesis by the Relations of Masses and Momenta', p. 65. The conversion factor from Moulton's unit to that of Darwin was almost exactly 150. About Moulton's unit Chamberlin merely said that it was based on 'convenient initial units' (p. 64). Chamberlin emphasised that in his computations he chose at every possible point the alternative which assured the largest and lowest angular momentum for the sun and for the nebula, respectively. Thus, he assumed the sun to be spherical and homogeneous (adopting the larger figure proposed by Darwin, or 0.679, for the sun, which, when added to the total orbital momentum of the planets, gave 22.766 for the whole system), and the nebula to have extended only to Neptune's orbit, although he considered more likely an extension surpassing it by 500 million miles, or by somewhat less than one-fifth of the orbit in question.

24. *Ibid.*

25. *Ibid.*, p. 67.

26. *Ibid.*, p. 69.

27. *Ibid.*, p. 68.

28. *Ibid.*, p. 72.

29. *Ibid.*, p. 73.

30. 'An Attempt to Test the Nebular Hypothesis', p. 73.
31. 'The Crossley Reflector of the Lick Observatory', *The Astrophysical Journal* 11 (1900): 325–49; for quotation, see p. 348. A collected edition of Keeler's best photographs of nebulae was published only in 1908.
32. 'On a Possible Function of Disruptive Approach in the Formation of Meteorites, Comets, and Nebulae', *The Astrophysical Journal* 14 (1901): 17–40.
33. *Ibid.*, p. 40.
34. *Ibid.*, p. 32.
35. *Ibid.*, p. 34.
36. *Ibid.*, pp. 35–36.
37. In his article, 'Lord Kelvin's Address on the Age of the Earth as an Abode Fitted for Life', *Science* 9 (1899): 889–901 and 10 (1899): 11–18.
38. 'On a Possible Function . . .', p. 34.
39. See on this Chapter VII, 'The Myth of One Island', in my *The Milky Way: An Elusive Road for Science* (New York: Science History Publications, 1972).
40. 'Fundamental Problems in Geology', in *Carnegie Institution of Washington: Year Book No. 3, 1904* (Washington, D. C.: Published by the Institution, 1905), pp. 195–258.
41. *Ibid.*, p. 200.
42. *Ibid.*, p. 207.
43. *Ibid.*
44. *Ibid.*, p. 208.
45. *Ibid.*
46. *Ibid.*, p. 210.
47. *Ibid.*, p. 208.
48. *Ibid.*, p. 210.
49. *Ibid.*
50. *Ibid.*, p. 215.
51. *Ibid.*
52. *Ibid.*, p. 213.
53. *Ibid.*, p. 215.
54. *Ibid.*, p. 217.
55. *Ibid.*
56. *Ibid.*, p. 219.
57. *Ibid.*, p. 218.
58. *Ibid.*, p. 225.
59. *Ibid.*, p. 228.
60. *Ibid.*, p. 229.
61. *Ibid.*, p. 231.
62. *Ibid.*, p. 232.
63. *Ibid.*, p. 230.
64. From Moulton's letter to Chamberlin, dated Sept. 29, 1904; *ibid.*, p. 255.
65. 'On the Evolution of the Solar System', *The Astrophysical Journal* 22 (1905): 165–81.
66. *Ibid.*, pp. 166–67.
67. *Ibid.*, p. 169.
68. *Ibid.*, p. 177.
69. Thomas C. Chamberlin and Rollin D. Salisbury, *Geology*, Vol. II, *Earth History: Genesis–Paleozoic* (New York: Henry Holt and Company, 1906), pp. 1–81.
70. *Ibid.*, p. 38 note.
71. F. R. Moulton, *An Introduction to Astronomy* (New York: The Macmillan Co., 1906).
72. *Ibid.*, p. 463 note. In the second, revised edition (New York: The Macmillan Co., 1916) the title of the corresponding section was changed to 'The Planetesimal Hypothesis', (pp. 421–46).
73. 'The Nebular Hypothesis', *Publications of the Astronomical Society of the Pacific* 18 (1906): 111–22; for quotation see p. 117. The theory of Chamberlin and Moulton was not discussed by Percival Lowell either in his *The Solar System*, the text of six lectures

delivered at the Massachusetts Institute of Technology in December 1902 (Boston: Houghton, Mifflin & Co., 1903), or in his course on cosmology given there in 1908–09 and published under the title, *The Evolution of Worlds* (New York: Macmillan, 1910). There were only three short references to Moulton (Chamberlin was not mentioned at all) in Agnes Clerke's *Modern Cosmogonies* (London: Adam & Charles Black, 1905), a collection of her short essays on cosmological topics. Clerke noted the arguments which Moulton marshaled against the Laplacian theory in his article of February 1900 in *The Astrophysical Journal* and also quoted (see p. 82) its conclusion about the spiral nebulae! But the planetesimal theory did not seem to exist to Clerke, although one chapter in her book was entitled 'Cosmogony in the Twentieth Century'.

74. 'The New Cosmogony', *Popular Astronomy* 14 (1906): 515–22; for quotation see p. 517.

75. *Ibid.*, p. 519.

76. See Chapter VI, 'End of the Sun – Origin of Nebulae', and Chapter VII, 'The Nebular and the Solar States', of the English translation, *Worlds in the Making* (translated by Dr. H. Borns; New York: Harper & Brothers, 1908) of his *Das Werden der Welten* (Leipzig: Akad. Verlagsgesellschaft, 1907). For *Illustration XXIX* see p. 141 of the German original. Arrhenius' reader could not form any specific idea about the objection raised by Chamberlin and Moulton to Laplace's theory. See *Worlds in the Making*, pp. 206–07. In his famous *Lehrbuch der kosmischen Physik* (Leipzig: Verlag von S. Hirzel, 1903) vol. 1, pp. 220–33, Arrhenius mentioned the supposedly high temperature of the original nebula as the 'greatest difficulty' in the Laplacian hypothesis, the basic truth of which he considered 'undeniable' (*ibid.*, p. 225).

77. *The Astrophysical Journal* 34 (1911): 251–60.

78. 'Researches on the Physical Constitution of the Heavenly Bodies', *Astronomische Nachrichten* Nr. 4053; 169 (1905), cols. 321–64.

79. 'Significance of the Spiral Nebulae', *Popular Astronomy* 14 (1906): 615.

80. Letter to the Editor of *Science* 29 (1909): 859. See tried to soften this remark by noting that much worthless material was being published in all leading scientific periodicals!

81. *Ibid.*

82. 'On the Cause of the Remarkable Circularity of the Orbits of the Planets and Satellites and on the Origin of the Planetary System', *Astronomische Nachrichten* Nr. 4308; 180 (1909), cols. 185–94.

83. *Ibid.*, col. 192.

84. *Ibid.*

85. *Ibid.*, col. 189.

86. *Ibid.*

87. *Ibid.*, col. 193. Nothing of importance was contained in See's further elaboration of his capture theory which appeared in mid-1909 in Nrs. 4341–43 of the *Astronomische Nachrichten*. See's massive *Researches on the Evolution of the Stellar Systems*, Volume II, *The Capture Theory of Cosmical Evolution* (Lynn, Mass.: Thos. P. Nichols & Sons, 1910) was a collection of many of his essays published since the printing of Volume I in 1896. See was also an early and sharp antagonist of Einstein's theory of special relativity.

88. 'A Theory of the Evolution of the Solar System', *Internationale Wochenschrift für Wissenschaft, Kunst und Technik* (Berlin), 3 (1909), cols. 921–34.

89. *Ibid.*, col. 930.

90. *Ibid.*

91. Articles and discussions on cosmogony which appeared in the *Bulletin* during that period failed even to consider the bearing of Fouché's objections against Laplace's theory. Undoubtedly, this might have been due to the dominating influence of Poincaré, one of the five-member editorial board of the *Bulletin*. In 1911 the *Bulletin* carried Poincaré's article, 'Remarques sur l'hypothèse de Laplace' (vol. 28, pp. 251–66), which was identical with a chapter to appear in the same year in his *Leçons* to be quoted below. In the popular *L'Astronomie*, monthly organ of the Société astronomique de France, there was a similar silence during the same period about Chamberlin and Moulton. The several and at times obscurantist papers on cosmogony that appeared in *L'Astronomie*

could easily distract from a report which tellingly enough came from Fouché. His report, given in connection with a discussion held on cosmogony on October 3, 1906, in the Headquarters of the Société, was a badly needed warning about the immense complexities facing cosmogonists, but was also a classic example of how both the past and present context of science could be misinterpreted. As to the past, Fouché claimed that Laplace's theory could appear satisfactory for the science of its time. Concerning the present, Fouché noted that recent developments were discouraging for making a new start in cosmogony. Among these developments he listed not only radioactivity and X-rays, but also, and in the same breath, the increase of mass with velocity and the decrease of velocity for a body moving through the ether. But was not that increase of mass a consequence of that special theory of relativity which meant the demise of the ether? Needless to say, Fouché did not seem to know of Chamberlin's and Moulton's efforts to overcome that very same difficulty of Laplace's theory which Fouché had pointed out twenty-two years earlier. See *L'Astronomie* 20 (1906): 481–82.

92. The substance of his numerous articles in the *Comptes rendus* and in the *Bulletin* was presented in 1911 in his *L'origine dualiste des mondes: Essai de cosmogonie tourbillonnaire* (Paris: Gauthier-Villars). In its revised form, published in 1924, under the title, *L'origine dualiste des mondes et la structure de notre univers* (Paris: Payot), Belot still kept ignoring Chamberlin and Moulton.
93. *Les planètes et leur origine* (Paris: Gauthier-Villars, 1909), p. 235.
94. *Ibid.*, p. 270.
95. *Ibid.*, p. 256.
96. *Leçons sur les hypothèses cosmogoniques*, professées à la Sorbonne par H. Poincaré, redigées par Henri Vergne (Paris: Libraire scientifique A. Hermann et Fils, 1911).
97. *Ibid.*, p. vi.
98. *Ibid.*, pp. 20–21.
99. *Ibid.*, p. 129. On Poincaré see also text to *Illustration XXXI*.
100. *Hypothèses cosmogoniques* (Paris: Hermann, 1914).
101. 'Sur l'origine des planètes et de leurs satellites', *Comptes rendus* 155 (1912): 892–95.
102. See, for instance, E. Saubert, 'Moultons neue Idee über die Planetenentstehung', *Die Natur* 49 (1900): 250–55; W. Foerster, 'Die Zweifel an der Kosmogonie von Kant und Laplace', *Mitteilungen der Vereinigung von Freunden der Astronomie und kosmischen Physik*, 12 (1902): 7–11; and G. Holzmüller, 'Die Kant-Laplace'sche Kosmogonie und ihre Kritik', *Sirius* 37 (1904): 73–82.
103. *Das Problem der Entwicklung unseres Planetensystems* (Berlin: Verlag von Julius Springer, 1908).
104. *Ibid.*, see especially pp. 2–10 and 124.
105. *Ibid.*, pp. 48–49. In addition to some errors in Nölke's account of the history of the topic, the organisation of the material in his book represented a not too helpful mixture of the historical and systematic viewpoints. He ignored the pre-Kantian part of the question as well as the overwhelming part of the 19th-century material. These shortcomings were not remedied in the second revised edition of the work, *Das Problem der Entwicklung unseres Planetensystems: Eine kritische Studie* (Berlin: Verlag von Julius Springer, 1919) in which Babinet and Fouché still found no place. In this edition the organisation of the material was completely subordinated to the systematic viewpoint.
106. See the first edition, pp. 72–74.
107. 'On Certain Possible Distributions of Meteoric Bodies in the Solar System', *Monthly Notices of the Royal Astronomical Society* 77 (1916): 84–112.
108. *Ibid.*, p. 104.
109. *Ibid.*
110. *Ibid.*, p. 105.
111. 'The Part Played by Rotation in Cosmic Evolution', *Monthly Notices* 77 (1917): 186–99.
112. 'Theories regarding the Origin of the Solar System', *Science Progress* 12 (July, 1917): 52–62.
113. 'The Motion of Tidally-distorted Masses with Special Reference to Theories of Cosmogony', *Memoirs of the Royal Astronomical Society* 62 pt. 1 (1917): 1–48.

114. 'On the Early History of the Solar System', *Monthly Notices* 78 (1918): 424–41.

115. *Problems of Cosmogony and Stellar Dynamics: Being an Essay to which the Adams Prize of the University of Cambridge for the Year 1917 was adjudged* (Cambridge: at the University Press, 1919).

116. *Ibid.*, p. 18.

117. *Ibid.*, p. 275.

118. *The Earth: Its Origin, History and Physical Constitution* (Cambridge: at the University Press, 1924).

119. *Ibid.*, pp. 34–35.

120. *Astronomy and Cosmogony* (Cambridge: at the University Press, 1928).

121. *The Earth: Its Origin, History and Physical Constitution* (2nd rev. ed.; Cambridge: University Press, 1929). Further editions of this work will be discussed in the next chapter.

122. *The Stars in Their Courses* (Cambridge: University Press, 1931), pp. 44–47.

123. *Problems of Cosmogony*, p. 284.

124. *Ibid.*, p. 15. Contrary to Jeans, Babinet did not consider the angular momentum of the whole planetary system. He merely offered calculations on the basis of the sun's present angular momentum about the equatorial speed of the solar nebula at its extension to various planetary orbits.

125. *Ibid.*, pp. 246–53.

126. *Ibid.*, pp. 128–31 and 276–78.

127. *Ibid.*, pp. 280–81.

128. 'Collision and the Origin of Rotation in the Solar System', *Monthly Notices* 89 (1929): 636–41.

129. *Ibid.*, p. 60.

130. Jeffreys did so in his Bakerian Lecture, 'The Origin of the Solar System', delivered on June 12, 1952. For its text, see *Proceedings of the Royal Society* (London) 214A (1952): 281–91. For quotation, see p. 282.

131. 'Versuch einer Entwicklungsgeschichte der Planeten', *Gerlands Beiträge zur Geophysik* 17 (1927): 1–68.

132. The reference to Chamberlin consisted in quoting his popularising booklet, *Origin of the Earth*, without any further identification. Of Darwin's works only his *Tides* was mentioned.

133. 'On the Electrostatic Field of the Sun due to its Corpuscular Rays', *Koninklijke Nederlandsche Akademie van Wetenschappen: Proceedings of the Section of Sciences* 33 (1930): 614–18 and 719–22; for quotation, see p. 720. Berlage's solution to the peculiar distribution of mass and angular momentum in the solar system was as follows: 'It happened that the mass of the disc was neglegible [*sic*] compared with the mass of the sun, the moment momentum of the sun, however, neglegible [*sic*] compared with the moment of momentum of the disc' (*ibid.*, p. 720). The same year saw the publication of Berlage's popularising cosmogony, *Het Onstaan en Vergaan der Werelden* (Amsterdam: Wereld-bibliothek, 1930).

134. 'Spontaneous development of a gaseous disc revolving round the sun into rings and planets', *Proc. Kon. Ned. Akad. v. Wetensch.*, 43 (1940): 532–40 and 557–66; for quotation, see p. 566.

135. *Ibid.*, p. 534.

136. 'The Early History of the Solar System on the Collision Theory', *Monthly Notices* 89 (1929): 731–38.

137. 'The Dissipation of Planetary Filaments', *The Astrophysical Journal* 90 (1939): 675–88; for quotation, see p. 687.

138. *The Solar System and Its Origin* (New York: The Macmillan Company, 1935).

139. *Ibid.*, p. 114.

140. *Ibid.*

141. *Nature*, 136 (1935): 933.

142. *Ibid.*

143. The *L'origine et formation des mondes* (Paris: O. Doin, 1922) by T. Moreux, director of the

Observatory of Bourges, closely reflected ideas already submitted by du Ligondès and Belot. B. Fessenkov's modification of the Laplacian theory ('L'évolution du système solaire', *Scientia* 40 [1926]: 9–16) consisted in the derivation of the gaseous nebula from a cloud of meteorites and in the assigning the formation of planets to convective turbu-lances inside the solar nebula. In his *Constitution et l'évolution de l'univers* (Paris: Librairie Octave Doin, 1927, pp. 378–411) A. Veronnet tried to give new respectability to old ideas about the spontaneous development of points of high density in the primordial cloud and of their 'slight' dissymmetry with respect to a central point of attraction. His bibliography contained neither the names of Babinet, Fouché, and Jeffreys, nor the original publications of Moulton and Chamberlin. These two, and especially Chamberlin, ignored to the end the various developments subsequent to the formulation of the planetesimal theory. In the almost 300-page-long monograph which Chamberlin published in 1927 under the title, *The Two Solar Families: The Sun's Children* (Chicago: University of Chicago Press) there were still no references to Jeans and to Jeffreys! Its section on the evolution of the system of planets was largely a reprint of the corresponding section in Chamberlin's *The Origin of the Earth* (Chicago: University of Chicago Press, 1916). Nölke's suggestion in his *Der Entwicklungsgang unseres Planeten-systems* (Berlin: Ferd. Dümmlers Verlag, 1930, see especially pp. 207–26) that the sun and the planets were born about the same time as condensations in a wisp of nebulosity was accompanied with the positing of a long list of rare occurrences somewhat anal-ogous to See's capture theory. For a discussion of Nölke's theory, see H. Jeffreys, 'On the Origin of the Solar System', *Monthly Notices* 92 (1932): 887–91, and Nölke's reply, *ibid.*, 93 (1933): 159–61.

144. *The Solar System and Its Origin*, pp. 135–36. Russell characterised the new approach a 'wild assumption', partly because the companion of the sun could not be greater than one percent of the sun's mass, a situation for which there was no astronomical evidence whatever.

145 'The Origin of the Solar System', *Monthly Notices* 96 (1936): 559–68. Lyttleton noted at the outset that Russell's new approach occurred to him independently 'before finishing reading Professor Russell's book'.

146. *Ibid.*, p. 568.

147. This detail was developed in Lyttleton's 'The Origin of Satellites', *Monthly Notices* 98 (1938): 633–45.

148. *The Solar System and Its Origin*, pp. 136–37.

149. *Ibid.*, p. 138.

150. 'A Few Remarks on the Dynamical Tidal Theory of the Solar System', *Astronomische Nachrichten* Nr. 6190; 258 (1936): cols. 381–84.

151. *Problems of Cosmogony . . .*, p. 17.

152. *Ibid.*, p. 288.

153. In his lecture 'The Origin of the Solar System', delivered at the Royal Institution on Feb. 15, 1924. The full text of the lecture was printed in *Nature* 113 (1924): 329–40; for quotation, see p. 340.

154. *Astronomy and Cosmogony*, p. 401.

155. *Ibid.*, p. 311.

156. *Ibid.*, p. 402.

157. *The Stars in Their Courses*, p. 47.

158. *The Nebular Hypothesis and Modern Cosmogony: Being the Halley Lecture Delivered on 23 May, 1922* (Oxford: Clarendon Press, 1923), p. 30.

159. *Eos or the Wider Aspects of Cosmogony* (London: Kegan Paul, Trench, Truebner & Co. Ltd., 1928), p. 86.

160. *The Universe around Us* (New York: Macmillan, 1929), p. 323.

161. *Ibid.* The last or fourth edition revised by Jeans carried both statements unchanged (Cambridge: University Press, 1945, p. 284).

162. *The Mysterious Universe* (New York: The Macmillan Company, 1930), pp. 1–2.

163. An afternoon lecture, entitled 'Is there Life on the Other Worlds?', given on Thursday,

November 20, 1941, at the Royal Institution. For its text, see *The Smithsonian Institution Report, 1942* (Washington D. C.: Smithsonian Institution, 1943), pp. 145–50.

164. *Life on Other Worlds* (New York: The Macmillan Company, 1940), p. 282.

165. *Ibid.*

166. *Earth, Moon and Planets* (Philadelphia: The Blakiston Company, 1941), p. 238.

167. 'The Origin of the Solar System', *Monthly Notices* 108 (1948): 101. Jeffreys seemed to note this trend with satisfaction.

CHAPTER EIGHT

THEATRE-IN-THE-ROUND

Dazzling props: stellar fission, triple stars, novae

In presenting, in 1959, to Raymond Arthur Lyttleton the Gold Medal of the Royal Astronomical Society, W. H. Steavenson, its President, pointedly remarked that in 1936 all theories on the origin of the solar system 'were completely on the rocks'.[1] The idiom was fitting in more than one sense. It conveyed vividly the real measure of a classic impasse and its plural form might have even served as a thinly veiled hint that in addition to the 'famous angular momentum difficulty', to which Steavenson restricted his remark, there was at least another rock, the nearly circular orbit of planets. Certainly, Steavenson did not overlook any of the areas of modern theoretical astronomy where Lyttleton played a decisive role in attracting attention to new avenues of investigation. One of those areas was the question of the origin of planetary systems. The novelty of the avenue opened up by Lyttleton was unquestionable though the same cannot be said of his appraisal of its merits. It rested in part of the support which it seemed to give in his estimate to the age-old wish to see in planetary systems a frequent, nay normal phenomenon of the realm of stars.

That 'more interest would attach to a theory implying the frequent occurrence of solar systems in space than [to] another theory' was an artful understatement on Lyttleton's part as he formally introduced in 1941 the idea that fission due to the rotational instability of a coalesced double star might be the process that can turn the trick.[2] For if it was true that the accretion of interstellar mass by stars would cause any star 'likely to become the member of a double system and later a member of a triple system, and so on to a more complex system still', then it readily followed that 'the number of stars capable of providing the requisite initial conditions [for the formation of planetary systems] must be regarded as gradually increasing and this, given sufficient time, to practically any extent'.[3] Interest was, indeed, very much in that direction. Witness the variety of theories which saw print since the publication of Lyttleton's paper. Different as they could be they all evidenced the same interest and the same intent. The intent was to rely on motion in-the-round, that is, to explore anew the possibilities connected with a rotating medium.

The rotation, if it was to be of any promise, had to be conceived in a mould different from the one which Laplace utilised. The failure of the Laplacian form was emphatically noted by Lyttleton in 1938 in a paper on the formation of satellite systems[4] which became, in retrospect at least, the starting point of the new trend in searching for an explanation of the origin of planetary systems.

Concern for the origin of satellite systems had always been a secondary matter in all theories and for understandable reasons. Compared with the planets, satellites appeared puny both in size and importance. Once a theory seemed to cope with the planets, satellites were readily thrown in for good measure. This was particularly true about the tidal theory. Yet, as Lyttleton remarked, whatever the merits of the tidal theory about the planets, one had to recognise at long last its basic inadequacy with respect to the satellites. These were most numerous around the outer planets which were too far from the sun to justify the assumption that originally they had passed close enough to it to undergo tidal fission. Furthermore, tidal fission of a planet passing near the massive sun was to partition the planet into two or more comparable masses. With this the case of the four great planets presented a striking contrast as the mass of Jupiter, for instance, exceeds by a factor of about ten thousand the mass of its four great satellites. Furthermore, those satellites have masses comparable to that of the moon. Herein lay a noticeable difference with respect to the planets, the masses of which vary through a range of more than three orders of magnitude. Most importantly, the distribution of angular momentum between planets and their satellites formed the very reverse of the case between the sun and the planets. The angular momentum of the orbital motion of satellites is negligible when compared with the angular momentum of the rotation of planets.

It was to cope with that special world of satellites that Lyttleton submitted a novel form of rotational mechanism. According to it the actual planets were the products of the rotational fission of a protoplanet. When such a body broke into two large fragments due to rotational instability, a filament of matter developed between the two fragments as they separated from one another. Condensations in the end parts of the filament became satellites of two actual great planets, the two principal fragments of the fission of the protoplanet, whereas condensations in the central part of the filament developed into the small, or terrestrial, planets. Lyttleton identified the two principal fragments with Jupiter and Saturn all the more readily as the ratio of their masses is very close to 3:1, the ratio which is the critical value for the desired break-up of a rotating body with the form of a Jacobian ellipsoid [*Illustration XXXI*]. In addition to the large number of small satellites around Jupiter and Saturn the theory seemed to account for their nearness to their respective planets. The sense of rotation of the latter could also be seen as the determining factor of the sense of orbiting of their satellites. Lyttleton then claimed, though without the aid of detailed mathematical analysis, that his suggestions were 'the first to be put forward that give any hope of explaining the peculiarities of density distribution of the terrestrial planets and the similarity of the Moon to a number of other satellites in the system'.[5] The former claim could appear reasonable, as the central part of the filament possibly came from the deeper layers of the protoplanet, but the justification of the claim about the moon was unclear. At any rate, the earth-moon system was the product of mere chance in Lyttleton's theory. Since the ratio of the masses of Uranus and Neptune is much less than 3:1, they could not have originated from the same protoplanet. The idea of their separate origin could be supported by the great disparity of the direction

of their axes of rotation, but it also implied the escape of their matching fragments from the solar system.

Within the aim of Lyttleton's paper the question of the origin of protoplanets could, strictly speaking, be ignored. By comparing the birth of satellites
to the emergence of planets from a filament Lyttleton did not wish to endorse
the tidal theory, aware as he was of the fatal objections to it. Actually, he called
the protoplanets 'autonomous bodies' with a reference to the incompetency
of 'all outside agencies' to produce them.[6] It took, however, several years,
before the meaning of these words could be seen in the context of the new
theory which Lyttleton offered on the origin of planetary systems. The road
to it led him through a penetrating re-examination[7] of the question of rotating
fluid masses and of the further interaction between the products of their fission.
His starting point was a result obtained much earlier by E. Cartan who found
that when a rotating fluid mass having the form of a Jacobian ellipsoid becomes
secularly unstable, it also becomes ordinarily unstable, that is, it will rapidly
fission under the influence of a small disturbance.[8] From this Lyttleton concluded that displacement of the smaller part produced by the fission would increase exponentially. In other words, it could not acquire the angular momentum needed by a circular orbit, but would recede along a hyperbolic path
from the larger part. While the fission process had to be recognised as impotent for the purpose of producing binary stars and for the purpose of producing planets, Lyttleton perceived that binaries presented an aspect until then
unexplored. It related to the interaction of their components provided they
were close to one another. Eager to seize on promising research topics over a
broad range of interest, Lyttleton was also working, in collaboration with F.
Hoyle, on the question of the increase of stellar mass through accretion of interstellar matter.[9] Increase of mass meant in the case of binaries their inevitable
rapprochment and coalescence. Analysis of the stages subsequent to this showed
that the new unit would, partly because of its increased rotation, break up into
two components with a filament extending between them.[10]

In most cases the ratio of masses resulting from the fission could be shown to
be such as to turn the two fragments and the filament into wholly independent
bodies. But, and this was the crucial step toward the new theory, what if the
double star going through this process formed part of a system, of which our
sun was a remote third member? In that case it was conceivable that the
filament, and especially its major condensations might be captured by the sun
as its protoplanets. The process was for Lyttleton not only conceivable, but also
sufficiently frequent, since, as he hastened to point out, triple stars consisting
'of a very close pair [the two parent stars] and a distant body [the sun]' were
'by no means rare'.[11] Thus the new mechanism seemed to secure to planetary
systems 'a much more frequent occurrence' than 'any process relying on an
encounter'.[12] As a further corroboration of this, Lyttleton added a remark,
already quoted, on the tendency of stars to group into larger and larger clusters,
again, in part, because of the accretion of their mass, which could only increase the chances for the formation of special triple stars capable of giving rise
to planetary systems.

To be sure, Lyttleton emphasised that only in a 'small proportion' of these triple stars would the distant body (the sun) be in a suitable position for capturing the filament containing the protoplanets.[13] Yet this 'small proportion' represented in his eyes a value well exceeding the one that could be obtained on the basis of close encounters. Analysis of the interplay of masses and distances in the triple system in question suggested to him that the limitation arising from the proper distance of the sun was 'not very serious'.[14] The same analysis showed, however, the distinct possibility that one of the main fragments produced from the fission of the coalesced double star would also be captured by the sun, a circumstance clearly incompatible with the formation of a stable planetary system around it. For, as Lyttleton himself noted, the formation of satellites and that of the planets, though very similar in many respects, had one crucial difference. In the case of satellites the filament and the two parent bodies (the fragments of the protoplanet) did not have to become gravitationally independent. In the case of the formation of planets it was absolutely necessary that the filament containing the protoplanets should be left completely behind by the fission products of the double star. The decisive role played by the filament was a matter which must have given second thoughts to anyone familiar with Spitzer's devastating conclusions concerning its role in the tidal theory.[15] The counter-arguments offered by Lyttleton[16] were not to generate confidence in his ingenious effort to explain the origin of the solar system on the basis of a rotational process within a very specific arrangement of three stars.

Partly because of Spitzers conclusions even less was the lasting value of the modification of Lyttleton's theory by Hoyle who replaced the triple star with a double star of which one was the sun and the other a nova.[17] According to Hoyle the planetary material was provided by an exploding nova whose two principal fragments left behind a filament in rotation around the sun. If the derivation of the filament from a nova appeared ingenious, it also pre-empted the chances for planetary condensations because the very high temperature of such a filament must have brought about its quick dissipation. There was a similar touch of contradictoriness in the manner in which Hoyle deplored the attitude of other proponents of planetary theories, each of whom, so Hoyle remarked, submitted his own theory as the only tenable one. In the same breath he referred to Lyttleton's theory, which was also his own, as the only one not 'untenable for obvious dynamical or physical reasons'.[18] To this he added that exploding novae formed the observational evidence on behalf of the theory advocated by Lyttleton and himself, and that the number of planetary systems had, therefore, to be comparable to the number of novae occurring in double stars.[19] When a year or so later he sought 'observational evidence' for the foregoing formation of planetary systems in the explosion of the supernova of 1054 (the Crab nebula),[20] he merely served evidence of the arbitrariness by which sophisticated observations can be manipulated to fit highly specific parameters, among them the amount, temperature, and velocity of the filament. He gave no indication of the observational traces of that long-exploded supernova from which our sun might have received the material needed for its planetary system. His was a dazzling but unconvincing performance in the theatre-in-the-round symbolised by some rotating medium or system taken as a starting point of explanation.

Further props: magnetism and ionisation

Another memorable recourse to a coupling of rotational dynamics with other physical factors, as a solution to the problem, came in late 1941 from Hannes Alfvén who was to receive twenty years later the Nobel Prize for his work in magneto-hydrodynamics. Reliance on the role of the sun's magnetic field was one of the principal characteristics of his theory,[21] the main aim of which was to explain the transference of much of the angular momentum of the sun to the planets. In that theory the sun was already formed and rapidly rotating when it passed through a gaseous cloud. Because of gravitational attraction the gas molecules would then be drawn toward the sun and undergo collisions resulting in their ionisation. Once ionised, they would be almost entirely under the influence of the sun's magnetic field with their courses presumably changed into a circular one in the equatorial plane of the sun. The work done in that process is at the expense of the sun's angular momentum.

Such was undoubtedly a novel approach to the problem (Alfvén was unaware of Birkeland's and Berlage's work), but represented only one step on a long and arduous road full of pitfalls. One of them was the distance from the sun at which molecules of various atomic weight supposedly became ionised as they fell 'from infinity' toward the sun. Since this distance was proportional to the atomic weight, the material of terrestrial planets mainly composed of heavier elements could not conceivably get, even by further spiralling toward the sun, into the present orbit of those planets. To extricate his theory from this obvious breakdown Alfvén postulated that following the formation of the big planets the sun passed through a suitable cloud of meteoric dust which in turn was volatilised and ionised in the sun's heat. Clearly, this was a recourse to a very fortuitous *ad hoc* factor, lessening considerably the chances for the formation of an orderly system of planets, a point which Alfvén did not make explicit. Some of his phrases were rather expressive of the age-old impasse: 'It is difficult to say how the volatilisation of the dust cloud would be effected. . . . It is difficult to understand why the gas should condense again, to form planets'.[22] There was in addition a grave inconsistency in the theory which escaped Alfvén's attention. His derivation of the respective distances for ionisation depended on the assumption that the gas molecules acquired the kinetic energy needed for that ionisation by falling toward the sun from infinity. This meant that they suffered no ionising collisions until they reached the vicinity of the critical distance where collisions suddenly became very frequent. In other words, the theory implied both an exceedingly long and an exceedingly short free mean path in order to secure the desired result. No wonder that as Alfvén proceeded with the application of his theory apparent successes were outnumbered by frustrating inconsistencies. As a result, in his second communication Alfvén had to raise from two to three the number of dust and gas clouds through which the sun was supposed to pass, and in a specific sequence at that, to acquire a planetary system of its own.[23] Finally, serious question marks emerged when a closer look was taken at the magneto-hydrodynamical waves effecting the transfer of angular momentum from the sun to the ring of ions occupying Jupiter's orbit. Could, for instance, the sun possess a dipole moment one million times larger than its present value when Jupiter was formed?[24] It was still fully true what Alfvén had earlier written that 'it is a very precarious task to develop a cosmogony'.[25]

The measure of Alfvén's dedication to that 'precarious task' was the monograph which he published in 1954 on the subject.[26] Its principal feature was the effort to utilise to the full the wealth of details which accumulated during the intervening years on the magnetic and electric properties of inter-stellar gas and dust, and on the early phase of stellar evolution. Alfvén thought that the presence of very dense and hot cloud layers around T Tauri stars might alone provide all the requirements which in the earlier form of his theory only the passage of the sun through three different clouds could secure. Alfvén also felt that the different processes of infall, orbital acceleration, and ionisation could now go on at the same time. His confidence in the transference of the sun's angular momentum to the planets by magnetohydrodynamical waves became now so strong as to let him state that 'all stars with slow rotation should possess planetary system'.[27]

Willingness to claim high frequency for planetary systems was once more an attitude shared by all other investigators for reasons of their own. Common was also to all theories the pattern which Alfvén described with a view on recent past: 'During the last few decades the investigations in the field have been very intense, but the main result has been a greater number of divergent hypotheses rather than real advance along one promising line'.[28] That Alfvén's own theory did not break this pattern derived also from the typical expectation that a particular physical process might do after all what it clearly could not. In the case of Alfvén's theory this point was unsparingly laid bare by Lyttleton. His remark concerned the very foundation of Alfvén's theory, the sun's mag-netic field. Did not Alfvén somehow assume the existence of a property of that field which it logically could not possess? Or as Lyttleton put it: 'It came as a surprise to find explicit reference to this kind of difficulty only in the closing pages of the book (p. 180) where it is suddenly remarked that *the lines of force must not go directly from the cloud to the central body*, whereas throughout most of the preceding 179 pages the reader has been allowed to think that the sun's dipole field, which has this inconvenient property, is suitable (at however enhanced a strength). Full attention to the difficulty might of course have involved a complete reconsideration of all that had gone before, and it is a moral problem for an author making such a discovery late in a piece of work to decide what he shall do about it'.[29] It should give a good glimpse of the true merits of some highly acclaimed scientific writings that not a hint was made in them to that aspect of Alfvén's theory in their glowing endorsement of it.[30]

The magic tenfold and five roller-bearings

The line of advance which was to show the highest promise was first jotted, though not without some ambiguities, by Bertrand M. Peek in late 1942.[31] He saw the principal merits of his theory in its great simplicity, in its method of coming to grips with the question of angular momentum which, in his words, 'sounded the knell of so many previous theories',[32] and in its implication that 'for a star to be accompanied by planets should be by no means a rare occurrence'.[33] The theory was certainly simple in the sense that it contained no mathematical analysis of any of its proposed stages of the evolution of the solar system. Its chief originality lay in tracing the angular momentum to random, turbulent motions and instabilities which broke up the original, diffuse, cosmic medium. Consequently, the primordial solar nebula, as pictured by Peek, had a dimension comparable to the average distance among stars and was not rotating, like its Laplacian counterpart, as a rigid body. Because of that lack of rigidity Peek assumed that much of the angular momentum was 'concentrated close to the rim of the invariable plane and, initially at any rate, may even be confined to a comparatively small arc of the circumference'.[34] This meant, in Peek's reasoning, that 'the bulk of mass will not therefore assume a lenticular form but will approximate more and more closely to a sphere as concentration proceeds and the effect of gas pressure increases'.[35] A planet then began to form when a small volume of matter near the rim detached itself from it. At this point Peek added the unclear phrase containing a fortunate detail: 'If the mass of all the particles moving in the original random current amounted to ten times the total mass of the planets, there would seem to be a good chance that enough of it would be attracted to our particular condensation S to furnish, after allowing for losses by diffusion and other causes, the material that would ultimately leave the rim and form the planets'.[36] Since condensation S, as shown in Peek's diagrams [*Illustration XXXII*] was the contracting sun, the rim mentioned in the context had to be its rim and not that of the original nebula with the angular momentum concentrated in *its* rim. It, therefore, still needed to be explained how the tenfold mass of planets had been first attracted to the contracting sun only to be detached from it later by the angular momentum.

The summary of the theory which appeared in early 1943 in the pages of *Nature*[37] did not mention the tenfold of planetary masses. It was rather emphasised that according to Peek the developing solar nebula was not lenticular or flat but spherical. That summary is worth noting because it was referred to by Carl Friedrich von Weizsäcker who was completing in Germany in mid-1943 a major theory on the origin of the solar system based in part on the idea of a flat, gaseous envelope (flache Gashülle) around the sun.[38] Curiously, Weizsäcker emphasised that the original substratum of planets had to possess the tenfold of planetary masses to permit a solution to the riddle of angular momentum. The similarity between Peek's and Weizsäcker's theories on this point was very superficial. Also, Peek's paper, or its summary, contained no reference to the question of the relative distances of planets, a chief target of Weizsäcker's paper dedicated to the 75th birthday of Arnold Sommerfeld who, as Weizsäcker noted, had best insisted on the analogy between the system of planets and the structure of atoms.[39] The remark was all the more fitting as in Weizsäcker's theory the radii of planetary orbits embodied a process of quantisation vaguely analogous to the one determining the set of electron orbits around the atomic nucleus.

The first step toward that crucial juncture in Weizsäcker's theory was a short survey of the historical background of his entire topic,[40] hardly the most attractive part of his essay. There was little merit in his claim that his own theory had closer ties with the ideas of Kant than with those of Laplace. Again, Weizsäcker gave undue importance to Nölke's proposal that the planetary system originated from an irregular nebula by presenting it as being one of the three main types of theories together with the condensation and collision, or close encounter, theories. Once more the history of the question suffered when Weizsäcker presented the second part of the dilemma facing the condensational theories. According to him, if the sun had already shrunk to its present size at the start of the formation of planets, the spreading out of the mass of planets would have yielded a medium of such low density out of which their condensation could not have taken place. This was undoubtedly true, but it was another matter to label it as 'Roche's condition'.[41]

The first part of the dilemma derived from the assumption that at the beginning of the formation of the planetary system the sun's mass was spread out to the present limits of the solar system. In that case one was faced with the classic problem of a hundredfold decrease in the sun's angular momentum. Weizsäcker's starting point for solving that problem was the latest research which indicated the overwhelming preponderance of hydrogen and helium in the sun's composition. Elements heavier than oxygen were found to constitute only one percent of the sun's mass, whereas much of the mass of the planets, and especially of the terrestrial planets, was due to the presence of those heavier elements. At this point Weizsäcker introduced the notion of a gaseous envelope surrounding the sun which he presumed to be in its early stage of development. He also postulated that the temperature in the gaseous envelope was at distances equivalent to present planetary distances the same as the surface temperature of the respective planets themselves. With this postulate Weizsäcker wanted to secure the role of condensation into dust particles and droplets during the process of planetary agglomeration inside the gaseous envelope which, of course, could not rotate in a stable condition. 'Thus, it is demanded by the principles of conservation of energy and of angular momentum that a part of the gas[eous envelope] should fall into the sun, and the energy thereby liberated enables the rest [of the gaseous envelope] to dissipate itself into cosmic spaces by carrying along all the angular momentum'.[42]

A detailed account of this process assumed the transformation of the rotating nebular envelope into a flat disk within which molecules had Keplerian orbits. The disk comprised about one-tenth of the sun's mass, extended to the orbit of Pluto and had a width of about one-hundredth of the radius of that orbit, that is, about forty times the sun's radius. The average density of matter within the disk was equal to one billionth of a gramme per cubic centimetre. Weizsäcker recognised that such a disk could form around the sun only during some earlier phase of galactic development. While this point could remain qualitative, the dissipation of much of the rotating gaseous disk demanded a quantitative analysis if the solution of the problem of angular momentum was to be convincing. Unfortunately, Weizsäcker did not carry his investigations in this respect beyond the conclusion that 0.71 of the material in the disk escaped because of viscous forces. These in turn were due to the variation of angular velocity with distance from the centre in agreement with Kepler's Third Law. Consequently much of the outer part of the disk became gradually detached from the system while much of the inner part was falling on the sun. The question was left unanswered whether this fall of additional (and rotating) material into the sun would indeed alter its rotation in the sense as to produce the great inequality of the distribution of angular momentum between the sun and the planets. Compared with the absence of specific information on this crucial point it was of relatively small advantage to learn that the dissipation of the disk required the same amount of time, about 100 million years, as did the accretion of planetary masses. Since the time in question was much shorter than the actual age of the solar system, questions about observational evidence of the primordial disk could readily be dispensed with.

The next phase in Weizsäcker's discussion concerned the question whether there was any possibility for the formation of systematic and quasi-stable patterns among the particles orbiting in the rotating disk. According to him this possibility could very well be substantiated if one assumed that inter-action between such particles was almost exclusively gravitational. While this assumption could be taken plausibly (Alfvén was only beginning to write on the subject), it was otherwise with Weizsäcker's next step which constituted the very core of his theory. The step consisted in the derivation of a set of concentric rings around the sun, each ring consisting of the same number of vortices. The diagram [*Illustration XXXIII*] of this set gained immense publicity in the pages of subsequent technical and popular discussions of planetary origins, but Weizsäcker gave only hints about the actual steps of information and intuitive insights which led him to conceive the figure as a clue to the formation of a planetary system like ours.[43] His interest was not that of the historian but that of the mathematical physicist concerned with the systematic justification of the final product. This he felt to have achieved by considering first only particles moving in Keplerian orbits of different eccentricities but having the same period and consequently the same major axis. Now, if a particle moving around the sun in a circle is viewed from a rotating co-ordinate system having the same period, the particle would appear motionless, or in other words its position would correspond to a point. If the particle moves along an ellipse of small eccentricity the picture of its path in the rotating frame of reference would be a small ellipse. This ellipse would not only become larger with the increase of eccentricity, but would turn into a bean-shaped figure. It is, therefore, conceivable, mathematically that is, to picture these elliptical or rather bean-shaped figures as smoothly enclosing one another and forming larger and larger concentric sets of bean-shaped vortices, provided that eccentricities of slightly varying values were grouped together.

The physics of the question, or the actual development of sets of vortices not destroying one another, demanded that an upper limit be set to the growth of each vortex. This upper limit was sought by Weizsäcker in a reasoning which was a most attractive though not a truly rigorous aspect of Weizsäcker's theory. By postulating a minimum of energy dissipation in the interaction of vortices in a ring he inferred that the sum of their angular widths must be equal to 2π, or a complete circle. Within the restrictions adopted by Weizsäcker concerning the eccentricities of Keplerian orbits (he excluded eccentricities larger than $1/3$), the foregoing 'quantum condition' implied five vortices within one ring. The chief consequence of this arrangement was a specific ratio of the radii of successive rings, a ratio of which the Titius–Bode law was a limiting case.[44]

This 'unveiling a so far hidden kind of integral number relation [Ganzzahligkeit]',[45] as Weizsäcker characterised his feat, had a subtle touch of the kind of apriorism which seeks one-to-one correspondence between a specific mathematical pattern and a very specific pattern of nature, in this case the actual distances of planets. As has always been the case with the application of the various refinements of the Titius–Bode law. Weizsäcker's theory too received some support from the facts. In connection with the distances of Mars, Jupiter, Saturn, and Uranus, he could proudly note that 'our values fit the observations somewhat better than do those of Titius and Bode'. But almost in the same breath he had to admit that 'as to the other planets our representation does not hold true'.[46] The partial success, or failure, of the theory was fully pictured in a Table[47] whose data were soon to cast doubt on the validity of the physical process underlying the determination of planetary orbits in Weizsäcker's theory. Its inability to cope with the task became further evident when the question of the formation of the bodies of planets was taken up. The place of the formation was between the vortex rings, and especially in the roundish spaces formed by three vortices touching one another. In those five 'roller-bearing' spots, unlike in the vortices themselves, the turbulence seemed to be sufficiently high to help produce globules considerably larger than dust particles. The emergence of those globules was indispensable for starting the growth of proto-planetary bodies, as Weizsäcker recognised that collision between two dust particles would most likely result in their volatilisation. On the other hand, collision of dust particles with globules could conceivably be pictured as a process leading to the emergence of bodies sufficiently large to acquire further material through gravitational attraction. During the time needed for the growth of globules to larger sizes the vortices were to retain their stability to assure both the continued growth of globules and their proper spacing from the centre. But, as Weizsäcker admitted with commendable frankness, the 'assertion of the respective stability of the vortex system is by far the most uncertain part of the theory'.[48] No more certainty exuded from Weizsäcker's speculations on the coagulation of the five protoplanetary bodies from the five 'roller-bearing' spaces into one single planet. Unfortunately,

Weizsäcker seemed to be unaware of previous investigations, especially of those of Moulton, which had established that a chain of bodies within one ring would not coalesce into one single body.

In view of all this it helped little that in Weizsäcker's theory the sense of the rotation of planets coincided with the direction of their orbital motion, since the rotation of vortices was opposite to the latter. The basic uncertainty about the formation and stability of vortices reflected also on the assumption that the very same process which brought about the planets could be effective in giving rise to systems of satellites around the planets themselves. Little if any assurance was carried in Weizsäcker's final remark that when one of the planets grew in size too rapidly a double star was the result and not a planetary system.[49]

Magic celebrated, discredited, and courted again

Weizsäcker's own trust in the role viscosity would play in a rotating medium, was well in evidence in his theory, and in particular in his non-technical summary of it published in June 1946.[50] A year later he even sought a derivation of the sizes, shapes, and dynamics of nebulae on the basis of turbulent motion in the primeval gaseous cloud.[51] His confidence for doing so could in part be ascribed to the fact that discussions of his theory were delayed by wartime conditions preventing a sufficiently broad circulation of the issue of the *Zeitschrift für Astrophysik* which carried Weizsäcker's article. At any rate, the first reactions to it were markedly enthusiastic in the United States. There G. Gamow and J. A. Hynek hastened to announce in *The Astrophysical Journal*, after receiving a reprint of Weizsäcker's paper in the Spring of 1945, that Weizsäcker 'has freed the "one-star" hypothesis of its classical difficulties and has allowed an interpretation of the Bode-Titius Law of planetary distances'.[52] In Gamow's and Hynek's estimate Weizsäcker's theory explained 'all the principal features of the solar system', namely, the common plane of revolution, the small eccentricities of orbits, the sense of rotation, and the lower densities of the larger planets.[53] The only point where Gamow and Hynek hinted at difficulty concerned Weizsäcker's own acknowledgment that the process of the emergence of a single planet from the five 'protoplanets' formed in the roller-bearings was 'difficult to visualise'.[54] Within such an optimistic outlook on the theory it was natural for Gamow and Hynek to assert that 'if the Weizsäcker theory holds, planetary systems of a wide variety of types must be the rule rather than the exception'.[55]

The growth of confidence in the truth of Weizsäcker's theory was further enhanced when a year later, in 1946, the noted astrophysicist at the University of Chicago, Subrahmanya Chandrasekhar, published, without adding critical comments, a lengthy account of it in the pages of *Reviews of Modern Physics*.[56] Two years later Thornton Page brought his article, 'The Origin of the Earth', in *Physics Today* to a conclusion with an outline of Weizsäcker's theory which, as Page put it, 'accounts for more of the observational data than any of the previous speculations', and implies 'that the formation of planets should be an extremely common occurrence'.[57] To this Page added words that should speak for themselves: 'Thus we might expect billions, if not hundreds of billions of planets in our galaxy, the strong likelihood that life has developed on a million or more of them, the high probability that there are other civilisations of mankind, and even the possibility that men on other planets are writing articles on the origins of their solar systems!'[58]

Implicit or explicit encomiums heaped on Weizsäcker's theory should have appeared all the more dubious by 1948 as two years earlier K. E. Edgeworth published his studies on certain aspects of stellar evolution.[59] Edgeworth took up the question of planetary evolution only as a sequel to his main theme and made no reference to Weizsäcker, yet his well-argued conclusions were in a sense a rebuttal of the latter's theory. According to Edgeworth, there appeared insurmountable difficulties against tracing the origin of stars either to millions of simultaneous condensations in a gaseous sphere as large as the galaxy, or to successive condensations within a rotating disk of similar dimensions. One of the difficulties was that in such cases stars would have acquired much larger angular momentum than they actually have. Edgeworth, therefore, proposed the break-up of the galactic disk into filaments within which condensations with low angular momentum could occur. If then such a condensation (protosun) captured some stray material from the filament, then one could have, in Edgeworth's words, a theory which would explain 'without any special or artificial assumptions, how the solar system may have acquired its remarkable distribution of angular momentum'.[60] The price of avoiding 'artificial assumptions' was, of course, the admission that the process of planetary formation was, like that of star formation, anything but an 'orderly' process. This in turn meant that our planetary system, contrary to the implication of Weizsäcker's theory, was not a typical feature of the stellar universe.

Weizsäcker's theory certainly did not appear promising once subjected to a close scrutiny. This was done independently by F. Nölke, whose paper appeared posthumously in 1948, and by Dirk ter Haar, who devoted to Weizsäcker's theory a considerable part of his doctoral dissertation which he defended in Leiden in June of the same year. Since Weizsäcker's ideas about the transformation of Keplerian orbits into rings of vortices were not precise enough with respect to the underlying physical process, three of Nölke's four objections[61] could not be strictly conclusive. It was otherwise with his fourth objection which concerned Weizsäcker's explanation of the distribution of angular momentum in the solar system. Nölke showed that if one considered only the formation of Jupiter from the original gaseous disk, exceeding by a factor of 100 the mass of Jupiter, then on the basis of the conservation of angular momentum 77 Jupiter mass had to fall back on the sun to permit the escape of 23 Jupiter mass carrying away the amount of angular momentum of which the system had to be liberated to conform to its present-day features. If the angular momenta of the other planets too were considered then one had to assume the fall into the sun of an additional amount of matter equal to 26 Jupiter mass to permit the escape from the system of an amount of matter equivalent to 14 Jupiter mass. The fall into the sun of 103 Jupiter mass meant, however, the slowing down of the sun's rotation by several hours,[62] making thereby the distribution of angular momentum in the solar system even more lopsided than it actually is.

That Weizsäcker's theory provided no satisfactory mechanism for the distribution of angular momentum was also pointedly brought out in ter Haar's dissertation.[63] According to him, if the present slow rotation of the sun was to be explained by the falling on it of part of the material of the original flat disk, then the difference between the actual value and the one predicted by the theory was of the order of one hundred thousand.[64] The discrepancy was only two magnitudes smaller when the explanation was sought on the basis of the transport of angular momentum from a rapidly rotating sun with the aid of waves of turbulence.[65] That Weizsäcker's set of vortices could be derived from hydrodynamical considerations was a chief claim of ter Haar, but it largely rested on his accepting the validity of a short essay by Jaakko Touminen.[66] The latter concluded, by using the formulas of Prandtl and von Kármán on transport of energy in eddies, that 'eddies caused by turbulence and those contemplated in von Weizsäcker's theory (combined with the law of Titius and Bode) *are of the same order of magnitude and that both increase linearly with the distance from the sun*'.[67] The italics attested the import Touminen saw in his finding.

Whatever the apparent promises offered by further investigations of the role of turbulence, ter Haar felt impelled to conclude that 'there seems at present to be no theory which can explain satisfactorily the various properties of our solar system'.[68] While describing a year later, in 1949, Weizsäcker's theory as the one that 'has had the most striking successes', Harold Spencer Jones also took the view in his popular survey of the latest on the topic that 'there is as yet no theory which can explain satisfactorily all the features of the system'.[69] The diagram [*Illustration XXXIV*], which the Astronomer Royal offered on 'the arrangement of vortices and roller-bearing eddies as pictured by Weizsäcker', was a good illustration of the furtive liberties that are often made with originals. Jeffreys, who a year earlier gave his own evaluation of the state of the art, spoke with pronounced reserve of Weizsäcker's theory, and urged further exploration of the role which turbulence and viscosity might play in planetary theories.[70]

Sober evaluations of a particular question did not necessarily convince all scientists working in the field. This could very well be seen in the manner in which Gamow presented in 1952 to uncounted readers of his *Creation of the Universe* the theory of Weizsäcker as a most satisfactory answer to the question of the evolution of planetary systems.[71] Even Jones seemed to become oblivious to the unsatisfactory points of Weizsäcker's theory, when he took up the topic in the same year in the second, revised edition of his *Life on Other Worlds*.[72] The only saving grace in his arguments in support of the high frequency of planetary systems consisted in his warning against visioning hominid civilisations around all neighbouring stars: 'It is unlikely', he reminded his readers of a basic facet of evolutionary theory, that 'evolution has followed a parallel course on any two worlds'.[73] Leaning heavily toward the likelihood of planetary systems everywhere in the universe Jones now had to put Weizsäcker's theory in the best possible light. According to him the theory afforded a 'plausible explanation' of all the principal features of the solar system,[74] and represented 'a workable scheme'[75] even for the explanation of the distribution of angular momentum between the planets and the sun. Such a benevolent evaluation was obviously motivated by the great frequency of planetary systems which Weizsäcker's theory seemed to secure. That it had been particularly influential in this regard was unwittingly recognised in the remark of the Astronomer Royal who noted with satisfaction 'the marked change in outlook in the last few years from that of twenty years ago'.[76]

That preferences and factual results kept pointing in opposite directions was well evidenced by work on the role of turbulences. The first to voice the small promise of turbulences for planetary theories was Chandrasekhar who devoted to the cosmogonical role of turbulence the Henry Norris Russell lecture of 1949. There he took the view that Weizsäcker's theory would prove no exception to the rule that 'it is the usual fate of cosmogonical theories not to survive'.[77] As further work was to show, turbulences and eddies could not be tamed for the purposes of an orderly and typical planetary evolution. Ironically, Weizsäcker provided ammunition against his theory with his study, made independently of D. Kolmogoroff, of the energy spectrum of eddies,[78] although he failed to perceive this, as did ter Haar and Chandrasekhar. These two claimed in a joint paper in 1950 that the Titius-Bode law was still the logical result when Kolmogoroff's conclusions were taken into account.[79] Quite contrary was the view expressed by Gerard P. Kuiper in a long essay which had been in a limited circulation for more than a year before its actual publication in 1951.[80] The revision of Weizsäcker's diagram provided by Kuiper[81] [Illustration XXXV] not only showed sufficiently large irregularities and discontinuities in the succession of vortex rings, but also the distinct possibility of the formation of 'interior eddies' within the vortices themselves. Clearly then, the regular pacing of planets could not reasonably be ascribed to vortices arising from turbulence. Or as Kuiper put it: 'It now appears established that the Kolmogoroff energy spectrum applies to the solar nebula, and we must look for an explanation of the Bode law elsewhere'.[82]

Gravitational instability, viscous disk, and ample confidence

Three other shortcomings of Weizsäcker's theory, which Kuiper pointed out, concerned the highly idealised process of accretion from dust particles to globules, the improbability of the persistence of large vortices for a sufficiently long time, and the relation of the shrinkage of the sun to the formation and dissipation of its discoidal nebula.[83] This last point could, as will be seen later, be turned against Kuiper himself who lost no time in submitting his own solution to the question of planetary origins. Without abandoning entirely the idea of vortices ('we see no escape from the conclusion', he wrote, 'that the planets derived from some kind of primary vortices'[84]) he proposed the idea of gravitational instability as the 'magic' factor in combination with the rotation of a nebula around the sun. In doing so, Kuiper set great store by the observational evidence that one in every two stars was part of a binary and that much smaller was the proportion of truly single stars. This implied the 'satisfying' situation, to quote Kuiper's very word, that 'the process of planetary formation is but a special case of the almost universal process of binary-star formation'.[85] He added forthwith that probably some 10^9 planetary systems existed in our galaxy alone and that his theory would make, therefore, the formation of planetary systems an occurrence 10^8 times more frequent than in the case if such systems depended on collision or near collision between two stars.

A hundred million planetary systems meant the presence of planets around one in every hundred stars in our galaxy, a ratio which closely corresponded to the restricted range in the density of a nebula, if its central part developing into a star was to have planets of its own. If that density was in various parts of the discoidal part of the nebula well above what Kuiper called the Roche-density[86] (the self gravitation of a cloud of gas equals at that density the force of solar tides), then binary stars developed, whereas innumerable small condensations were the results when the density was well below the Roche-density. The development of seven or eight large condensations was, therefore, in Kuiper's theory a somewhat fortuitous occurrence, as was the emergence of an equal number of large vortices. The same held true of their relative sizes and Kuiper did, indeed, remark that the 'Bode law' was not really a law but merely a reflection of the initial density distribution in the solar nebula. Within the context of his theory this distribution could take on many different forms leading to as many types of planetary systems. It also followed from Kuiper's theory that the solar nebula broke up under gravitational instability into as many parts as there are planets and that the original, still gaseous, protoplanets were so large as to touch one another.

A consequence of such a configuration was the strong impact of solar tides on the protoplanets, bringing about a synchronisation between their rotational and orbital periods. With their further contraction this synchronisation yielded to much shorter rotational periods. Thus the actual angular momentum per unit mass in each planet had to be much lower than the maximum angular momentum allowable for each protoplanet immediately after its formation. This reduction of angular momentum in the planets was assigned by Kuiper to the action of solar tides and to the escape of gases (largely hydrogen) from the periphery of planets partly under the influence of solar winds. Such was hardly a satisfactory handling of a thorny question. Furthermore, Kuiper's ideas on the effectiveness of the Roche-density implied a protosun which developed simultaneously with the planetary system. But as ter Haar noted in an unsparing criticism of Kuiper's theory, 'in that case one meets again with the vexing problem of the slow rotation of the sun'.[87]

Weizsäcker's theory showed keen awareness of the fact that the presence of a rotating gaseous disk around the sun needed some justification before one could introduce the seminal factor (viscosity) responsible for the actual formation of planets. From what was already mentioned in the preceding chapter about the work of Berlage it should be clear that he in a sense had anticipated Weizsäcker. The latter's essay was a prompting for Berlage to pursue his own investigations on the action of viscosity in a disk developing around the sun.

His communication of 1948 to the Academy of Sciences of the Netherlands began, in fact, with a reference to Weizsäcker.[88] The same initial remark also reflected the age-old pattern of dismissing former theories and of presenting a new one in full confidence of its correctness. Or as Berlage spoke of himself: 'He admits that in his odyssey through several attempts to attain a rational picture of the evolution of the planets, he was many times led astray, but is now in the position to formulate a rather concise theory as a working basis'.[89] Berlage's new theory claimed the evolution of a concentric set of 'toruslike structure' in the gaseous disk [*Illustration XXXVI*]. While the mutual separation of these toruses could be shown, under very suitable assumptions, to follow a law similar to that of Bode, nothing specific was offered by Berlage on the coalescence of each of those toruses into one planetary body. In general, his developments could readily give the impression of arbitrariness and eclecticism to anyone capable of seeing through the facade of impeccable and extensive mathematics. An example of that arbitrariness was Berlage's facile handling of the question of angular momentum with a short reference to the alleged vanishing of much of the hydrogen and helium 'from the nebula by evaporation into space before the planets were formed'.[90] Oversight of principal features of important past theories supported the unabashed eclecticism of his concluding words. They expressed his belief that his considerations 'show that CHAMBERLIN and MOULTON's planetesimal theory may be essentially right, as the theories of DESCARTES, KANT, and LAPLACE proved also essentially right'.[91] The one really right was Berlage himself, at least in his own eyes. A decade and several papers later, which evoked no significant comment, he characterised his theory that planets as well as satellites developed from concentric rings 'as one almost impossible to be averted'.[92]

Games with dust clouds

It was shortly afterwards that wider access was secured to O. Y. Schmidt's disk theory, which he developed in the 1940's and early 1950's, through the translation into English of the third edition of his work, *A Theory of Earth's Origin*, a set of four lectures.[93] The real merit and originality of Schmidt's speculations lay not, however, in details of the development of the disk into planets. In that respect Schmidt offered only the dubious device of 'the averaging of the *angular momenta of many bodies*'[94] moving originally in orbits of most varied directions, planes and eccentricities to explain the motion of planets in the same plane and direction. Again, he offered nothing original, or convincing for that matter, with respect to the agglomeration of matter into bodies of planetary sizes. The touch of originality in his derivation of the Titius–Bode law was, however, predicated on the unproved assumption that the angular momentum per unit mass increased linearly with distance in the still undifferentiated disk. On such a basis it was possible to introduce into the disk zones whose sequence was similar in form to the law in question.[95] Schmidt's persistent attention to the question of angular momentum paid some dividend as he took up the question of the 'ultimate origin' of the net angular momentum in the disk. He started with the appropriate remark that although the validity of the law of the conrvati n of angular momentum 'had been tested and proved millions of times', it 'seemed to have penetrated very slowly into the depths of astronomy'.[96] His general solution to that 'ultimate origin' was that the disk was captured by the sun as it passed through a dust-cloud section of the galaxy. In other words, the angular momentum of the solar system derived 'from a redistribution of the angular momentum in the Galaxy; part of the angular momentum possessed by the cloud in respect to the passing Sun would be retained by the part of the cloud captured by the Sun'.[97] The special solution was based on the work of O. A. Sizova, who found that three bodies can meet in such a way that two of them would stay together as a binary system with the third leaving forever the scene after making its dynamic contribution. More specifically, Schmidt's theory rested on the assumption that the meeting of the sun with a wisp of dust cloud (the future disk) was accompanied by a very specific encounter with another star. The diagram [*Illustration XXXVII*] showed the process of detaching a planet P_1 from its elliptical orbit around the sun P_0 under the influence of star P_2 passing by. One, therefore, only needed to replace P_1 with a wisp of dust cloud and picture the process in reverse order to see that a proper capture was indeed a distinct possibility.

While all this could appear as a *tour de force*, it did not explain the distribution of angular momentum in the actual planetary system, and it certainly implied an extremely low frequency for the existence of such systems. The highly specific encounter between two suns and a dust cloud could only be seen as a very rare occurrence, regardless of the great number of stars and of dust clouds. That these latter too were very numerous became gradually evident as observational data accumulated during the 1940's about the presence in our galaxy of small dark nebulae, possibly the stage preceding the actual formation of stars. Estimates of their number had just about been completed,[98] when they were seized upon as a clue for planetary origins by Fred L. Whipple. He did so in the concluding section of his study of the kinetics of cosmic clouds, which he presented at the Centennial Symposia of Harvard Observatory in December, 1946.[99] It was these ideas that he later called the dust-cloud hypothesis.

Whipple's starting point was the postulate that 'a small group of partially condensed clouds were captured by, or developed in the original cloud, and that they then spiralled inward by accretion at the proper rate to be left in approximately the orbits of the present planets when the main cloud underwent its final rapid collapse'.[100] On that basis Whipple expected the first three of seven requirements to be satisfied by a viable theory, namely, the existence of planets, their direct motion in a circular orbit, and their motion close to a common plane. The fourth requirement, or the direct rotation of planets, was in turn satisfied by Whipple by assuming that the density gradient was negative in outward direction from the centre of the main cloud. As to the fifth requirement, the slow rotation of the sun, Whipple claimed that it 'might easily be established by chaotic matter from the small-cloud group or from one or more of the small clouds that suffered mishap and failed to develop'.[101] This was a somewhat cryptic reasoning, motivated perhaps by Whipple's full realisation of the gravity of the problem posed by the sun's slow rotation. To eliminate that problem, he postulated at the very outset that the original cloud could only possess an 'extremely small net angular momentum'.[102]

Such a drastic restriction of a basic parameter could only enhance the rarity of the process envisaged by Whipple who tried to satisfy the sixth requirement, the formation of satellite systems, from assuming that the planets themselves were condensed 'from clouds of much larger dimensions'.[103] An even sharper glimpse of that rarity was revealed when Whipple analysed in detail what he called the 'basic process' of his theory, the spiralling of nascent planets toward the centre of the main cloud as it began to collapse rapidly. The spiralling, since it meant a longer route for the interior planets, heated them up more effectively through collision with particles of the remnants of the outer part of the main cloud than was the case with the other planets. Whipple, therefore, argued that the interior planets lost much of their primordial atmosphere and part of their angular momentum, and consequently had to rotate more slowly than the outer planets.

It now remained to take a closer look at the mechanism and probability of the 'basic process', which required that 'condensations in a captured smaller cloud, or in one moving stream of the large cloud, maintain their identities and grow by accretion; this latter process also causes them to spiral inwards'.[104] His careful mathematical presentation of the process did not offset his admission that 'presumably a specified narrow range of conditions, perhaps in surface-area/mass, coupled with a lucky avoidance of disrupting encounters, would be necessary for the successful culmination of a small cloud as a planet'.[10] Unfortunately, Whipple offered no proof concerning the dynamically most 'basic' aspect of his theory, the turning of spiralling motion into a circular one. The indirect proof he advanced in this connection was the collapse of the main cloud at the moment, which he should have characterized as an even 'luckier' aspect of his theory than was the suitable capture of a set of dust-condensations by the main cloud.

That the theory embodied a long chain of 'lucky' events was again overlooked by Whipple, when he presented a year or so later a popular version of it which, unlike his former publication, reached a very large audience.[106] Whipple now claimed that his theory possessed 'a plausibility that other theories about the origin of stars and planets lacked'.[107] About the transition of spiralling into a circular orbit he merely asserted with the aid of a diagram [*Illustration XXXVIII*] that it 'can be demonstrated mathematically beyond question'.[108] Actually, what seemed to be beyond question was the inevitable rush of those spiralling bodies into the sun. The only feature of the solar system which the dust-cloud theory could not explain was, according to its proponent, the regular spacing of the planets. He also tried to put in better light his own theory by reviewing the failure of some previous theories.

If the dust-cloud hypothesis was satisfactory in all but one respect, why was there then any need to state in the end that 'all of the current theories about the birth of the stars and planets leave much to be desired', that 'evolution of the solar system remains foggy', and that 'perhaps an entirely new advance in science will be required to light our way'?[109] Clearly, if such was the case, why was it then justified to claim that because of the dust-cloud hypothesis a goodly frequency of planetary systems with 'human or intelligent life' no longer belonged to the realm of science fiction?[110] The inconsistency wholly escaped Whipple who submitted in the same breath: 'If intelligent beings exist on other planets, we may some day establish radio contact with them. Conceivably we may be able to send ships into space to cruise among planetary systems belonging to other stars in our neighbourhood. There our descendants may find strange types of intelligent beings – or at least settle the argument'.[111]

Debt to chemists and momentary balance sheet

The most creditable part in Whipple's arguments concerned his emphasis on a hitherto largely neglected detail, the chemical processes at work in the formation and growth of dust particles. The first major step toward unfolding the chemistry of planetary origins was embodied in a monograph by Harold C. Urey,[112] whose awareness of the novelty of his own approach was all too evident as he concluded a short survey of previous theories: 'During the three hundred years since Copernicus again discovered the heliocentric solar system no progress seems to have been made in explaining its origin. *Ad hoc* assumptions have been made which have all been questioned or appear highly improbable. The subject seems always beset with miracles, especially when considered by others than serious scientists. . . . But even scientists have a marked tendency to call for miracles whenever some phase of the subject moves outside their own specialities. Under these circumstances it is well to look for more data and for new approaches'.[113]

Unlike most proponents of other theories Urey did not present his investigations as a definitive solution to the question. Compared with the emphasis in previous theories on the dynamics of planetary evolution, little attention was given in his discussion to such questions as the distribution of angular momentum and the gravitational instability in the primordial cloud. He simply assumed that an effective formation of protoplanets took place in the interstellar dust cloud along the lines specified by Kuiper.[114] Urey's two main conclusions, (1) the relatively low temperature at which the earth and the planets formed, (2) the existence of a more uniform distribution of iron throughout the silicate phases of the earth's development than is the case now,[115] should have appeared a trifle in comparison with the sweeping tone of cosmological theories. Again, there was a great disparity between the simplifying moulds of other cosmologies and Urey's painstaking compilation of data. A brief look at Table 19 in his book[116] could make it abundantly clear that data on the cosmic abundance of elements, on the interaction among chemical processes, on the composition of meteorites, and on the surface of the earth, moon, and planets, set stringent limits to cosmological theorising. His analysis of meteorites proved to be of lasting value concerning the cosmic abundance of elements. Whether it substantiated his claim that the core of all protoplanets was a moon-sized accretion of solid particles cannot be maintained with the same assurance. He could not suspect that within two decades actual samples from the moon's surface would make largely irrelevant all previous assertions about it and would pose fresh problems to speculations on planetary origins. But he certainly was well ahead of his time by putting so much emphasis in this connection on the chemistry and physics of the moon's surface.[117]

The year 1952, when Urey's work appeared, was made doubly memorable in the history of recent speculations on planetary origins by Harold Jeffreys' Bakerian Lecture in which a most competent balance sheet was drawn between real gains and still unsolved problems. The lecture ended with the hopeful assertion that 'we are approaching a stage where we can ask the right questions, and a problem correctly stated is usually three-quarters solved'.[118] Advancing to that stage was not, however, something to be quickly accomplished. In Jeffreys' words the problem of planetary origins was 'extremely difficult', and among the many features of the solar system he could single out hardly one 'as satisfactorily explained'.[119] Latest research of particulars seemed to turn up more puzzles than to secure definitive solutions. Among the latter he noted the process in which small bodies could form without first going through the gaseous state.[120] Among the former was the possible necessity to attribute separate origins to Venus and Mercury.[121] Compared with the tidal and collisional theories he saw distinct advantage in the disk theories, but he emphasised the difficulty of the coalescence of the material of the disk into the actual planets. He also noted that considerations about gravitational instability in matter in pure rotation could not be applied to the case when rotation was combined with turbulence.[122] His most incisive remark was reserved to the supposed escape of the overwhelming portion of the disk's material. The process in question was, in Jeffreys' words, not only 'not very clearly explained', but 'definitely impossible'. His succinct reasoning deserves to be quoted in full: 'As the angular momentum per unit mass in such a disk varies like $r^{\frac{1}{2}}$, no internal reaction, whether due to gravity or viscosity, could expel a positive mass beyond some definite distance. Hence if the excess matter was to be removed it could only be by absorption into the sun. . . . If several times the mass of Jupiter had been absorbed into the sun we are back against the old problem of seeing how the sun comes to rotate so slowly'.[123]

In making his evaluation of the state of the art Jeffreys noted that 'it is difficult enough to find an explanation of one solar system without having the extra responsibility of explaining millions'.[124] The remark evidenced his impatience with arguments made against the collisional or close-encounter theories on the ground that within their context the formation of planetary systems was a very rare event. In fact, the appeal of a particular theory had all too often been in direct proportion to the degree of frequency of planetary systems it promised. This in turn implied the danger that in order to assure a high measure of that frequency, the weak points of the theory were not weighed in their true gravity. Whether for this reason or not, these two symptoms appeared together in an important address given by Kuiper in 1955 on the state of development of theories on planetary origins.[125] In his own admission, he derived comfort from the fact that he could predict the presence of planets around 1 to 10 percent of the main-sequence stars, which meant the existence of a billion or more planetary systems in our galaxy alone.[126] Introducing this view of his was his confident claim that the development of a solar nebula as part of the stellar evolution had been sufficiently clarified, and that all four stages of the development (in all of which the angular momentum played a principal role[127]) could be studied empirically.[128] In concluding he took the view that 'perhaps it is correct to say that one now has a beginning of a theory of the origin of the solar system, presenting no known obstacles to further development'.[129] Indeed, he submitted explicitly about the two main difficulties of the nebular hypothesis, of which his own was a modern version, that they were shown 'to be only apparent'.[130]

Dashing but futile tries

One of the two main difficulties was, of course, the distribution of angular momentum. That it was far from being only 'apparent' was amply demonstrated by the continued efforts to find a convincing solution to it. Some of these efforts were extraordinary in more than one sense. Thus, Arthur W. Titherley, dean of the Faculty of Science at the University of Liverpool, pictured in his *Origin of the Solar System*[131] the sun, to secure its slow rotation, as one of three fragments (the other two were identified as α Centauri A and B) of a broken up Cepheid. The planets, in turn, were traced by him to the early pulsations of the sun, which supposedly emitted a succession of at least ten, but perhaps fifteen clouds, ready to coalesce into protoplanets (Ur-planets). Titherley attributed their actual orbit to the damping of their spiralling return toward the sun by the resisting medium. According to him, to quote only one of the not-too-extravagant passages from his book, 'each cloud is envisaged as, at first, a heated balloon which on leaving the Sun received a counter-clockwise rotary impulse, and the resulting rotation, as with other large masses of gas, would more or less flatten it to a lens shape. After it had become cold by expansion it would hold in suspension the solidified chondritic drops of iron, silicates, etc., which in about 1,300 years, when a distance of 10^{16} cm had been reached (17 times Pluto's present distance), began to gravitate together and form a nucleus at the centre of the disc; but gas resistance would prevent their direct inward fall. Instead, as noted, a whorl would develop in which particles, including cooled gases, followed sub-spiral paths, similar (on a smaller scale) to that of each entire Ur-planet itself towards the Sun after it had reached its limiting distance'.[132] Such was Titherley's own summary of the formation of planets and satellites. About it he could rightly say that 'the complexity of the process baffles [exact] solution'.[133]

The situation about the angular momentum remained no less baffling when sober speculations were combined with exact mathematical treatment and calculations by W. H. McCrea, in 1959, in a paper presented before the Ninth Astrophysical Symposium in Liège.[134] His theory became associated with the word 'floccules', that is, patches of interstellar material moving at random among themselves and composed mainly of molecular hydrogen. While McCrea made much of the evidence on behalf of floccules and by the invariable connection of sun-like stars with clusters, his principal aim was the derivation of the slow rotation of the sun as a result of the formation of planets. He felt this could be done by making very specific and even 'to a considerable extent arbitrary' assumptions about the process leading to the formation of the protosun.[135] The starting point in his theory was a 'cloud' with a mass of several hundred times that of the sun, and consisting of a large number of 'cloudlets' or 'floccules' moving randomly within it. The next step was the emergence, through head-on collision of floccules, of ever larger floccules, until further condensation by gravitational attraction could come into play. In this steadily 'favourable' chain of collisions among randomly moving floccules the angular momentum within a given region fluctuated and, therefore, statistical analysis could be applied. McCrea showed that the most probable net angular momentum would be very small in any condensation. His analysis concerned a configuration of N floccules within a radius R equal to Pluto's distance, with each floccule moving with a velocity V of about 5 km/sec and constituting a total mass M equal to that of the sun. By taking the total angular momentum within that sphere $(MVR/N^{\frac{1}{3}})$ equal to the sun's angular momentum he obtained numerical values for the number and size of floccules. With the additional assumption that only head-on collisions of floccules with the proto-sun would result in its growth he carried his calculations back to the actual value of the angular momentum of the sun.

Whatever the value of showing that there may be a purely mechanical explanation of the problem of the sun's angular momentum, the gain, though dazzling in execution, was not free of a touch of circularity and of some highly precarious assumptions. It was still to be proved that floccules were relatively permanent entities and that their collisions would not result in complete volatilisation. Certainly, the theory was not geared to give a consistent explanation of the formation of terrestrial planets and of the satellites of great planets, to say nothing of the nemesis of all theories, the moon. In McCrea's candid admission, the random collision of floccules was 'not likely to reproduce Bode's law'.[136]

The same law was only one of the many features of the solar system that found not even a semblance of explanation in the theory which Hoyle proposed in 1960.[137] In a typical Hoylean fashion the theory was a mixture of bold exploration of novel possibilities, of masterful though at times wishful correlation of quantitative parameters, and of some patently unsound utterances. Among the latter belonged Hoyle's opening remark that 'at the symposium on the Origin of Planets, held at the Moscow meeting of the I.A.U. [International Astronomical Union], it became clear that there is a very widespread disagreement among astronomers on this whole problem'.[138] Publications and statements on the matter had already made that disagreement abundantly clear during the dozen or so years prior to 1952, and Hoyle's own theory hardly diminished it in any noticeable degree, in spite of the fact that, as will be seen shortly, it prompted further efforts along the same line. The theory represented a detailed, quantitative presentation of the argument which Hoyle had qualitatively described and illustrated with a diagram [*Illustration XXXIX*] six years earlier in his well known popularisation of modern astrophysics and cosmology, *Frontiers of Astronomy*,[139] the expanded form of his lecture series given at Princeton University in 1954.

Tellingly enough, the starting point of the detailed form of Hoyle's theory was a reference to the fact that 'the angular momentum per unit mass about the centre of gravity of the solar system is on the average some 50,000 times greater for planetary material than it is for solar material'.[140] Hoyle then suggested that instead of asking the traditional question how did the planetary material acquire so much angular momentum one should invert the problem by asking how did so little angular momentum go into the sun. Actually, he asked the traditional question after dismissing two possibilities for the sun's slow rotation. One was that the condensing volume was filamentary rather than spherical, the other was that the condensation took place in regions of a cloud where the local differential rotation was abnormally small. It was then necessary to retain the third possibility, namely, that, in Hoyle's strange phrasing, 'angular momentum may not be conserved during condensation'.[141] By this he meant that the angular momentum of a fast-spinning sun was transferred to a disk surrounding it in the equatorial plane.

The mechanism supposed to do this was the magnetic torque, which was also suggested in Germany by R. Lüst and A. Schlüter,[142] independently of Hoyle, about the time when the latter's Princeton lectures were being published. Hoyle defined the mechanism as follows: 'It is generally believed that a magnetic field pervades the interstellar gas. Thus a condensing volume of gas will be magnetically connected to its surroundings. The possibility then arises that a torque is conveyed through the magnetic field from the condensation to the interstellar gas, and that angular momentum is transferred outwards through the agency of this torque'.[143] Hoyle had, however, to admit in the immediate context that L. Mestel and L. Spitzer had already shown[144] that as the contraction of stars goes on the condensing gas must be able to slip readily across the magnetic lines of force. To cope with this problem Hoyle disregarded all but the final phase of star formation for the emergence of a planetary system with the known distribution of angular momentum. This phase was, in Hoyle's words, 'a final slow contraction to the main sequence, the slip across the lines of force being inappreciable in this phase'.[145]

With hardly more than one-tenth of his discussion completed Hoyle now rushed to his grand conclusion. To make it appear somehow well founded he offered a calculation about the angular momentum of the contracting solar nebula just before it entered its 'final slow contraction to the main sequence'. The value was 8×10^{51} gm cm^2 sec^{-1} with the radius being about 500 times the radius of Mercury's orbit, or about 5 times the radius of Pluto's orbit. Much the same value could be obtained if one assumed in close analogy to Weizsäcker's argument that in view of the densities of Uranus and Neptune these planets originally contained 100 times more mass, largely in form of hydrogen. From this consideration Hoyle jumped to the startling inference that because of the excellent agreement of the two calculations 'a process occurred whereby the angular momentum of the primitive solar condensation was transferred from the Sun to the planetary material'.[146]

This was a momentous *non sequitur* within the framework outlined by Hoyle. On the basis of his reasoning he could have only written, for sake of clarity and logic, that one therefore had at hand a problem of explaining by some mechanism involving the loss of much of the sun's angular momentum the dissipation of so much material from the planetary disk. But Hoyle's mind was already preoccupied with much broader vistas unfolded with a similar neglect of rigorous logic. Immediately afterwards he asserted: 'It is implicit that a similar process has been operative in the great majority of dwarf stars, and hence that such stars possess planetary systems'.[147] What really asserted itself once more was the inner logic of an urge to vindicate the claim about the high frequency of planetary systems.

It was now Hoyle's task to offer more than the analogy outlined in his Princeton lectures of elastic strings holding together a fast spinning hub and a rim falling behind, a process which suggested that the hub would slow down and the rim would speed up. The magnetic coupling between the sun and the disk was a far more delicate mechanism, and Hoyle's own calculations showed that it was fraught with grave uncertainties. One of them derived from the necessity of assigning to the protosun a magnetic field thousands of times larger than the present value. The other concerned the degree of ionisation needed in the disk, or at least in its inner edge, so that the magnetic lines of the sun might be firmly attached to the disk to ensure the working of the torque [*Illustration XL*]. He calculated that at least one in ten million hydrogen atoms near the inner edge of the disk had to be ionised. Yet at temperatures of about $1,000°$ K only a much smaller ratio of ions could be expected. Here Hoyle defensively remarked: 'In view of the non-thermodynamic activity of the present-day Sun it would be rash to argue on these grounds that the requisite ionisation could not be maintained'.[148] With respect to rash arguments the shoe was on the other foot.

Impasse with essentials and return to details

Further difficulties of Hoyle's theory were pointed out only years later.[149] The crucial question of angular momentum was considered essentially solved at the Conference on the Origin of Planetary Systems, held in early 1962 at the Goddard Institute for Space Studies in New York, at which Hoyle himself outlined his theory.[150] This cavalier approach to the question of angular momentum was a blessing in disguise, because it seemed to help focus attention on three grave problems for which, in T. Gold's admission, there were no answers in sight.[151] One had to do with the question of the agglomeration of solid particles, especially the growth of centimetre-sized bodies to sizes of a few kilometres, so that growth by gravitation might start. The other related to the stability and co-ordination of those spiralling, almost circular paths in the close neighbourhood of the sun produced by its magnetic torque. The problem here was all the more acute as evidence was lacking that the foregoing situation existed in any degree around the sun today. In Gold's words, 'if it was a process that is very different from anything that is going on now, then we are in a much poorer position to find out'.[152] The third problem which loomed large in Gold's estimate concerned the escape of hydrogen from the nebular disk and the simultaneous staying behind of oxygen, nitrogen, and carbon. Diffusive separation, the most efficient process one could think of in that connection, could achieve the desired effect only in a time several times longer than a few billion years, the age of the solar system. About one of the alternatives, radiant solar flux, Gold remarked that 'one needs to strain one's self to say the least' to see it work.[153]

The impasse in the field was also a chief lesson of the historical survey[154] which introduced the text of the dozen or so papers delivered at the Conference. At the end of that survey answer was sought to the question: 'Why are there so many theories that all claim to have solved the problem?'[155] Behind claims of this kind lay the spectre of persistent failure which could rightfully be traced to the fact that even the most quantitative theories were such only in part. The lack of observational data on many a decisive point prevented rigorous treatment and left each theory open to almost immediate rebuttal. No wonder that the Conference reflected the emergence of a new thinking according to which broad speculations on planetary origins had to yield, for a while at least, to careful work on details. Acquisition of new data meant, however, not only the possibility of a more solid basis for a future planetary theory, but also increased the difficulty of its formulation, since an ever larger set of boundary conditions had to be satisfied. Consequently, confident claims about planetary systems as 'normal features' of the starry realm were more than ever to be buttressed by quantitatively strong arguments, if they were to deserve scientific respectability. In sum, the need became more evident for a complete rehauling of the theatre-in-the-round to meet more exacting standards for future performances. Such seemed to be the main lesson of an era of sleights-of-hand producing planetary systems in every nook and cranny of the universe.

The last ten or so years have, indeed, been mostly a period of work on details among which the transfer of angular momentum by magnetic torque received special attention. Once more the old pattern was in evidence as new 'solutions' proved invariably to be the source of new problems. Thus explanation of the very slow rotation of some stars, such as the sun, by the passing of sunflares and jets through the stellar magnetic field, as proposed by E. Schatzman,[156] could only be achieved by sacrificing the idea of the steady transference of angular momentum to the surrounding disk through permanent magnetic coupling. Further applications of the Schatzman-mechanism implied, therefore, a readiness to ignore serious difficulties.[157] The same attitude was very much the case when, in 1960, M. M. Woolfson reintroduced the tidal theory.[158] His original idea was that the passing star produced a travelling tidal wave on the sun's surface with the result that in every 80 minutes a huge bulge was detached from the point of the sun's surface facing the passing star [Illustration XLI]. The formation of ten planets was achieved in 12 hours in a regular sequence, which, according to Woolfson, accounted for the 'well-known but hitherto unexplained Bode's law'.[159] That he made no reference to the question of angular momentum was doubly ironical, because shortly afterwards there appeared a

paper by Lyttleton,[160] the originality of which lay in part in the use of computers. Previously simplified calculations of the dynamics implied in tidal
theories were now replaced by detailed analysis of hitherto intractable details.
The results showed, among other things, that the difficulty first outlined by
Russell in 1934 was, in fact, so serious that a primitive planet 'would simply
fall back to the solar surface almost if not quite where it was detached'.[161] It
seemed to escape Woolfson that the problem of angular momentum remained
the same for the case when the material of the planets was captured from a
light, diffuse star passing close to the sun, a mechanism which he submitted
in 1964 in mathematical detail.[162]

    That such efforts stirred no significant echo seemed to be due to the
increased awareness of some difficulties resisting all ingenuity and to the dismay
one could feel in the presence of theories, the number of which, as Alfvén
remarked in his Harold Jeffreys lecture in 1967, increased 'by perhaps an order
of magnitude' since Poincaré had listed in 1911 half a dozen or so.[163] The net
result of all these efforts spanning some half a century was, in Alfvén's estimate,
that a state had been reached which was no longer that of speculation, but one
permitting a systematic scientific analysis of problems. According to him this
justified the expectation that 'it is likely that we shall be able to understand
at least the main features of the cosmogonic process'.[164] He warned at the same
time that to many questions connected with the origin and evolution of the
solar system 'our answer will be *ignorabimus*'.[165] As a principal reason for this
he called attention to the fundamental elusiveness of the 'initial conditions' of
the process leading to the formation of our planetary system. Alfvén pictured
the cosmogonist not so much as a theoretical astrophysicist boldly unfolding
the consequences of initial assumptions, but as an archeologist cautiously
pushing farther back into the past on the basis of fragmentary evidence.
Accordingly, in Alfvén's words, 'we should not look for *the* cosmogonic
theory . . . which solves the problem at once, but for a number of theories
clarifying a great multitude of detailed questions of which the total cosmogonic
problem consists'.[166]

    Behind these words of caution there lay a quarter of a century of intensive
attention to the problem on Alfvén's part. Lyttleton, who had already been
working on the topic for three decades, displayed even greater reserve when
he gave, in 1968, about the true status of the art an evaluation the theme of
which was that 'almost everywhere we look in the solar system we are faced by
unsolved problems'.[167] Characteristically enough, he drew up the balance with
an eye on the question of the frequency of planetary systems: 'If we had a
reliable theory of the origin of planets, if we knew of some mechanism consistent with the laws of physics, so that we *understood* how planets form, then
clearly we could make use of it to estimate the probability that other stars have
attendant planets'.[168] The same concern seemed to prompt his remark that
'scientists cannot really quite relax' until the mystery is entirely cleared up and
that they can feel sure that 'no "special", super-natural, event has to be postulated' to account for the solar system.[169] While such a proviso should seem
most legitimate, one cannot read without puzzlement Lyttleton's statement that

the possibility of the extreme rarity of the solar system on the basis of the encounter theories 'ran counter to the widespread uncritical assumption of many that the world is here for the purpose of providing an arena for man's activities'.[170] Actually, the ones who greeted this possibility with some sympathy were those for whom nothing in the universe, however rare, was ever the outcome of mere chance. Whether in this case, too, they had put undue emphasis on the gaps in scientific knowledge for support of some metaphysical or religious outlook about the uniqueness of mankind is another matter. They certainly should remind themselves that gaps in scientific knowledge have an uncanny way of being filled. On the other hand, those who really found the possibility in question 'an emotionally distasteful hypothesis'[171] should have kept in mind an old truth which Lyttleton himself put in the following words: 'The whole problem is far from concluded, and even if the general hypotheses . . . are correct the working out of them to show what processes result can scarcely be said to have been more than begun'.[172]

What was true in the mid-1960's has been still true in the mid-1970's. Revealingly, the reason for this lies in the massive accumulation of new data about the planets and the sun as well as about the more distant cosmic regions and objects. This new wealth of information not only helps clarify questions of detail, but makes increasingly greater demand on theories with respect to exactness and specificity. Gone are the days when simple references to nebulae and dust clouds sufficed as a starting point for one's story of the solar system. The physical and chemical states of nebulae and interstellar dust are now known in such detail as to impose stringent limits on speculations concerning any step in the development of a stellar nebula into a star with planets. As was acknowledged in a recent survey of the state of the art, 'the first and last steps are still as vague as ever'.[173] What this means is that the classic difficulties of the theories on the origin of planetary systems are still unsolved. To secure solutions, assumptions of many kinds will be made, but in this respect too the classic lesson of our story will remain as valid as ever. It was put in that survey in words that deserve to be quoted in full: 'Judging from the diversity of assumptions, models, and predispositions among those hardy scientists who venture to try to outguess the course of evolution of the nebula that presumably predated us all, more constraint is precisely what is needed'.[174]

A historian, whether of politics or of science, mindful of Hegel's famous remark that the only thing man learns from history is that man never learns from history, will entertain no illusion about the coming of an age sympathetic to constraint, intellectual or political. His only satisfaction will be his success of setting forth the vast evidence supporting the principal lessons of history for those relatively few ready to learn uneasy lessons. They will be far outnumbered by those whose chief interest is in that stance which invariably makes much of recent progress by unduly extrapolating its relevance[175] and by those who simply want to hear that 'we are in fact very near to a solution of this age-old problem'.[176] A careful reading of the record, both recent and remote, is not compatible with such hopeful utterances. The same record should also cast strong doubt on the efforts of those who, in order to see planetary systems with thriving civilisations and supercivilisations everywhere in the universe, systematically gloss over the chronic failure to find the precise clues to the development of the solar system. It is hardly the sign of serious scholarship to mention in such a connection only the dust-cloud hypothesis and suggest that it is a satisfactory explanation.[177] Again, the transfer of angular momentum by magnetic torque is not free of serious difficulties to such extent as to be used as a basis for stating that 'most single stars in the solar neighbourhood are expected to possess planetary systems'.[178] This type of handling patently unresolved questions reveals itself also in the readiness to seize on any slender evidence, such as the indirect observation of 'companions' (planets) around several nearby stars.[179]

The 'presence' of Jupiter-sized 'planets' around neighbouring stars can exercise a decisive influence even on the thinking of those who otherwise admit, on the basis of the inability to explain the origin of our planetary system, the 'excessive rarity' of such systems in the universe. Thus, Z. Kopal in his book, *The Solar System*, concluded that since 'we know that at least 1 percent (and probably more) of the stars in our neighbourhood possess companions whose mass is of planetary, rather than stellar, order of magnitude, . . . planetary systems would seem to be reasonably common, not rare, in the realm of the stars that we know best'.[180] To see the weakness of such reasoning it is enough to recall the extremely narrow margin for the sizes of planets comparable to Jupiter, if the stability of a system like ours is to be secured. It was the study of this detail that forced Shiv S. Kumar to conclude that the 'total number of planetary systems in the Galaxy is much less than the numbers generally quoted in the literature'.[181] According to him, for the million million (10^{12}) or more stars in the Galaxy there are at most a million with a planetary system, but the actual number may very well be closer to bare unity. Since a planetary system does not necessarily mean a proper abode for the evolution of higher forms of life, he felt impelled to add by way of final remark: 'The probability of intelligent life in our Galaxy, other than our own, must unfortunately be considered negligible!'[182]

All this attention to the serious shortcomings of explanations of the origin of the system of planets (and to the *non sequiturs* of extrapolations to a high frequency of planetary systems in the stellar universe) should not be construed as a disagreement with the claim that the fascinating orderliness and structure embodied in the solar system is the outcome of the interplay of physical forces and factors. The disagreement concerns the chronic unwillingness to consider the solar system the product of an extremely complicated chain of events, the outcome of the interaction of a very wide range of physical factors. Great as may be the value and attractiveness of the principle of simplicity in the method of exact science, it may very well be that explanation of the solar system demands a very different approach, which in turn could forcefully indicate its excessive rarity if not uniqueness. In particular, the disagreement concerns the facile marshalling of each and every 'explanation' having but a modicum of truth in support of a grand conclusion which is still to be demonstrated. For the truth of the matter is that, in spite of countless and at times ingenious efforts making use of every *tour de force* of mathematical physics and observational lore, no theoretician has yet even remotely succeeded in turning our planetary system into a self-explaining unit, let alone into a very typical phenomenon. Those efforts deserve the highest admiration, for truth, as Francis Bacon once remarked, shall much sooner come out from saying something than from saying nothing. Yet efforts, undaunted and admirable as they may be, do not necessarily mean success.

Wilful efforts and sanguine vistas

Keen awareness of the difference between efforts and success has always been the hallmark of sound scientific reasoning. The same reasoning implies also the recognition of the point that in a universe of radical chance, if it is indeed such, all configurations, all results, however rare, are 'self-explanatory'. This holds true even of the efforts of the new breed of exobiologists who, by throwing consistency to the wind, ignore the ground rules of random evolution while asserting on its basis the multiple duplication of technological civilisations on an uncounted number of planets.[183] They handle the lessons of the history of planetary origins with the same inconsistency. For one is merely reassuming an old but hardly creditable attitude when one seizes on the latest 'insight' and claims that 'magnetohydrodynamic considerations appear to eliminate the angular momentum difficulty in the nebular hypothesis'.[184] The 'demonstration' of this and similar claims is inevitably studded with conditional subclauses. Their frequency is proportional to the blithely assumed high frequency of planetary systems and in particular to the true value of the dictum burdened with a subtle touch of contradictoriness: 'Before the problem of the origin of planetary systems is definitively solved, much more work in theoretical physics and observational astronomy must be performed. But a beginning has been made, and the contours of a well structured theory have emerged'.[185]

The true cement keeping such a statement together is something not too different from wishful thinking, a factor even more questionable when it translates itself into such programmes as the listening by huge radiotelescopes to possible broadcasting of 'universally comprehensible propositions', like Pythagoras' theorem, from planets around some neighbouring stars. It has even been argued that computers and not men should be put on extrasolar space probes, since man would hardly be intelligent enough to cope with the superior acumen of his cosmic relatives. Actually, Pioneer 10, destined to escape from the solar system, is carrying a plaque (*Illustration XLII*) with a message etched into it for the benefit of other civilisations, unfamiliar though they could be with 'human' logical processes. The promoters of the venture admitted, mindful of the 80,000 years needed by Pioneer 10 to reach the vicinity of the nearest star, that the message may never be found or decoded. At the same time they claimed that the inclusion of the plaque on Pioneer 10 seems to be 'a hopeful symbol of a vigorous civilisation on earth'.[186] Vigour is, however, of more than one kind, and can hardly be called such when used in detriment of sober reasoning. It took indeed but sober calculus (the rather elementary feat was done by a rightfully indignant Nobel-laureate), to show that plans to travel to other planetary systems should remain where mostly advertised, on the sides of cereal boxes.[187]

This sober calculus and reasoning is not the best side of those who keep giving brave discourses on the presence of planetary systems everywhere in the universe. Their bravery in conjuring up such systems is matched by their sanguine vistas of planetarians outside the solar system and of some forms of life on at least some of its members other than the earth. The failure of lunar rocks to reveal the faintest trace of life dampened not a whit their hopeful expectations about finding life on Venus and especially on Mars. Yet some time before the first Viking artificial satellite touched on Martian soil, they had already insisted on the inconclusiveness of a possibly negative outcome of that multibillion-dollar space venture. Life on Mars, it was hastily noted, might be locked there in a deeper layer than what can be sampled by a space-robot.[188] A similar proviso against the possible necessity to admit failure is lurking behind remarks that life on Mars might after all be totally different from the kind we know on earth and that, as Carl Sagan put it, in searching for life there 'we simply may be asking the wrong questions'.[189] Yet should it be blotted out of memory that the principal boost to the expensive Viking project had come from that Space Board of the National Academy of Science which threw its full scientific authority enhanced by Nobel Prize medals behind the claim that the likelihood of finding vegetation on Mars was considerable? Does not the bleak Martian soil seem to justify the suspicion that 'proponents of this project have conveyed a far greater probability for life on Mars than they know in their hearts exists'?[190]

This suspicion, voiced more than ten years ago, will not be weighed in its true significance even if the Viking venture justified the same scientist who took the view in 1965 that 'in looking for life on Mars we could establish for ourselves the reputation of being the greatest Simple Simons of all time'.[191] No lesson, however sharp, will make a dent on the conviction of those who claim that man ought to be considered a freak in the universe if he failed to find traces of life elsewhere.[192] Such is a Mephistophelic metaphysics typical of a methodically irresponsibile journalism and is a most telling example of the destructive interference which undue concern for planetarians can have not only with thinking about planets but with thinking itself. That interference, rich as it may appear in speculation, has only a superficial resemblance to the precept of that great explorer of the realm of nebulae, William Herschel, who felt that it was preferable to err by theorising too much than too little.[193] Regrettably, many of those, who in offering theories on the origin of the system of planets did not encumber their discourse with speculations on denizens on other planets and on their number, still acted as if they were ready to 'swallow the universe like a pill'.[194] It is, therefore, well to recall that while man's grasp of the universe gained enormously during this century, a tiny but peculiar portion of the universe, the solar system, is still to be digested. The end of an already long road, the one leading to the solution of the emergence of the puzzling world of planets, is not in sight yet, brave and sanguine statements notwithstanding. On that road we are still 'planets', that is, wanderers.

NOTES TO CHAPTER EIGHT

1. 'The President's Address on the Award of the Gold Medal to Dr. Raymond Arthur Lyttleton', *Monthly Notices of the Royal Astronomical Society* 119 (1959): 445.
2. 'On the Origin of the Solar System', *Monthly Notices* 101 (1941): 216.
3. *Ibid.*, p. 223.
4. 'The Origin of Satellites', *Monthly Notices* 98 (1938): 633–45.
5. *Ibid.*, p. 642.
6. *Ibid.*, p. 637.
7. 'On the Origin of Binary Stars', *Monthly Notices* 98 (1938): 646–50.
8. 'Sur la stabilité ordinaire des ellipsoides de Jacobi', in Elie Cartan, *Oeuvres complètes, Partie III, Volume I, Divers, géométrie différentielle* (Paris: Gauthier-Villars, 1955), pp. 103–11. The work was originally presented in 1924 and first published in 1928.
9. F. Hoyle and R. A. Lyttleton, 'On the Accretion Theory of Stellar Evolution', *Monthly Notices* 101 (1941): 227–36.
10. 'On the Origin of the Solar System', p. 223.
11. *Ibid.*
12. *Ibid.*
13. *Ibid.*
14. *Ibid.*
15. As discussed in Chapter Seven, p. 209.
16. 'On the Origin of the Solar System', pp. 225–26.
17. F. Hoyle, 'On the Origin of the Solar System', *Proceedings of the Cambridge Philosophical Society* 40 (1944): 256–58.
18. *Ibid.*, p. 256.
19. *Ibid.*, p. 258.
20. 'Note on the Origin of the Solar System', *Monthly Notices* 45 (1945): 175–78. While Lyttleton's and Hoyle's collisional theories could seem extravagant they evidenced full awareness of the problem of angular momentum. It was noticeably absent in A. Dauvillier's massive monograph, *Genèse, nature et évolution des planètes: Cosmogonie du système solaire – Geogénie – Genèse de la vie* (Paris: Hermann et Cie), which although published in 1947 represented a stage where theories of planetary origins stood before Russell's criticism of the tidal theory, a criticism of which Dauvillier seemed to be unaware in spite of his familiarity with Lyttleton's work. Dauvillier's overconfident handling of the question of life's origin in the context of planetary evolution was typical of the ease by which he passed over some of the classic problems of that evolution. The chief characteristic of his theory lay in the proposition that planets formed in pairs. Russell was mentioned in 1963 in Dauvillier's *Les hypothèses cosmogoniques: Théories des cycles cosmiques et des planètes jumelles* (Paris: Masson et Cie, p. 197) but only as one who proposed a variant of Bickerton's theory which still was for Dauvillier the key to the origin of the system of planets!
21. 'On the Cosmogony of the Solar System', communicated on Nov. 12, 1941, in *Stockholms Observatoriums Annaler* Band 14, No. 2 (1942).
22. *Ibid.*, p. 18.
23. 'On the Cosmogony of the Solar System II', communicated on Oct. 14, 1942, *ibid.*, Band 14, No. 5 (1943).
24. 'On the Cosmogony of the Solar System III', communicated on Dec. 5, 1945, *ibid.*, Band 14, No. 9 (1946), p. 26.
25. See the second communication, p. 3.
26. *On the Origin of the Solar System* (Oxford: Clarendon Press, 1954).
27. *Ibid.*, p. 187.
28. *Ibid.*, p. 1. In this book Alfvén recognised (pp. 128–30) that the putting of ionised atoms into orbit around the sun by its magnetic field implies a selection of orbits expressed in the Titius-Bode law.

29. See Lyttleton's review of Alfvén's work in *Proceedings of the Physical Society* (London) B68 (1954): 694.

30. See, for instance, F. Hoyle, *Frontiers of Astronomy* (New York: Harper & Brothers, 1955), p. 91, and its paperback reprint (New York: The New American Library of World Literature, 1957), p. 89.

31. 'The Formation of Planetary Systems', *Journal of the British Astronomical Association* 53 (1942–1943): 23–30.

32. *Ibid.*, p. 23.

33. *Ibid.*

34. *Ibid.*, p. 24.

35. *Ibid.*

36. *Ibid.*, p. 26.

37. *Nature* 151 (1943): 200.

38. 'Über die Entstehung des Planetensystems', *Zeitschrift für Astrophysik* 22 (1943): 319–55. For Weizsäcker's reference to Peek, see p. 328.

39. *Ibid.*, p. 355.

40. *Ibid.*, pp. 320–21.

41. *Ibid.*, p. 321.

42. *Ibid.*, p. 322.

43. In this connection (pp. 335–36) Weizsäcker acknowledged the collaboration of Dr. Höcker, who computed and drew the figure, and of Prof. Hellerich, who called his attention to the Tables for the computation of the radius vector in elliptical orbits with eccentricity-angle 0 to 26 published by the Astronomisches Recheninstitut (Berlin-Dahlem) in 1933 as an enlarged version of the first edition (1912). Neither edition contains any diagram.

44. The ratio (*ibid.*, p. 339) was expressed in the equality, $r_n/r_{n-1} = 1 + e_{max}/1 - e_{max}$. Its value is 2 because the maximum eccentricity (e_{max}) was taken as $1/3$. Since e_{max} was by assumption independent of r (distance from the sun), the ratio could be simplified to r_n/r^2, a close approximation of the Titius-Bode law. On the role of that law in cosmogonical speculations much interesting material was brought together by M. M. Nieto in his *The Titius-Bode Law of Planetary Distances: its History and Theory* (New York: Pergamon Press, 1972).

45. 'Über die Entstehung des Planetensystems', p. 342.

46. *Ibid.*

47. From Weizsäcker's Table (*ibid.*) it should suffice here to reproduce the figures from the last two of its five columns. The figures give from Mercury to Pluto (including the asteroid belt) the computed number of bean-shaped vortices in each ring between two neighbouring planets together with the integer, in parentheses, closest to that number: Mercury-Venus 4.91 (5), Venus-Earth 9.7 (10), Earth-Mars 6.46 (6–7), Mars-Asteroids 5 (5), Asteroids-Jupiter 5 (5), Jupiter-Saturn 5.23 (5), Saturn-Uranus 4.6 (5), Uranus-Neptune 7.03 (7), Neptune-Pluto 11.5 (11–12). The figures evidenced both a striking agreement with the expected number 5, and a striking discrepancy from it.

48. *Ibid.*, p. 349.

49. *Ibid.*, p. 354.

50. 'Die Entstehung des Planetensystems', *Die Naturwissenschaften* 33 (1946): 8–14.

51. 'Zur Kosmogonie', *Zeitschrift für Astrophysik* 24 (1948): 181–206.

52. 'A New Theory by C. F. von Weizsäcker of the Origin of the Planetary System', *Astrophysical Journal* 101 (1945): 249–54. For quotation, see p. 254.

53. *Ibid.*, p. 249.

54. *Ibid.*, p. 252.

55. *Ibid.*, p. 250.

56. 'On a New Theory of Weizsäcker on the Origin of the Solar System', *Reviews of Modern Physics* 18 (1946): 94–102.

57. 'The Origin of the Earth', *Physics Today* 1 (Oct. 1948): 21.

58. *Ibid.* The chief value of Page's article lay in its diagrams of theories of Kant, Laplace,

Chamberlin-Moulton, Jeans-Jeffreys, Lyttleton (1936), Hoyle, Berlage (1930), Alfvén, and Weizsäcker.

59. 'Some Aspects of Stellar Evolution', *Monthly Notices* 106 (1946): 470–75, 476–83, 484–90.

60. *Ibid.*, p. 490.

61. 'Zu C. F. v. Weizsäckers Hypothese über die Entstehung des Planetensystems', *Zeitschrift für Astrophysik* 25 (1948): 58–69. These four objections related to the proper functioning of the vortices. Prior to that Nölke expressed strong doubts about the development of the vortices themselves.

62. *Ibid.*, pp. 67–68.

63. *Studies on the Origin of the Solar System* (Assen: Van Gorcum & Comp., 1948). Published in identical form also as Bind XXV, Nr. 3 of *Det Kgl. Danske Videnskabernes Selskab, Matematisk-Fysiske Meddelelser* (Copenhagen: Ejnar Munksgaard, 1948).

64. *Ibid.*, p. 16.

65. *Ibid.*, p. 41.

66. 'Weizsäcker's Theory of the Origin of the Solar System and the Theory of Turbulence', *Annales d'Astrophysique* 10 (1947): 179–80.

67. *Ibid.*, p. 180.

68. *Studies on the Origin of the Solar System*, p. 20.

69. 'The Origin of the Solar System', in *Science News, Number 14*, edited by J. L. Crammer (Harmondsworth-Middlesex: Penguin Books, 1949), pp. 74–93; for quotation see p. 92. Jones' historical survey of the question contained such phrases as 'the oldest hypothesis which is deserving of mention was suggested independently by the theologian, Swedenborg, and the philosopher, Kant' (*ibid.*, p. 75).

70. 'The Origin of the Solar System', *Monthly Notices* 108 (1948): 103.

71. *The Creation of the Universe* (New York: Viking Press, 1952), pp. 104–11. As late as 1961 Gamow found it unimportant to rephrase ever so slightly his evaluation of Weizsäcker's theory in the 2d revised edition of the work (New York: Viking Press, 1961), pp. 104–11.

72. *Life on Other Worlds* (2d rev. ed.; New York: Macmillan, 1951), p. 145.

73. *Ibid.*, p. 153.

74. *Ibid.*, p. 147.

75. *Ibid.*, p. 148.

76. *Ibid.*

77. 'Turbulence – A Physical Theory of Astrophysical Interest', *Astrophysical Journal* 110 (1949): 338.

78. 'Das Spektrum der Turbulenz bei grossen Reynoldsschen Zahlen', *Zeitschrift für Physik* 124 (1947): 614–27.

79. 'The Scale of Turbulence in a Differentially Rotating Gaseous Medium', *Astrophysical Journal* 111 (1950): 187–90.

80. 'On the Origin of the Solar System', in J. A. Hynek (ed.), *Astrophysics: A Topical Symposium* (New York: McGraw-Hill, 1951), pp. 357–424. Parts of that essay appeared with minor modifications in *Proceedings of the National Academy of Sciences* 37 (1951): 1–14, 383–93, 717–20.

81. *Ibid.*, p. 374.

82. *Ibid.*, p. 375.

83. *Ibid.*, p. 372.

84. *Ibid.*, p. 377.

85. *Ibid.*, p. 416.

86. *Ibid.*, p. 417.

87. 'On Kuiper's Theory of the Origin of the Solar System', *Proceedings of the Royal Society of Edinburgh* 64 (1953): 1–7; for quotation see p. 7.

88. 'The Disc Theory of the Origin of the Solar System', *Koninklijke Nederlandsche Akademie van Wetenschappen – Proceedings of the Section of Sciences* 51 (1948): 796–806, and its sequel, 'Types of Satellite Systems and the Disc Theory of the Origin of the Planetary System', *ibid.*, pp. 965–68.

89. 'The Disc Theory of the Origin of the Solar System', p. 796.

90. *Ibid.*, p. 799.

91. *Ibid.*, p. 806.

92. 'The Basic Scheme of Any Planetary or Satellite System', *Kon. Nederl. Akad. Wetenschap. Proceedings of the Section of Sciences* B 60 (1957): 75–87; for quotation see p. 86.

93. *A Theory of Earth's Origin: Four Lectures* (Moscow: Foreign Languages Publishing House, 1958). Of the four lectures the first deals with history and method, the second with the differentiation within the disk, the third with the encounter of the sun with a dust cloud and another star, the fourth with geophysics. The work shows ample traces also of that ideological straitjacket that hampered Soviet scientists particularly during the 1940's and 1950's.

94. *Ibid.*, p. 43.

95. *Ibid.*, p. 49. Schmidt let λ become zero in the function cq^λ, an approximation of $f(q)$, where q is the angular momentum per unit mass in the still undifferentiated disk. Then the equation $q_n = \frac{q_{n+1}+q_{n-1}}{2}$, where n and $n+1$ define zones between two neighbouring planets, could be given a general solution of the form $q_n = A + B_n$, similar in form to the Titius-Bode law.

96. *Ibid.*, p. 82.

97. *Ibid.*, p. 84.

98. B. J. Bok and E. F. Reilly, 'Small Dark Nebulae', *Astrophysical Journal* 105 (1947): 255–57.

99. 'Kinetics of Cosmic Clouds', in *Harvard Observatory Monographs No. 7* (Cambridge, Mass.: Harvard Observatory, 1948), pp. 109–42. For the outline of the theory, see section 3, 'Possibilities of Planetary Evolution in a Collapsing Interstellar Cloud', pp. 133–42.

100. *Ibid.*, p. 134.

101. *Ibid.*, p. 135.

102. *Ibid.*, p. 134.

103. *Ibid.*, p. 135.

104. *Ibid.*, p. 137.

105. *Ibid.*

106. 'The Dust Cloud Hypothesis', *Scientific American* 178 (May 1948): 34–45.

107. *Ibid.*, p. 35.

108. *Ibid.*, p. 41.

109. *Ibid.*, p. 44.

110. *Ibid.*

111. *Ibid.*

112. *The Planets: Their Origin and Development* (New Haven: Yale University Press, 1952).

113. *Ibid.*, p. 5.

114. *Ibid.*, p. 105.

115. *Ibid.*, p. 223.

116. *Ibid.*, p. 217.

117. Almost one fourth of the work, its lengthy Chapter 2, dealt with the moon.

118. 'The Origin of the Solar System', *Proceedings of the Royal Society* (London) 214 A (1952): 281–91; for quotation see p. 290.

119. *Ibid.*, p. 290.

120. *Ibid.*, p. 187. Jeffreys referred to the work of A. L. Parson, 'The Vapour Pressures of Planetary Constituents at Low Temperatures, and their Bearing on the Question of the Origin of the Planets', *Monthly Notices* 105 (1945): 244–45.

121. *Ibid.*, p. 288. Here Jeffreys emphasised the importance of the findings of E. Rabe, 'Derivations of Fundamental Astronomical Constants from the Observations of Eros', *The Astronomical Journal* 55 (1950): 112–26.

122. *Ibid.*, p. 285. In this connection Jeffreys called attention to the work of Edgeworth.

123. *Ibid.*, p. 284.

124. *Ibid.*, p. 283.

125. 'The Formation of the Planets', *Journal of the Royal Astronomical Society of Canada* 50 (1956): 57–68, 105–21, 158–76.

126. 'The Formation of the Planets', p. 167.
127. *Ibid.*, p. 164. The four stages were in Kuiper's words, '(1) gravitational collapse until angular momentum arrests process; (2) quiescent period of slow internal rearrangement of nebular matter, which by now is disk-shaped and in fairly rapid rotation; cooling of nebula by radiation to space; onset of gravitational instability in densest parts; (3) central star becomes luminous; interior of nebula becomes ionised; angular velocity of star reduced by magnetic coupling to ionised nebula in contact with it; (4) central star attains full brightness and nebula is dissipated'.
128. *Ibid.*, p. 164.
129. *Ibid.*, p. 173.
130. *Ibid.*, p. 108.
131. Winchester: Warren & Son Ltd, 1957.
132. *Ibid.*, p. 38.
133. *Ibid.*
134. The following discussion of McCrea's theory is based on its more detailed form, 'The Origin of the Solar System', *Proceedings of the Royal Society* 256 A (1960): 245–66.
135. *Ibid.*, p. 249.
136. *Ibid.*, p. 264.
137. 'On the Origin of the Solar Nebula', *Quarterly Journal of the Royal Astronomical Society* 1 (1960): 28–55.
138. *Ibid.*, p. 28.
139. New York: Harper & Brothers, 1955. See Chapter VI, 'The Origin of the Planets'. It is worth noting that the first subdivision of that Chapter is introduced with the caption, 'Planets by the Billion'.
140. 'On the Origin of the Solar Nebula', p. 29.
141. *Ibid.*
142. 'Drehimpulstransport durch Magnetfelder und die Abbremsung rotierender Sterne', *Zeitschrift für Astrophysik* 38 (1955): 190–211. While stressing the adequacy of magnetic torque for the slowing down of even very fast rotating stars, Lüst and Schlüter admitted that the 'exact evaluation of its effectiveness is tied to a series of assumptions of small certainty' (*ibid.*, p. 208).
143. 'On the Origin of the Solar Nebula', pp. 29–30.
144. 'Star Formation in Magnetic Dust Clouds', *Monthly Notices* 116 (1956): 503–14.
145. 'On the Origin of the Solar Nebula', p. 30.
146. *Ibid.*, p. 32.
147. *Ibid.*
148. *Ibid.*, p. 45.
149. Lyttleton, in his *Mysteries of the Solar System* (Oxford: Clarendon Press, 1968, pp. 33–35), noted that in addition to the high magnetic field postulated in Hoyle's theory, it also implied a condensation process for stars with too large initial radius. Also, the theory seemed to make all stars rotate very rapidly without explaining why only certain classes, such as stars similar to the sun, slowed down subsequently. According to F. L. Whipple, while the magnetic torque seemed to be an efficient mechanism to drive gaseous ions out of the sun's domain, it was impotent to propel small solid condensations as far as the orbital region of the great planets. See his *Earth, Moon and Planets* (3d ed.; Cambridge, Mass.: Harvard University Press, 1968), p. 252.
150. 'Formation of the Planets', in Robert Jastrow and A. G. W. Cameron (eds.), *Origin of the Solar System. Proceedings of a Conference Held at the Goddard Institute for Space Studies, New York, January 23–24, 1962* (New York: Academic Press, 1963), pp. 63–72.
151. 'Problems Requiring Solutions', *ibid.*, pp. 171–74.
152. *Ibid.*, p. 173.
153. *Ibid.*
154. D. ter Haar and A. G. W. Cameron, 'Historical Review of Theories of the Origin of the Solar System', *ibid.*, pp. 1–37.
155. *Ibid.*, p. 34.

156. 'A Theory of the Role of Magnetic Activity during Star Formation', *Annales d'Astrophysique* vol. 25 (1962), Nr. 1.

157. As illustrated by the papers of I. Okamoto, 'On the Loss of Angular Momentum from the Protosun and the Formation of the Solar System', *Publications of the Astronomical Society of Japan* 21 (1969): 25–53 and of T. Nakano, 'Origin of the Solar System', *Progress of Theoretical Physics* 44 (1970): 77–98.

158. 'Origin of the Solar System', *Nature* 187 (1960): 47–48.

159. *Ibid.*, p. 47.

160. 'Dynamical Calculations Relating to the Origin of the Solar System', *Monthly Notices* 121 (1960): 551–69.

161. *Ibid.*, p. 558.

162. 'A Capture Theory of the Origin of the Solar System', *Proceedings of the Royal Society* 282 A (1964): 485–507.

163. 'On the Origin of the Solar System', *Quarterly Journal of the Royal Astronomical Society* 8 (1967): 215–26; for quotation see p. 215.

164. *Ibid.*, p. 225.

165. *Ibid.*

166. *Ibid.*, p. 223.

167. *Mysteries of the Solar System* (Oxford: Clarendon Press, 1968), p. 1.

168. *Ibid.*, p. 4.

169. *Ibid.*, p. 5.

170. *Ibid.*, p. 23.

171. *Ibid.*, p. 24.

172. *Ibid.*, p. 44.

173. W. D. Metz, 'Exploring the Solar System (II): Models of the Origin', *Science* 186 (1974): 814–18; for quotation, see p. 818.

174. *Ibid.*

175. See, for instance, I. P. Williams and A. W. Cremin, 'A Survey of Theories Relating to the Origin of the Solar System', *Quarterly Journal of the Royal Astronomical Society* 9 (1968): 40–62. According to William and Cremin, the most satisfactory theories were those proposed by Hoyle and McCrea, and they confidently stated that 'the correct theory for the origin of the solar system will be somewhat similar to one or other of these' (*ibid.*, p. 60).

176. H. P. Berlage, *The Origin of the Solar System* (Oxford: Pergamon Press, 1968), p. 2. A rather strange claim on the part of one who discarded half a dozen theories of his own construction. This book of Berlage is the expanded English version of his *Het ontstaan van het zonnestelsel* (Volksuniversiteits Bibliotheek, Twede Reeks, No. 58; Haarlem: De Erven F. Bohn N.V., 1956).

177. Su-Shu Huang, 'Life outside the Solar System', *Scientific American* 202 (April, 1960): 55.

178. Su-Shu Huang, 'Occurrence of Planetary Systems in the Universe as a Problem in Stellar Astronomy', in Arthur Beer (ed.), *Vistas in Astronomy*, vol. II (New York: Pergamon Press, 1968), pp. 217–63; for quotation see p. 259.

179. *Ibid.*, p. 238, where observations by P. Van de Kamp and others of periodic variations in the position and path of several nearby stars are listed.

180. London: Oxford University Press, 1972, p. 128.

181. 'Planetary Systems', in W. C. Saslaw and K. C. Jacobs (eds.), *The Emerging Universe* (Charlottesville: University Press of Virginia, 1972), pp. 25–34; for quotation see p. 33.

182. *Ibid.*, p. 34.

183. A recent major example of this trend is the work of Josif S. Shklovskii and Carl Sagan, *Intelligent Life in the Universe*, translated by Paula Fern (San Francisco: Holden-Day, 1966).

184. *Ibid.*, p. 167.

185. *Ibid.*, p. 178.

186. See the report by Carl Sagan, Linda Salzman Sagan, and Frank Drake in *Science* 175 (1972): 884.

187. Edward Purcell, 'Radioastronomy and Communication Through Space', Brookhaven Lecture Series, Number 1, Nov. 16, 1960, p. 11.

188. TIME, July 5, 1976, p. 90.

189. TIME, August 2, 1976, p. 23.

190. Philip Abelson, editor of *Science*, quoted in NEWSWEEK, July 26, 1965, p. 57. NEWS-WEEK sided with Colin Pittendrigh, who remarked: 'It seems worth the risk. The whole point of science is to investigate the unknown, isn't it?' Science investigates, it is well to remember, the partially known.

191. Abelson in his editorial, 'The Martian Environment', *Science* 147 (1965): 681.

192. As claimed by TIME, July 5, 1976, p. 90. Mars provided its own plainly non-Mephistophelic answer to search for life there by not even yielding evidence of decomposed organisms in its soil, a fact which prompted one of the investigators of data transmitted from Mars to remark with tongue in cheek: 'There's every sign of life except death'. See TIME, September 20, 1976, p. 87.

193. In his famous memoir 'On the Construction of the Heavens', *Philosophical Transactions* 75 (1785): 213–14.

194. The phrase was coined by Robert Louis Stevenson. See his essay, 'Crabbed Age and Youth' (1878), in *Essays by Robert Louis Stevenson*, with an Introduction by William Lyon Phelps (New York: Charles Scribner's Sons, 1918), p. 136. As these pages go to press, the public is once more being treated to excited reports about an imminent solution of the riddle of the origin and formation of planetary systems. Photographs showing a ring around the rapidly contracting star MWC 349 are certainly worthy of attention, but so is the strange silence (see, for instance, TIME June 27, 1977, p. 77) about the fact that, as Moulton had shown more than seventy years ago, a ring can at best coalesce into several planets. These being at the same distance from the star can hardly form a stable system.

ILLUSTRATIONS

XV   Planets with their satellites and satellites around the sun (p. 78);
        ibid., Tab. XXVII facing p. 396; reduced by half.

XVI   God's arm behind the formation of planets in a filament detached
        from the sun by a passing comet (p. 87); from Buffon, *Histoire
        naturelle, générale et particulière, avec la description du Cabinet du Roi.
        Tome Premier* (Paris: de l'Imprimerie Royale, 1749), facing p. 127.

XVII   The earth's encounter with a comet (p. 90); from W. Whiston,
        *A New Theory of the Earth, from its Original, to the Consummation of
        All Things* (London: printed by R. Roberts for Benj. Tooke,
        1696); Fig. 1. on end-foldout.

XVIII   The earth's passing through the atmosphere of a comet (p. 90);
        ibid.; Fig. 7. on end-foldout.

XIX   Putting a projectile into orbit around the earth (p. 98); from I.
        Newton, *De mundi systemate liber Isaaci Newtoni* (London: impensis
        J. Tonson, J. Osborn, & T. Longman, 1728), Fig. I. on end-foldout.

XX   Transformation by rotation of a spherical lump of oil into a ring
        (p. 139); from J. Plateau, 'Mémoire sur les phénomènes que pré-
        sente une masse liquide libre et soustraite à l'action de la pésanteur',
        in *Nouveaux Mémoires de l'Académie Royale des Sciences et Belles-
        lettres de Bruxelles*, vol. XVI (Bruxelles: M. Hayez, 1843), p. 135.

XXI   Transformation of the Laplacian solar nebula into rings (p. 141);
        from J. P. Nichol, *Views of the Architecture of the Heavens. In a Series
        of Letters to a Lady* (Edinburgh: William Tait, 1837), Plate XXI
        facing p. 170.

XXII   The difference between inner and outer orbital velocities in a
        segment of gaseous ring as basis for the forward rotation of planets
        (p. 141); from J. P. Nichol, *The Planetary System: Its Origin and
        Physical Structure* (London: H. Baillière, 1850), p. 239.

XXIII   Shearing collision of two stars leading to the formation of a spiral
        nebula (p. 168); from A. W. Bickerton, 'Partial Impact (Paper
        No 3): On the Origin of the Visible Universe', *Transactions and
        Proceedings of the New Zealand Institute* 12 (1879), facing p. 184.

XXIV   Development of a spiral nebula into a Laplacian nebula leaving
        behind a ring (p. 168); *ibid.*

XXV   The forward and backward rotation of planets in the solar system
        (p. 173); from H. Faye, *Sur l'origine du monde: Théories cosmogoniques
        des anciens et des modernes* (Paris: Gauthier-Villars, 1884), p. 191.

XXVI   Cross section of circles of low density in rotating nebula as places
        of its rupturing into rings (p. 179); from R. du Ligondès, *Formation
        mécanique du système du monde* (Paris: Gauthier-Villars, 1897), p. xiii.

XXVII   The progressive elongation and rotation of two suns (p. 192);
        from T. C. Chamberlin, 'On a Possible Function of Disruptive
        Approach in the Formation of Meteorites, Comets, and Nebulae',
        The Astrophysical Journal 14 (1901), p. 35.

XXVIII Formation of spiral arms in the solar nebula shortly after its close
        encounter with a passing star (p. 198); from F. R. Moulton, 'On

magnetic torque in the solar cloud (p. 242); from F. Hoyle, *Frontiers of Astronomy* (New York: Harper and Brothers, 1955), p. 92.

XL   Lines of magnetic field from the sun to its disk ensuring the effectiveness of transfer of angular momentum (p. 244); from F. Hoyle, 'On the Origin of the Solar Nebula', *Quarterly Journal of the Royal Astronomical Society* 1 (1960), p. 38.

XLI   Formation of planets through rhythmic tidal wave touched off by passing star on the sun's surface (p. 245); from M. M. Woolfson, 'Origin of the Solar System', *Nature* 187 (1960), p. 47.

XLII Partly coded message engraved on a plaque attached to Pioneer 10 for the information of extrasolar civilisations about earthlings (p. 249); from C. Sagan, L. S. Sagan, and F. Drake, 'A Message from Earth', *Science* 175 (1972), p. 883.

Special acknowledgement is hereby registered for permission to use illustrative material: to Corpus Christi College, Oxford (*I*, *II*, *III*, *IV*), to Universiteits-Bibliotheek van Amsterdam (*V*, *VI*, *VII*, *VIII*, *IX*, *X*, *XI*), to Seton Hall University (*XVI*), to Bodleian Library, Oxford University (*XVII*, *XVIII*, *XIX*), to Scottish National Library (*XXI*), to Firestone Library, Princeton University (*XX*, *XXII*, *XXIII*, *XXIV*, *XXXI*), to University of Chicago Press (*XXVII*, *XXVIII*), to British Astronomical Association (*XXXII*), to Springer Verlag (*XXXIII*), to Penguin Books (*XXXIV*), to McGraw-Hill Book Company (*XXXV*), to W. H. Freeman and Company '(*XXXVIII*), to Koninklijke Nederlandse Akademie von Wetenschappen (*XXXVI*), to Harper and Brothers (*XXXIX*), to Royal Astronomical Association (*XL*), to Macmillan Journals (*XLI*), to American Association for the Advancement of Science (*XLII*).

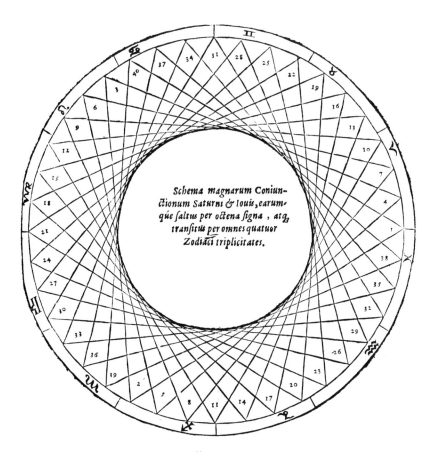

Schema magnarum Coniun-
ctionum Saturni & Iouis, earum-
que saltus per octena signa, atq́
transitús per omnes quatuor
Zodíací triplicitates.

Illustration I

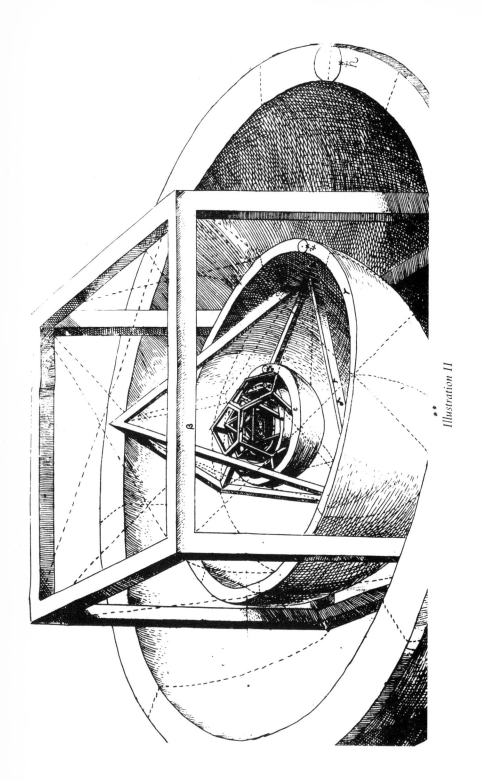

Illustration II

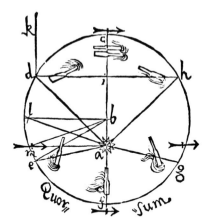

Illustration III

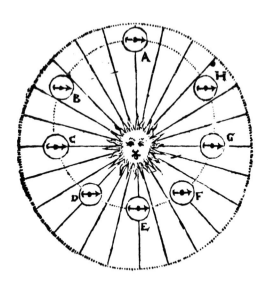

Illustration IV

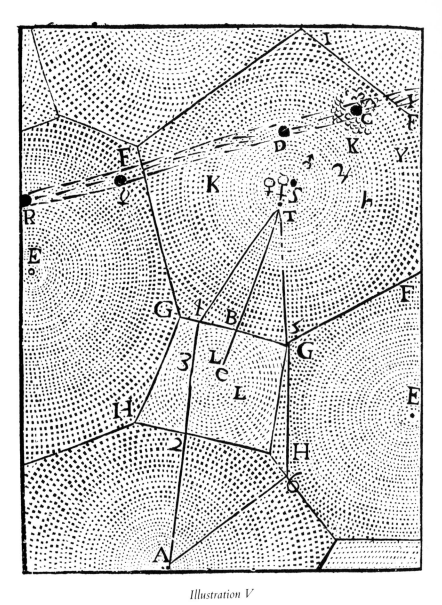

Illustration V

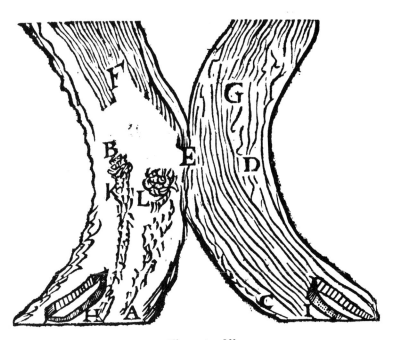

Illustration VI

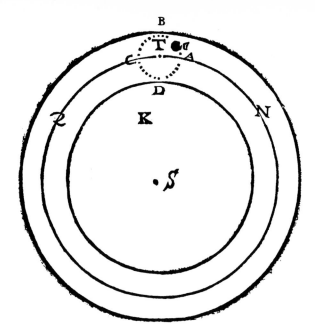

Illustration VII

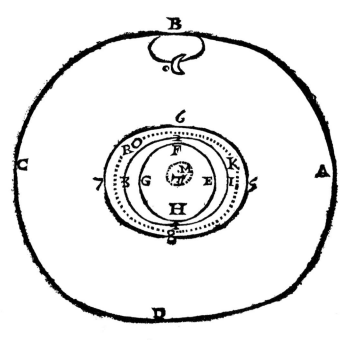

Illustration VIII

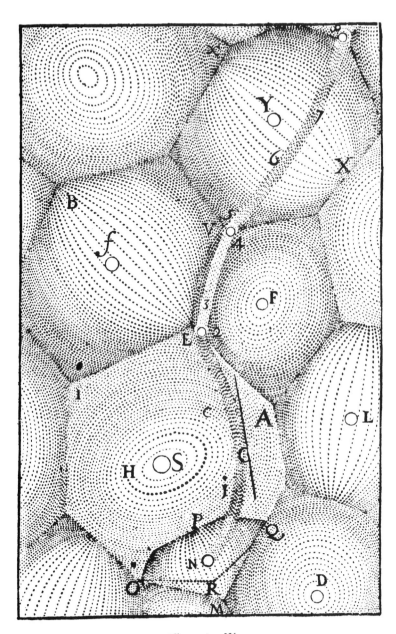

Illustration IX

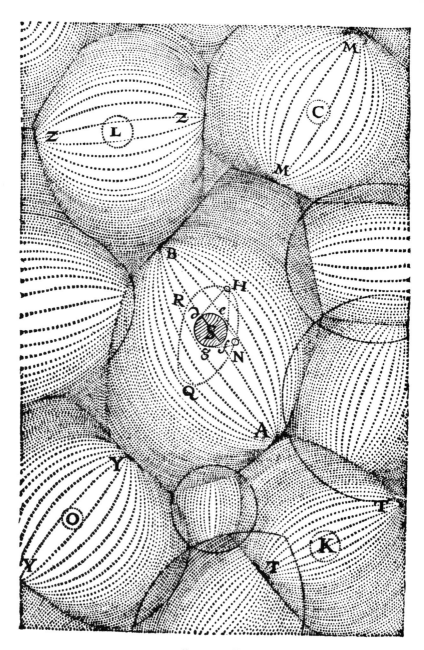

Illustration X

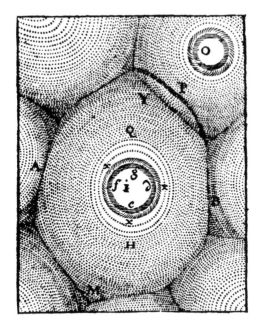

Illustration XI

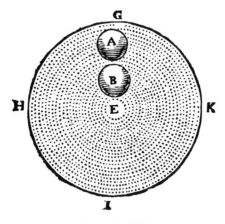

Illustration XII

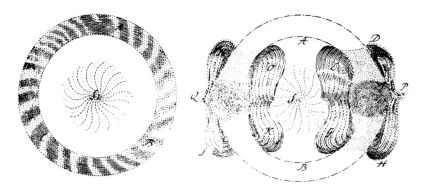

Illustration XIII

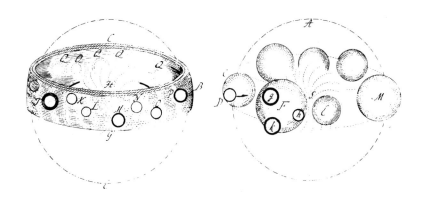

Illustration XIV

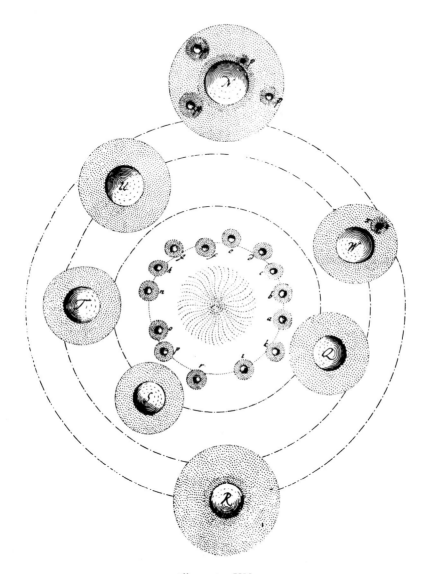

Illustration XV

Illustration XVI

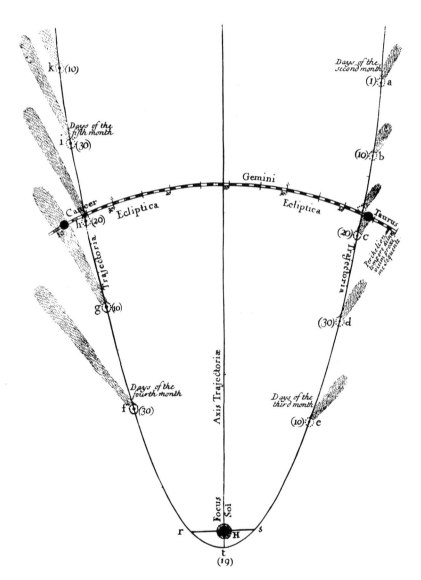

Illustration XVII

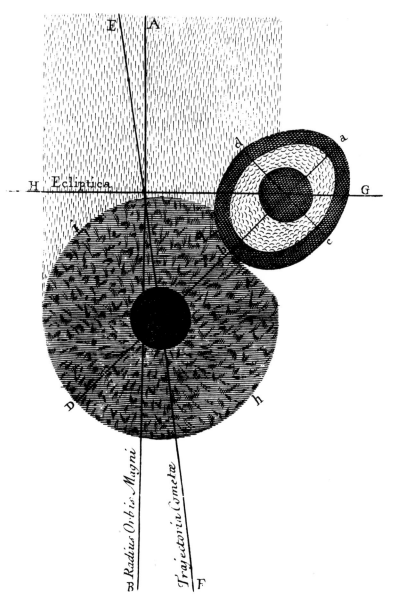

Illustration XVIII

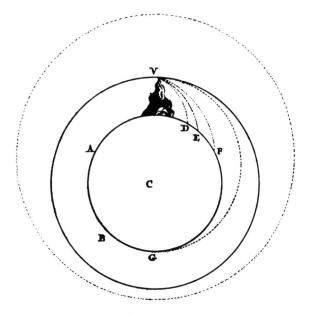

Illustration XIX

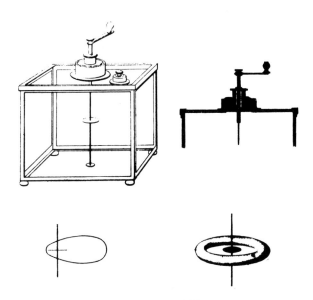

Illustration XX

Illustration XXI

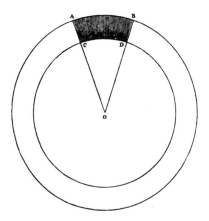

Illustration XXII

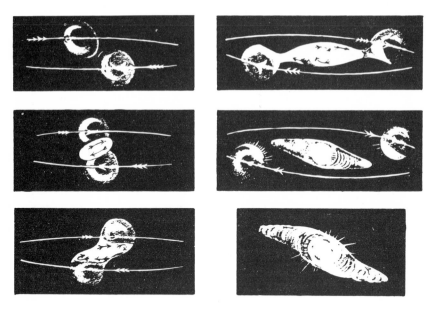

Illustration XXIII

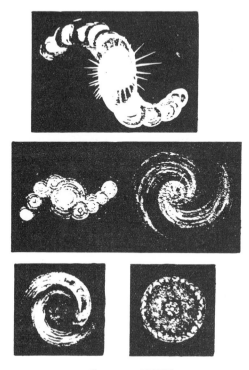

Illustration XXIV

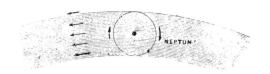

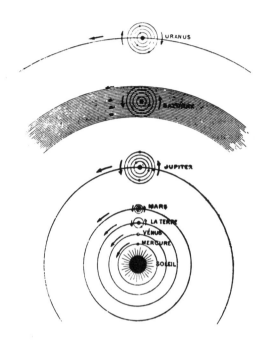

Illustration XXV

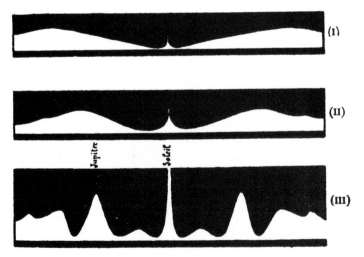

Illustration XXVI

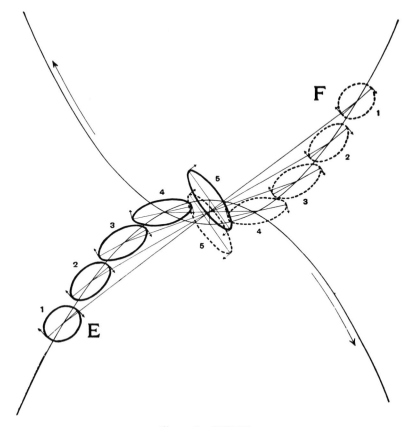

Illustration XXVII

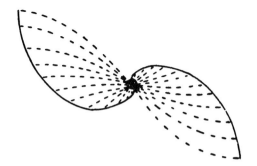

Illustration XXVIII

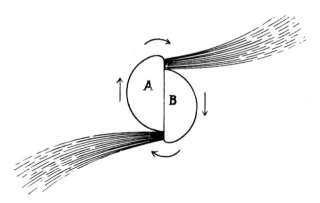

Illustration XXIX

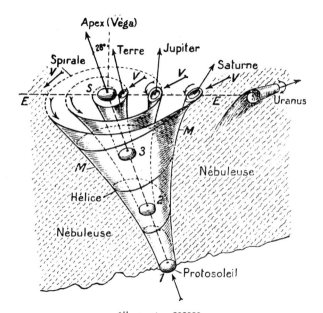

Illustration XXX

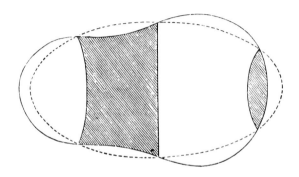

Illustration XXXI

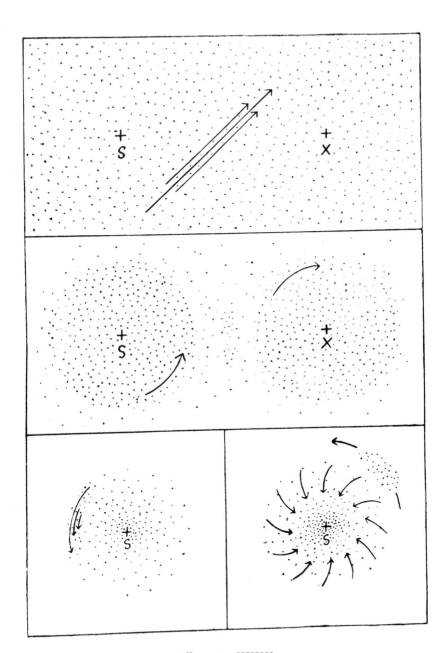

Illustration XXXII

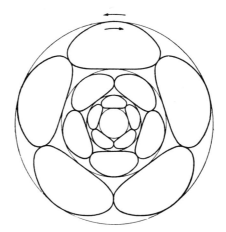

Illustration XXXIII

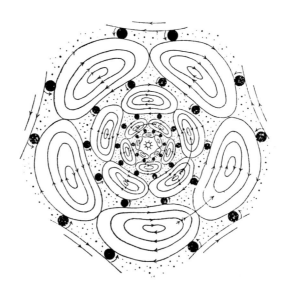

Illustration XXXIV

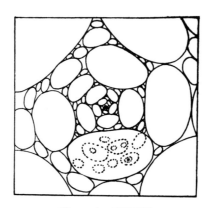

Illustration XXXV

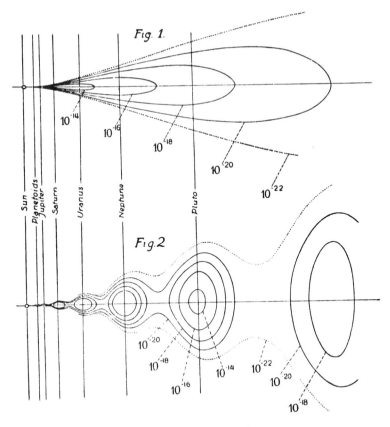

Fig. 1.
10^-14
10^-16
10^-18
10^-20
10^-22
Sun
Planetoids
Jupiter
Saturn
Uranus
Neptune
Pluto
Fig. 2
10^-20
10^-18
10^-16
10^-14
10^-22
10^-20
10^-18

Illustration XXXVI

P2
P1
P0
300
200
100
-100
-200
-100
0
100
200
300
400
500
700
P2

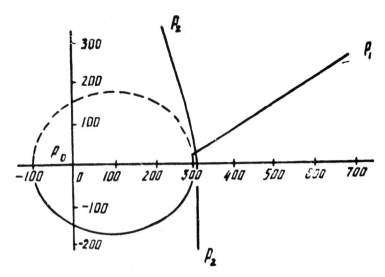

Illustration XXXVII

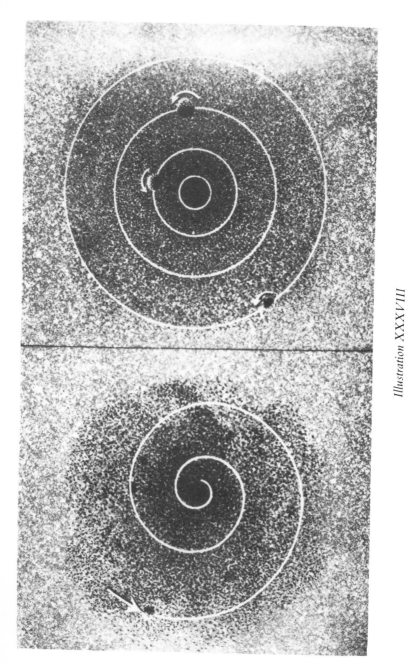

Illustration XXXVIII

Rotation

Illustration XXXIX

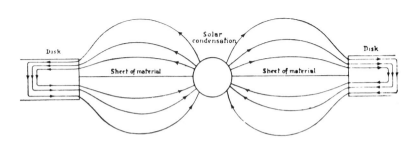

Illustration XL

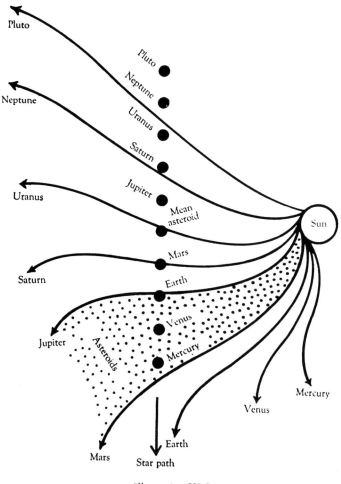

Illustration XLI

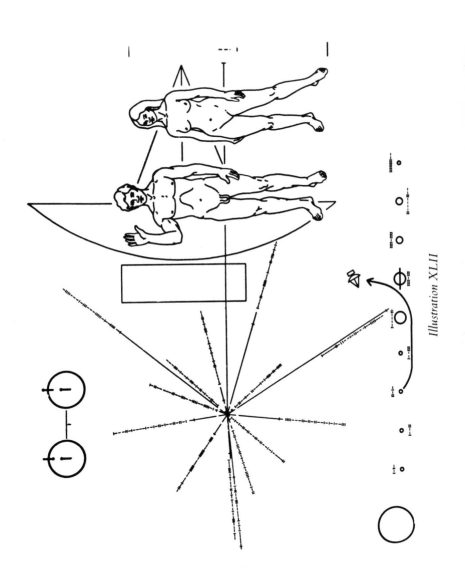

Illustration XLII

INDEX OF NAMES